Microorganisms for Sustainability

Volume 7

Series editor

Naveen Kumar Arora, Environmental Microbiology, School for Environmental Science, Babasaheb Bhimrao Ambedkar University, Lucknow, Uttar Pradesh, India

More information about this series at http://www.springer.com/series/14379

Deepak G. Panpatte · Yogeshvari K. Jhala
Harsha N. Shelat · Rajababu V. Vyas
Editors

Microorganisms for Green Revolution

Volume 2 : Microbes for Sustainable Agro-ecosystem

Springer

Editors
Deepak G. Panpatte
Department of Agricultural Microbiology,
B. A. College of Agriculture
Anand Agricultural University
Anand, Gujarat, India

Yogeshvari K. Jhala
Department of Agricultural Microbiology,
B. A. College of Agriculture
Anand Agricultural University
Anand, Gujarat, India

Harsha N. Shelat
Department of Agricultural Microbiology,
B. A. College of Agriculture
Anand Agricultural University
Anand, Gujarat, India

Rajababu V. Vyas
Department of Agricultural Microbiology,
B. A. College of Agriculture
Anand Agricultural University
Anand, Gujarat, India

ISSN 2512-1901 ISSN 2512-1898 (electronic)
Microorganisms for Sustainability
ISBN 978-981-10-7145-4 ISBN 978-981-10-7146-1 (eBook)
https://doi.org/10.1007/978-981-10-7146-1

Library of Congress Control Number: 2017958135

Printed on acid-free paper

This Springer imprint is published by Springer Nature
The registered company is Springer Nature Singapore Pte Ltd.
The registered company address is: 152 Beach Road, #21-01/04 Gateway East, Singapore 189721, Singapore

Foreword

Agroecosystem is presently disturbed due to human interference in agriculture by impacting on soil, water, microorganism, plant, and animal resources to produce their own requirements. Agroecosystem supports the crops, but due to continuous uncontrolled exploitation of natural resources, imbalance in natural biogeochemical cycles has occurred. This ignorance is now intensified in the era of green revolution, wherein concentration was given only on increasing productivity without worrying for the cost to be paid by agroecosystem. As a result, presently, we are witnessing reduced soil fertility and added ecological factors like climate change which are adversely affecting agricultural production. Promoting the healthy functioning of ecosystem ensures the resilience of agriculture through microorganisms which can set soil cycles properly. Climate change and related stresses are noticeably affecting basic functions of agroecosystem. The key to sustain the agroecosystem lies in the fact that the linkages between soil, plant, water, and microorganisms in agroecosystem should be strengthened. Microorganisms are in the nexus being able to mitigate the climate impact.

In view of above, the book *Microbes for Sustainable Agroecosystem – Volume II* contributes to sustainable agriculture. It is an attempt to explore knowledge of various national and international researchers and will be of great interest to students, teachers, and scientists working in the area of environmental, soil, and agricultural microbiology. The team of microbiologists from Anand Agricultural University have tried to compile and provide knowledge of experts at a single platform. I am sure that this book will serve as a resource for developing microorganism-based strategies for sustainable agroecosystem.

Anand Agricultural University
Anand, Gujarat, India

N.C. Patel,
Vice-Chancellor

Preface

Agroecosystem comprises of air, water, and soil which support the growth of all living things and total food web; however, numerous microorganisms critically involved with all living things are the driving forces of its functioning. It is a need to determine beneficial aspects of plant and microbial interactions in agroecosystem in light of the present challenge like drought, soil fertility loss, climate change, and soil contamination. In order to enhance the food security through sustainable agricultural practices, the role of soil microbes is certainly a priority component for productivity and preservation of soil. Environment and resource conservation for assuring food quality for the quickly growing human population is the major challenge in agriculture during the last few decades. Microbe-based ecological research plays a key role in developing technological approaches to protect the environment and overall agroecosystem for sustainable development as well as for future policies.

This book contains chapters from distinguished experts and dedicated researchers. Efforts have been made to compile the latest information on the present status of microorganism-based mitigation and key strategies for sustainable agroecosystem development and promotion for welfare of mankind. Some chapters deal with bioremediation of pesticides and toxic metals from soil, agro-waste management, and mitigation of effects of climate change on agriculture. The book also provides in-depth insights and scientific linkages as well as evidences to show the importance of microorganisms to mitigate climate change and enable plants to fight with abiotic stresses. We hope that this book *Microbes for Sustainable Agroecosystem – Volume II* will help the researchers in planning their future line of research and the policy makers in taking rational decisions on sustainable agroecosystem for the benefit of farmers as well as other stakeholders and will serve as a treasure of knowledge to students. It must be mentioned here that the scholarly chapters included in this volume do help to enrich the readers' understanding on the issues related to sustainability of agroecosystem with mitigation strategies keeping microbial world of soil in focus. The views expressed by the authors in their respective papers are based on their vast experience in the agricultural and allied research areas. We thank all the contributors of this volume, and we are grateful for their valuable contributions.

Anand, Gujarat, India

Deepak G. Panpatte
Yogeshvari K. Jhala
Harsha N. Shelat
Rajababu V. Vyas

Contents

About the Editors

Deepak G. Panpatte is a research scholar working for the past 7 years. His research interests include agriculturally beneficial microorganisms, viz., biofertilizers, biopesticides, and biodegraders. He has done pioneering work in the development of fortified biocontrol bacterial consortium with phyto-extracts for the management of phytopathogenic nematodes and fungi. He has received six awards, three for presentation of research outcomes and three for his remarkable role in agriculture sector. His publication profile includes 13 research papers, 1 book and 9 book chapters with Springer publishing house, 1 practical manual, 26 popular articles, and 2 editorial pages.

Yogeshvari K. Jhala is an assistant professor having 10 years of teaching and research experience. Her field of interest includes agriculturally beneficial microorganisms, viz., biofertilizers, biopesticides, and biodegraders. She has world over first time reported five unique strains of methanotrophic bacteria. For her outstanding research work of methanotrophic bacteria, she was honored with All India Best Research Award and Young Faculty Award. Her publications include 17 research papers, 1 book and 9 book chapters with Springer publishing house, 2 teaching manuals, 18 popular articles, and 2 editorial pages.

Harsha N. Shelat is an associate research scientist and award-winning researcher at Anand Agricultural University, Anand, India. She has 33 years of experience in the field of agriculturally advantageous microbes chiefly biofertilizers, biodegraders, and microbial pesticides. Her contribution in society is the development and dissemination of liquid biofertilizer technology for the benefit of farming community. She has written 3 books and 7 book chapters and has published more than 35 research papers in national as well as international journals.

Rajababu V. Vyas MSc (microbiology), PhD, serves as research scientist and head of the Department of Agricultural Microbiology, AAU, Anand. For the past 31 years, he has worked on agriculturally beneficial microorganisms on isolation and characterization, the development of mass production technique, and laboratory and field testing of biofertilizers for crop production. He has also developed native microbial agents for biological control of insect pests and plant parasitic nematodes for crop protection and PGPR for bioremediation of methane and agro-waste to support organic farming approach. His publications include 112 research papers, 2 review papers, 5 books and manuals, 5 training manuals, and 11 book chapters, 2 in CAB International and Michigan State University Press, USA. He is a recipient of six prestigious awards and instrumental for technology patenting, commercialization, licensing, and services.

Introduction

In recent years, we have witnessed fast growth in developing countries with concurrent progress in industrialization and agriculture. These developments have many environmental side effects in the form of pollution. If such type of unrestricted disposal of various wastes to agricultural land is continued, the liveliness of soil and its ability to sustain other ecosystems will be at risk. As climate change and environmental degradation increase, it will challenge sustainable crop production. Climate change has the prospective to intensify abiotic stress on crops while shifting the environmental niches for pathogens and weeds. Consecutively, enhancing food production using current technologies increases fuel requirements, but depleted fuel reserves drive demand for biofuels. Such increase in demand of biofuels creates problem of food vs fuel. Each of these events is closely related to parallel change in environmental quality.

This book entitled *Microbes for Sustainable Agroecosystem – Volume II* focuses on ideas of environmental security by microorganisms in sustainable manner. Microbes can be considered as natural cleaners utilizing garbage of earth. These tiny creatures can utilize almost everything including agrochemicals, heavy metals, and gaseous pollutants. The book focuses on the role of microorganisms in mitigation of environmental stress on crops as well as reduction of greenhouse gases in agriculture. It also contains reviews and original research of reputed scientists to highlight the latest developments in microbiological research, to cope up with the problem of changing climate along with remediation strategies practiced at various stages for improvement of agroecological health. The book will be a valuable resource for scientists working to develop mitigation strategies for the development of sustainable agroecosystem and also serves as inspiration and ready reckoner for the students who want to pursue studies pertaining to bioremediation of environment making them ready for future challenges.

Department of Agricultural Microbiology
Anand Agricultural University
Anand, Gujarat, India

Deepak G. Panpatte

1 Effect of Agrichemicals on Biocontrol Agents of Plant Disease Control

A. K. Patibanda and M. Ranganathswamy

Abstract

Plant diseases need to be controlled to maintain the quality and abundance of food, feed, and fiber produced by growers around the world. Different approaches may be used to prevent, mitigate, or control plant diseases. Beyond good agronomic and horticultural practices, growers often rely heavily on chemical fertilizers and pesticides. Such inputs to agriculture have contributed significantly to the spectacular improvements in crop productivity and quality over the past 100 years. However, the environmental pollution caused by excessive use and misuse of agrochemicals, as well as fear-mongering by some opponents of pesticides, has led to considerable changes in people's attitudes toward the use of pesticides in agriculture. Consequently, some pest management researchers have focused their efforts on developing alternative inputs to synthetic chemicals for controlling pests and diseases. Among these alternatives biological control is the major emerging tool for disease control in sustainable agriculture. Biological control offers a novel approach when applied either alone or in combination with other management practices. Biocontrol and chemicals are important tools in IPM strategy wherein agrochemicals applied by any of the methods (seed treatment, soil application, and foliar spray) will ultimately reach soil and interact with the biocontrol agents. These agrochemicals drastically affect the growth, reproduction, and survivability of biocontrol agents in a particular cropping system and thereby on their biocontrol efficacy. The available literature on the effect

A. K. Patibanda (✉)
Department of Plant Pathology, Agriculture College, ANGRAU, Bapatla, Andhra Pradesh, India
e-mail: anilpatibanda@yahoo.co.in

M. Ranganathswamy
Department of Plant Pathology, Agriculture College, A.A.U, Jabugam, Gujarat, India

D. G. Panpatte et al. (eds.), *Microorganisms for Green Revolution*, Microorganisms for Sustainability 7,
https://doi.org/10.1007/978-981-10-7146-1_1

of agrochemicals on biocontrol agents revealed that most of the agrochemicals (fungicides, insecticides, fertilizers, and herbicides) are reported to have adverse effect on the soil microflora including biocontrol agents. The need-based and judicious application of agrochemicals, checking the compatibility between the agrochemicals and biocontrol agents and developing the new pesticide resistant strains of biocontrol agents using biotechnology tool, will help to address the above Problems.

Keywords
Biocontrol agents · Environment · Fertilizers · Fungicides · Herbicides · Insecticides · Toxicity

1.1 Introduction

Plant diseases are of significant concern due to the intimate relationship between plant health and the welfare of people, animals, and the environment. The ability to provide adequate food and fiber is becoming increasingly strained, and continued improvement in sustainable plant disease management is required to help meet these demands. Plant diseases often substantially reduce quality and quantity of agricultural commodities (Cramer 1967). Also, infestation by microorganisms in the field or in postharvest storage can affect the health of humans and livestock, especially if the contaminating organism produces toxic residues in or on consumable products (Sinha and Bhatnagar 1998). Control of plant disease is vital in protecting the plants from diseases, reducing yield losses, and finally achieving the food security. In this concern agrochemicals (pesticides and fertilizers) are looked upon as a vehicle for improved crop production technology though it is a costly input. Balanced use, optimum doses, correct method, and right time of application of agrochemicals ensure increased crop production. Yet, chemical method of disease management, while effective in the control of plant pathogens, may negatively impact nontarget organisms; hence, it is time to look upon sustainable, eco-friendly approaches for disease management. At this juncture biological control plays a vital role in reducing the excess use of chemical pesticides and thereby prevents the adverse effect of pesticides. This chapter will highlight the role of environment on disease occurrence and on biocontrol agents, pros and cons of agrochemicals, adverse effect of agrochemicals on biocontrol agents, and strategies to mitigate adverse effect of agrochemicals.

1.2 Abiotic Environment

For a disease to occur and to develop optimally, a combination of three factors must be present: susceptible plant, infective pathogen, and favorable environment. However, although plant susceptibility and pathogen infectivity remain essentially unchanged in the same plant for at least several days, and sometimes for weeks or months, the environmental conditions may change more or less suddenly and to various degrees. Such changes may drastically influence the development of diseases in progress or the initiation of new diseases. Of course, a change in any environmental factor may favor the host, the pathogen, or both, or it may be more favorable to one than it is to the other. As a result, the expression of disease will be affected accordingly. Plant diseases generally occur over a fairly wide range of the various environmental conditions. The environmental factors that affect the initiation and development of infectious plant diseases most seriously are temperature and moisture on the plant surface. Soil nutrients also play an important role in some diseases and, to a lesser extent, light and soil pH. Environment and edaphic factors affect disease development through their influence on the growth and susceptibility of the host and on the multiplication and activity of the pathogen and biocontrol agents which finally relates to the severity of symptom development.

1.2.1 Abiotic Environment on Biocontrol Agents

As the crops are affected by abiotic stresses such as soil moisture-deficit stress, high temperature, soil salinity, and so forth, microbes are also known to be affected by these conditions. Reports from Madhya Pradesh and Chhattisgarh of India indicated that the free-living rhizobial population declines to lower than the minimum threshold levels required for nodulation due to high soil temperatures requiring inoculation every year (Gupta et al. 2000; Rao 1996–2000). Widden and Hsu (1987) observed that the ability of different species of *Trichoderma* to colonize pine or maple litter differed with temperatures. Mukherjee and Raghu (1997) reported that as temperature increases (>30° C), biocontrol efficacy of *Trichoderma* decreases drastically. Abiotic stresses greatly influence the performance of biocontrol agents. Sankaranarayanan and Rajeswari (2001) studied the effect of moisture and pH on the efficacy of VAM (*Glomus mosseae*) on *Meloidogyne incognita* in black gram. Among the moisture levels tested 70% moisture and among the pH levels, pH 7 was found suitable for the better mycorrhizal colonization, spore production, and well establishment of VAM to control root knot nematode. Higher moisture level (80–100%) was found detrimental to VAM fungus. Leo Daniel et al. (2011) evaluated *T. viride* for its in vitro abiotic stress tolerance ability and its field bioefficacy against root rot disease in black gram, and it was reported that growth of *T. viride* decreased with increase in salinity, temperature, and drought. Reetha et al. (2014) studied the effect of temperature and pH on growth of *Trichoderma harzianum* and reported that *T. harzianum* was grown better at 25–30 °C and very slowly grown at 37 °C and not grown at 45 °C.

1.3 Agrochemicals

Agrochemicals cover a wide range of compounds including insecticides, fungicides, herbicides, rodenticides, molluscicides, nematicides, plant growth regulators, and others. The introduction of synthetic insecticides – organophosphate (OP) insecticides in the 1960s, carbamates in the 1970s, and pyrethroids in the 1980s – and the introduction of herbicides and fungicides in the 1970s–1980s contributed greatly to pest control and agricultural output. Ideally a pesticide must be lethal to the targeted pests, but not to nontarget species, including man. Unfortunately, this is not the case, so the controversy of use and abuse of pesticides has surfaced. The rampant use of these chemicals, under the adage, "if little is good, a lot more will be better," has played havoc with human and other life forms.

1.3.1 Production and Usage of Pesticides

Modern agriculture depends on the four main factors, viz., water, fertilizers, seed, and pesticides. Pesticides are the integral part of modern agriculture. The production of pesticides started in India in 1952 with the establishment of a plant for the production of BHC near Calcutta, and India is now the second largest manufacturer of pesticides in Asia after China and ranks fourth globally (FICCI 2016). There has been a steady growth in the production of technical grade pesticides in India, from 5,000 metric tonnes in 1958 to 102,240 metric tonnes in 1998. In 1996–1997 the demand for pesticides in terms of value was estimated to be around Rs. 22 billion (USD 0.5 billion), which is about 2% of the total world market. The pattern of pesticide usage in India is different from that for the world in general. In India 76% of the pesticide used is insecticide, as against 44% globally (Mathur 1999). The use of herbicides and fungicides is correspondingly less heavy. The main use of pesticides in India is for cotton crops (45%), followed by paddy and wheat. The Indian domestic demand is growing at the rate of 8–9% and export demand at 15–16%. The per capita consumption of pesticides in India is 0.6 Kg/ha. The per capita pesticide consumption in China and the USA is 13 Kg/ha and 7 Kg/ha, respectively. India is the second biggest consumer of fertilizer in the world next only to China. The total consumption of fertilizer (NPK) in India is nearly about 255.4 thousand tonnes.

1.3.2 Positive Effects (Benefits) of Agrochemicals

1.3.2.1 Improving Productivity

A UN study on global population trends predicts that India will surpass China to become the most populous nation in the world by 2022. With a present size of 1.32 billion, India currently supports nearly 17.84% of the world population, with 2.4% land resources and 4 % of water resources (FICCI 2016). It is also noted that about 30–35% potential crop production is lost due to pests, weeds, and diseases.

Keeping pace with these growing numbers, the country will not only have to raise its agricultural production but also the productivity to ensure food and nutrition security of the nation. Crop protection and crop enhancement solutions, based on best global practices and the latest technologies available, are the answer. Tremendous benefits have been derived from the use of pesticides in agriculture. Food grain production, which stood at a mere 50 million tons in 1948–1949, had increased almost fourfold to 198 million tonnes in 1996–1997. This result has been achieved by the use of high-yield varieties of seeds, advanced irrigation technologies, and agricultural chemicals (Employment Information: Indian Labour Statistics 1997). Similarly outputs and productivity have increased dramatically in most countries, for example, wheat yields in the UK and corn yields in the USA. Increases in productivity have been due to several factors including use of fertilizer, better varieties, and use of machinery. Pesticides have been an integral part of the process by reducing losses from the weeds, diseases, and insect pests that can markedly reduce the amount of harvestable produce.

1.3.2.2 Protection of Crop Losses

About 30–35 % crop production is lost due to insects, weeds, and diseases. The total number of pests attacking major crops has increased significantly from the 1940s. For instance, the number of pests which are harmful for crops such as rice has increased from 10 to 17, whereas for wheat it has increased from 2 to 19, respectively. The increased damage to crops from pests and subsequent losses pose a serious threat to food security and further underscore the importance of agrochemicals. The use of pesticides helped to manage the weeds, insects, and diseases in agriculture resulting in scaling down the loss from these pests. Webster et al. (1999) stated that "considerable economic losses" would be suffered without pesticide use and quantified the significant increases in yield and economic margin that result from pesticide use.

1.3.2.3 Vector Disease Control

Most of the agriculture crops are prone to be infected by several viral diseases. These viral diseases are mainly transmitted by insect vectors (aphids, whitefly, leafhopper, etc.). The development of systemic insecticides has helped to control the insect vectors, thereby addressing the problem of plant viral disease management.

1.3.3 Negative Effects (Hazards of Pesticides)

1.3.3.1 Direct Impact on Humans

If the credits of pesticides include enhanced economic potential in terms of increased production of food and fiber, and amelioration of vector-borne diseases, then their debits have resulted in serious health implications to man and his environment.

There is now overwhelming evidence that some of these chemicals do pose a potential risk to humans and other life forms and unwanted side effects to the environment (Forget 1993; Igbedioh 1991; Jeyaratnam 1981). No segment of the population is completely protected against exposure to pesticides, and the potentially serious health effects, though a disproportionate burden, are shouldered by the people of developing countries and by high-risk groups in each country (WHO 1990). The high-risk groups exposed to pesticides include production workers, formulators, sprayers, mixers, loaders, and agricultural farmworkers.

1.3.3.2 Pesticide Residue in Food Commodities

The indiscriminate use of pesticides for controlling the insect pests and diseases resulted in pesticide residue problems in food commodities. Pesticide residue refers to the pesticides that may remain on or in food after they are applied to food crops. The maximum allowable levels of these residues in foods are often stipulated by regulatory bodies in many countries. In India the first report of poisoning due to pesticides was from Kerala in 1958, where over 100 people died after consuming wheat flour contaminated with parathion (Karunakaran 1958). This prompted the Special Committee on Harmful Effects of Pesticides constituted by the ICAR to focus attention on the problem (Report of the Special Committee of ICAR 1972). In 2006, the Ministry of Agriculture initiated a program to monitor pesticide residues at the national level. Working through the Indian Council of Agricultural Research, the Ministry selected 20 laboratories to collect and analyze samples of vegetables, fruits, spices, pulses, cereals, milk, fish, tea, honey, meat, animal feed, and groundwater. In 2011–2012, 16948 food commodities were analyzed for pesticides, in which 1668 (9.8%) commodities were found positive, while 290 (1.7%) had pesticides above MRL (AINP Annual report 2013).

1.3.3.3 Impact on Environment

Pesticide sprays can directly hit nontarget vegetation or can drift or volatilize from the treated area and contaminate air, soil, and nontarget plants. Some pesticide drift occurs during every application, even from ground equipment (Glotfelty and Schomburg 1989). Drift can account for a loss of 2–25% of the chemical being applied, which can spread over a distance of a few yards to several hundred miles. As much as 80–90% of an applied pesticide can be volatilized within a few days of application (Majewski and Capel 1995). In addition to killing insects or weeds, pesticides can be toxic to a host of other organisms including birds, fish, beneficial insects, and nontarget plants.

1.3.3.4 Surface Water Contamination

Pesticides can reach surface water through runoff from treated plants and soil. Contamination of water by pesticides is widespread. The results of a comprehensive set of studies done by the US Geological Survey (1998) on major river basins across the country in the early to mid-90s yielded startling results. More than 90% of water and fish samples from all streams contained one or, more often, several pesticides (Kole et al. 2001). Pesticides were found in all samples from major rivers with mixed agricultural land use (Bortleson and Davis 1987–1995).

1.3.3.5 Groundwater Contamination

Groundwater pollution due to pesticides is a worldwide problem. Pesticides enter surface and groundwater primarily as runoff from crops and are most prevalent in agricultural areas. A decadal assessment by the National Water-Quality Assessment (NAWQA) Program of the US Geological Survey (1998) showed that pesticides are frequently present in streams and groundwater at concentrations that may have effects on aquatic life or fish-eating wildlife and human beings. According to the USGS, at least 143 different pesticides and 21 transformation products have been found in groundwater, including pesticides from every major chemical class.

1.3.3.6 Soil Contamination

Agrochemicals sprayed aerially reach the soil (by means of air currents or are washed off the plant surface due to rain) and contaminate soil. A large number of transformation products (TPs) from a wide range of pesticides have been documented (Barcelo and Hennion 1997; Roberts 1998; Roberts and Hutson 1999). Persistency and movement of these pesticides and their TPs are determined by some parameters, such as water solubility, soil-sorption constant (Koc), water partition coefficient (K_ow), and half-life in soil (DT_{50}). Pesticides and TPs could be grouped into hydrophobic, persistent, and bioaccumulable pesticides that are strongly bound to soil. Pesticides that exhibit such behavior include the organochlorine DDT, endosulfan, endrin, heptachlor, lindane, and their TPs. Polar pesticides are represented by herbicides, carbamates, fungicides, and some organophosphorus insecticides and their TPs. They can be moved to soil and thereby contaminate soil.

1.3.3.7 Pesticide Resistance in Pest Species

Problems associated with reliance on chemical control include the development of pesticide resistance in important pest species. This encourages an increase in dosage and number of pesticide applications which magnifies the adverse effects on natural enemies. Pests may also resurge because of the destruction of predators and parasitoids, breeding without restraint from natural enemies. If pesticides eliminate natural enemies, populations of pests may increase dramatically. Melander (1914) first time reported the resistance of San Jose scale (*Quadraspidiotus perniciosus* L.) insect to lime sulfur in the USA. In India Pradhan (1960) reported resistance of singhara beetle (*Galerucella birmanica* (Jacoby)) to DDT and BHC. The problem of fungicide resistance became apparent following the registration and widespread use of the systemic fungicides. Fungicide resistance is now a widespread problem in global agriculture. Fungicide resistance problems in the field have been documented for nearly 150 diseases (crop-pathogen combinations) and within about half of the known fungicide groups. Resistance is now found in more than 500 insect and mite pests, in over 100 weeds, and in about 150 plant pathogens (WRI 1994).

1.3.3.8 Effect on Beneficial Organisms

Heavy treatment of soil with pesticides can cause populations of beneficial soil microorganisms to decline. According to the soil scientist Dr. Elaine Ingham, "If we lose both bacteria and fungi, then the soil degrades. Overuse of chemical fertilizers and pesticides have effects on the soil organisms that are similar to human overuse of antibiotics. Indiscriminate use of chemicals might work for a few years, but after awhile, there aren't enough beneficial soil organisms to hold onto the nutrients" (Savonen 1997). For example, plants depend on a variety of soil microorganisms to transform atmospheric nitrogen into nitrates, which plants can use. Glyphosate reduces the growth and activity of free-living nitrogen-fixing bacteria in soil (Santos and Flores 1995). 2,4-D reduces nitrogen fixation by the bacteria that live on the roots of bean plants (Arias and Fabra 1993; Fabra et al. 1997). 2,4-D reduces the growth and activity of nitrogen-fixing blue-green algae (Singh and Singh 1989). Mycorrhizal fungi grow with the roots of many plants and aid in nutrient uptake. These fungi can also be damaged by herbicides in the soil. One study found that oryzalin and trifluralin both inhibited the growth of certain species of mycorrhizal fungi (Kelley and South 1978). Roundup has been shown to be toxic to mycorrhizal fungi in laboratory studies (Chakravarty and Sidhu 1987; Estok et al. 1989).

1.4 Biocontrol

Biological control, the use of specific microorganisms that interfere with plant pathogens and pests, is a nature-friendly, ecological approach to overcome the problems caused by standard chemical methods of plant protection (Harman et al. 2004). Most biological control agents, including predators and parasitoids, at work in our agricultural environment are naturally occurring ones, which provide excellent regulation of many insect pests and diseases with little or no assistance from humans. The existence of naturally occurring biological control agents is one reason that many plant-feeding insects and plant pathogens do not ordinarily become economic pests. The importance of such agents often becomes quite apparent when pesticides applied to control one pest cause an outbreak of other pests because of the chemical destruction of important natural enemies. There is great potential for increasing the benefits derived from naturally occurring biological controls, through the elimination or reduction in the use of pesticides toxic to natural enemies. Biological control offers a novel approach when applied either alone or in combination with other management practices (Papavizas 1985; Mukhopadhyay 1987). Several advantages of using biological control agents have been reported by different workers.

1. Eco-friendly (Bohra et al. 2006 and Gaur et al. 2005)
2. Effective in managing diseases caused by soilborne plant pathogens which cannot be easily controlled by chemicals (Harman 2000; Howell 2003 and Manoranjitham et al. 2000)
3. Ease of multiplying antagonists with less cost of production (Gaur et al. 2005 and Das et al. 2006)
4. Growth-promoting effect (Mukhopadhyay 1987; Jang et al. 1993; Das et al. 2006; Pan and Bhagat 2007)
5. Long-lasting effective disease management (Harman 2000; Howell 2003)

1.4.1 Effect of Agrochemicals on Biocontrol Agents

Understanding the effect of agrochemicals on the beneficial activities of microorganisms is important to assess the hazards associated with agrochemicals used in agriculture. Crop productivity and economic returns will be maximized with the use of products controlling pests but preserving beneficial organisms. *Trichoderma* spp. are common soil inhabitants and have been extensively studied as biocontrol agents in the management of soilborne plant pathogens (Elad et al. 1980; Papavizas 1985; Upadhyay and Mukhopadhyay 1986; Adams 1990; Muthamilan and Jeyarajan 1992; Manoranjitham et al. 2000; Whipps 2001; Mohanan 2007 and RakeshKumar and Indra Hooda 2007). *Trichoderma* spp. offer biological control against soilborne plant pathogens by utilizing mechanisms such as antibiosis (Sivan et al. 1984; Shanmugam and Varma 1999; Hazarika et al. 2000; and Das et al. 2006), competition (Khan and Sinha 2007), and mycoparasitism (Upadhyay and Mukhopadhyay 1986 and Pandey et al. 2005). *Trichoderma* spp. are used either in seed treatment or soil application. In both the cases, the antagonist will be continuously exposed to different agrochemicals such as fungicides, insecticides, fertilizers, and herbicides applied to the field either in soil or as foliar sprays. Agrochemicals sprayed aerially reach the soil (by means of air currents or are washed off the plant surface due to rain) and influence the efficacy of native or applied biocontrol agents. Effects of agrochemicals on antagonists were reported by several workers (Pandey and Upadhyay 1998; Vijayaraghavan and Abraham 2004; Rai Ajay Kumar et al. 2005; Surajit et al. 2006 and Lal and Maharshi 2007). However, a great deal of the work done was based on arbitrary concentrations that were less than the field concentrations. Further, as variation exists between strains and species within a genus isolated from different climatic conditions, a thorough knowledge on sensitivity of applied biocontrol agent is necessary so as to exploit the full potential of the biocontrol agent. In this chapter an attempt has been made to summarize the effect of agrochemicals on biocontrol agent based on our experimentations (Ranganathswamy 2009) as well as reports of previous workers.

1.4.1.1 Effect of Fungicides

Plant pathogens cause significant loss (14.1%) in agricultural crops (Agrios 2005). Use of fungicides and bactericides is an age-old practice and has been proven effective in controlling the plant pathogens and thereby the loss caused by them. Fungicides are applied in various methods such as seed treatment and soil and foliar application. Fungicides applied by either of any methods will ultimately reach soil and interact with the biocontrol agents. These fungicides may drastically affect the growth, reproduction, and survivability of biocontrol agents in a particular cropping system and thereby on their biocontrol efficacy. Several workers (Karpagavalli 1997; Upadhyay et al. 2000; Akbari and Parakhia 2001; Vijayaraghavan and Abraham 2004; Tiwari and Singh 2004; Bhattiprolu 2007 and Madhavi et al. 2008) reported that fungicides were toxic to the biocontrol agents. We have (Ranganathswamy et al. 2013) evaluated the effect of the most commonly used 18 fungicides on assimilative (radial growth) and spore phases of the antagonist (*T. harzianum* and *T. virens*). Comparison was made between the two isolates selected, between the growth phases (assimilative and spore phase) and between different agrochemicals. Further, EC_{50} and EC_{90} values of test chemicals were calculated. Based on these observations, the chemicals were grouped into dangerous, cautious, and safe to antagonist.

1.4.1.1.1 Copper Group

Copper fungicides still form a major group of fungicides especially in the management of diseases caused by soilborne lower fungi such as *Pythium* and *Phytophthora*. Being soilborne, their management essentially deals with treating the soil through pre- or post-planting applications such as drenching. At this juncture, there is every reason to believe that the soil-inhabiting *Trichoderma* (employed through inundative or augmentative measures) is under continuous direct exposure to the applied field concentration. Several workers indicated that in general copper fungicides especially Bordeaux mixture and copper oxychloride have their impact on the growth of biocontrol agents (*Trichoderma* spp. and *Pseudomonas fluorescens*). Several reports published based on the in vitro assay indicated that copper fungicides are in general less toxic to *Trichoderma* spp. (Karpagavalli 1997; Upadhyay et al. 2000; Akbari and Parakhia 2001; Vijayaraghavan and Abraham 2004; Tiwari and Singh 2004; Bhattiprolu 2007 and Madhavi et al. 2008). However, variation existed in the type of chemical under study, i.e., Bordeaux mixture, copper oxychloride, and cuprous hydroxide.

Srinivasulu et al. (2002) reported 100% inhibition of *T. harzianum*, *T. viride*, and *T. hamatum* in Bordeaux mixture amended medium at 1% concentration. Similar report of highly incompatible nature of Bordeaux mixture with *Trichoderma* spp. was reported by Vijayaraghavan and Abraham (2004). Ranganathswamy et al. (2011) reported copper oxychloride had higher inhibitory effect on *Trichoderma* spp. compared to Bordeaux mixture on assimilative phase (mycelium), while it was reverse in the case of spore phase where Bordeaux mixture showed cent percent inhibition on spore germination.

1.4.1.1.2 Dithiocarbamates

Among the dithiocarbamates, mancozeb and thiram are the most commonly used in plant disease management – mancozeb mainly as foliar spray and thiram as seed treatment. Thiram will have a direct effect on the antagonist on the efficacy of *Trichoderma* spp. especially when applied as seed treatment in integration. Desai et al. (2002) reported that the radial growth inhibition of *T. harzianum* in mancozeb amended medium was 5.7% and 4.9% at 500ppm and 1000 ppm concentrations, respectively, which was found compatible. Moderate compatibility of mancozeb with *T. harzianum* was also reported by Vijayaraghavan and Abraham (2004), Tiwari and Singh (2004), Kumar et al. (2005), and Surajit et al. (2006). Madhavi et al. (2008) reported isolate variation in sensitivity to mancozeb when two mutants ThM_1 and TvM_1 were assessed. ThM_1 was compatible up to 0.125%, while TvM_1 was not compatible at the same concentrations.

Several reports suggested the incompatibility of thiram with *Trichoderma*. Complete inhibition of *Trichoderma* growth on thiram amended medium was reported by Sharma and Mishra (1995), Karpagavalli (1997), Pandey and Upadhyay (1998), Desai and Kulkarni (2004), and Upadhyay et al. (2004). In our study also thiram was highly toxic to *Trichoderma* spp., while mancozeb and wettable sulfur were least toxic (Ranganathswamy et al. 2011).

1.4.1.1.3 Heterocyclic Nitrogenous Compounds

Captan, a heterocyclic nitrogenous compound and another most commonly used seed dressing fungicide, was found to be less toxic with little inhibition of *Trichoderma* growth (Sharma and Mishra 1995; Gupta 2004; Desai and Kulkarni 2004 and Pandey et al. 2006). Contradictory to the above reports, a more toxic nature of captan to *Trichoderma* was reported by Vijayaraghavan and Abraham (2004), Tiwari and Singh (2004), Upadhyay et al. (2004), and Surajit et al. (2006).

1.4.1.1.4 Systemic Fungicides

Most of the systemic fungicides such as benzimidazoles, triazoles, acylalanines, strobilurins, etc. were reported toxic to biocontrol agents. Metalaxyl, an acylalanine group member highly effective against lower fungi, was reported to be less toxic to *Trichoderma* (Sharma and Mishra 1995). Indu and Mukhopadhyay (1990) reported zero inhibition at 50 ppm and 30% inhibition at 1000 ppm concentration in metalaxyl-insensitive isolate of *Trichoderma* (IMI No. 238493).

Carbendazim and benomyl, two most commonly used systemic fungicides belonging to benzimidazoles, were found to be extremely toxic to *Trichoderma* isolates in all the reports published so far. Carboxin, an oxathiin derivative highly effective against smuts and *Sclerotium rolfsii*, was reported to be toxic to *Trichoderma* spp. (Desai et al. 2002 and Tiwari and Singh 2004). Mondal et al. (1995) reported higher inhibitory effect of carboxin on *T. harzianum*. Oxycarboxin, another oxathiin derivative effective against rusts, was reported to have no effect on the growth of *T. harzianum* (Tiwari and Singh 2004).

All the triazole compounds such as tebuconazole, hexaconazole, propiconazole, tricyclazole, and epoxiconazole were reported to be highly toxic to biocontrol agents (Tiwari and Singh 2004; Pandey et al. 2006). Tridemorph, member of morpholine group, showed complete inhibition (100%) of assimilative and spore phases of both the test *Trichoderma* isolates at field concentration (Srinivasulu et al. 2002).

Ranganathswamy et al. (2011) evaluated ten systemic fungicides, viz., carbendazim, benomyl, metalaxyl, carboxin, propiconazole, hexaconazole, tricyclazole, tridemorph, fosetyl-Al, and azoxystrobin, which are commonly used to control plant pathogens in different agricultural and horticultural crops. It was reported that all the ten fungicides were toxic to biocontrol agents (*Trichoderma* spp.) affecting the vegetative (mycelium) as well as reproductive growth (spore) at their filed concentration.

1.4.1.2 Effect of Insecticides

1.4.1.2.1 Old-Generation Insecticides

Several kinds of insecticides were commonly used for the management of insect pests in agriculture crops. These insecticides were reported to affect the growth and survivability of biocontrol agents.

Endosulfan, a synthetic hydrocarbon belonging to cyclodiene compounds, constitutes a major insecticide used for the management of foliage-eating insects in vegetables. Endosulfan being a contact insecticide has retarding effect on the growth and survivability of concerned organism when comes in direct contact with particular organism. Several reports published based on in vitro assay indicated toxicity of endosulfan on *Trichoderma* spp. (Sushir and Pandey 2001; Vijayaraghavan and Abraham 2004; Tiwari et al. 2004 and Lal and Maharshi 2007). However, variation existed with respect to the concentration evaluated and isolate studied. Sushir and Pandey (2001) reported 55.5% inhibition of radial growth of *T. harzianum* at 50ppm concentration, while Vijayaraghavan and Abraham (2004) observed more than 83% inhibition at 400 ppm concentration. Tiwari et al. (2004) reported that endosulfan, in dust as well as EC formulation, was found toxic on the radial growth of *T. harzianum*.

Organophosphate (OP) compounds constitute a major group of chemicals used for the management of sucking pests. Several workers reported toxicity of OP compounds to *Trichoderma* and other biocontrol agents; however, variation was reported with respect to the concentration, isolates, and formulation of compound. Vijayaraghavan and Abraham (2004) reported toxicity of chlorpyriphos (0.02%), quinalphos (0.04%), and dimethoate (0.04%) to *Trichoderma* spp. Cent percent inhibition of radial growth of *Trichoderma* spp. at all the concentrations (500, 1000, 1500, and 2000 ppm) of chlorpyriphos tested was reported by Desai and Kulkarni (2005). However, non-inhibitory effect of chlorpyriphos on *T. harzianum* even up to 2000 ppm was reported by Sharma and Mishra (1995). Tiwari et al. (2004) reported that dust formulations of chlorpyriphos and quinalphos were nontoxic, while EC formulation showed toxic effect on *T. harzianum*. Ranganathswamy et al. (2011) reported the toxicity effect of chlorpyriphos, quinalphos, and dimethoate on *Trichoderma* spp. It may be noted here that all the OP compounds are having long residual effect in soil indicating their long-term interaction with test *Trichoderma* isolates, thereby reducing growth and survivability of test antagonists in such soil environment.

Tiwari et al. (2004) reported inhibitory effect of monocrotophos and triazophos on *T. harzianum* at 500–2000 ppm concentration. The least inhibitory effect of acephate on *Trichoderma* spp. was reported by Desai and Kulkarni (2005).

Several workers reported non-inhibitory effect of phorate and carbamate insecticides, on *Trichoderma* (Sharma and Mishra 1995; Gupta 2004; Vijayaraghavan and Abraham 2004 and Tiwari et al. 2004). However, Jayaraj and Ramabadran (1996) reported inhibitory effect of phorate, who observed almost nil growth (5 mm) of *T. harzianum* at 500 ppm concentration of phorate. Sharma and Mishra (1995) reported least toxicity of carbofuran and aldicarb on *T. harzianum*. Tiwari et al. (2004) reported inhibitory effect of carbofuran and fenobucarb on the radial growth of *Trichoderma harzianum* at all the concentrations tested (500–2000 ppm).

1.4.1.2.2 New-Generation Insecticides

New-generation insecticides, viz., indoxacarb, imidacloprid, thiamethoxam, fipronil, emamectin benzoate, and spinosad, were reported nontoxic or least toxic to biocontrol agents. Tiwari et al. (2004) reported non-inhibitory effect of indoxacarb on *T. harzianum* at all the concentrations (500, 1000, 1500, and 2000 ppm). Prasanna et al. (2002) reported the compatibility of *Trichoderma harzianum* to thiamethoxam even at 1% concentration. Lal and Maharshi (2007) reported the highly sensitive nature of *Trichoderma* spp. to imidacloprid at 500 ppm. Ranganathswamy et al. (2012a, b, c) reported the least toxicity of thiamethoxam, emamectin benzoate, and fipronil and nontoxicity of indoxacarb and spinosad. Tiwari et al. (2004) reported inhibitory effect of cartap hydrochloride on *T. harzianum* at all the concentrations tested (500–2000 ppm). Most of the synthetic pyrethroids were reported to be inhibitory on *Trichoderma*. Incompatibility of *Trichoderma* spp. to cypermethrin at 0.04% was reported by Vijayaraghavan and Abraham (2004). The toxic nature of alfamethrin, cypermethrin, and fenvalerate on *T. harzianum* at 500–2000 ppm was reported by Tiwari et al. (2004).

1.4.1.3 Effect of Fertilizers

Fertilizer is an important basic input in modern agriculture. Use of fertilizer gained momentum in the early 1960s when the use of dwarf, short-duration, and high-yielding varieties of wheat and paddy was brought into cultivation. Now, the chemical fertilizers become an integral part of agriculture in most of the developing countries. As the fertilizers were directly applied into the soil as a basal application or top dressing, they will interact with the biocontrol agents and affect their growth, reproduction, and survivability.

Several workers like Sharma and Mishra (1995); Jayaraj and Ramabadran (1998); Monga (2001); Reshmy Vijayaraghavan and Koshy Abraham (2004); Sharma and Mathur (2008); Gade et al. (2009) and Ranganathswamy et al. (2012a, b, c) reported the toxic as well as non-toxic effect of chemical fertilizers on soil beneficial microflora.

Among amide forms of nitrogenous fertilizers, urea was the most commonly used for almost all the crops. Sharma and Mishra (1995) reported the stimulatory effect of urea on growth and sporulation of *T. harzianum*; however, Vijayaraghavan and Abraham (2004) reported the compatibility of *Trichoderma* spp. to urea up to 1%, while concentration exceeding 1% resulted in toxicity. The least growth and sporulation of *T. harzianum* in urea amended medium were reported by Jayaraj and Ramabadran (1998) and Gade et al. (2009).

Most of the ammoniacal forms of nitrogenous fertilizers were reported to have no inhibitory effect on growth of *Trichoderma*. Good growth and sporulation of *T. harzianum* in ammonium sulfate and ammonium nitrate while poor growth in calcium nitrate were reported by Jayaraj and Ramabadran (1998). Among nitrate forms, sodium nitrate was found to support good growth and sporulation of *T.harzianum* (Jayaraj and Ramabadran 1998; Gade et al. 2009).

Vijayaraghavan and Abraham (2004) reported nontoxicity of phosphate fertilizer on *Trichoderma* spp. at 2% and 2.5% and even at 3% concentrations.

Potassium fertilizers irrespective of forms were reported nontoxic and compatible to *Trichoderma*. Compatible nature of *Trichoderma* spp. to muriate of potash was reported by Sharma and Mishra (1995) and Vijayaraghavan and Abraham (2004). Monga (2001) reported that potassium nitrate was proved to be the best for the growth of all the *Trichoderma* spp. Ranganathswamy et al. (2012a, b, c) reported toxicity of zinc sulfate, DAP, and urea, while nontoxicity of muriate of potash, ammonium sulfate, single superphosphate, and potassium nitrate on *Trichoderma* spp. Sharma and Mathur (2008) reported compatibility of *T. harzianum* to magnesium sulfate. Sharma and Mishra (1995) reported toxic effect of zinc sulfate on *T. harzianum* at 2%.

1.4.1.4 Effect of Weedicides (Herbicides)

Most of the herbicides, viz., pendimethalin, glyphosate, alachlor, butachlor, paraquat, atrazine, trifluralin, fluchloralin, oxadizin, 2,4-D, etc., were reported toxic on biocontrol agents. Acetamide compounds, viz., alachlor and butachlor, were reported toxic on *Trichoderma* spp. by Desai and Kulkarni (2004) and Subhalakshmi et al. (2006). Madhavi et al. (2008) reported toxicity of alachlor and butachlor even on *Trichoderma* mutants. Nil or least toxicity of 2,4-D on *Trichoderma* was reported by Jayaraj and Radhakrishnan (2000). Nitroaniline compounds, viz., fluchloralin and pendimethalin, were found toxic even on mutant isolates of *Trichoderma* spp. (Madhavi et al. 2008). Non inhibitory effect of pendimethalin, fluchloralin and oxazdiazon on *Trichoderma* spp. was reported by Parakhia and Akbari (2001). Desai and Kulkarni (2004) reported that glyphosate inhibited growth of *T. harzianum* at all the concentrations tested (500, 1000, and 2000 ppm). Triazine compound atrazine was moderately compatible with *T. harzianum* up to 2000 ppm concentration (Desai and Kulkarni 2004). Rai Ajay Kumar et al. (2005) reported high compatible nature of *Trichoderma harzianum* to isoproturon at all the concentrations tested (100–10000 ppm). Ranganathswamy et al. (2012a, b, c) evaluated the effect of four herbicides, viz., alachlor, glyphosate, pendimethalin, and 2,4-D, on *Trichoderma* spp. and reported toxicity of all the four herbicides on biocontrol agents.

Based on our experimental result (Ranganathswamy et al. 2013), all the agrochemicals were grouped as dangerous, cautious, and safe to antagonist. As the formulations of biocontrol agents are mainly based on spore preparations (with more shelf life), inhibition of spores by test agrochemicals was given higher weightage compared to mycelial inhibition while grouping into ten different categories of compatibility.

I Category: 100% inhibition of spore germination and 100 % inhibition of radial growth

II Category: 100% inhibition of spore germination and >50% inhibition of radial growth

III Category: 100% inhibition of spore germination and <50% inhibition of radial growth

IV Category: >50% inhibition of spore germination and 100% inhibition of radial growth

V Category: <50% inhibition of spore germination and 100% inhibition of radial growth

All the above five categories were placed in Group I, i.e., "dangerous" chemicals to antagonists.

VI Category: >50% inhibition of spore germination and >50% inhibition of radial growth

VII Category: >50% inhibition of spore germination and <50% inhibition of radial growth

VIII Category: <50% inhibition of spore germination and >50% inhibition of radial growth

All the above three categories were placed in Group II, i.e., "cautious" chemicals to antagonists. Further, chemicals with EC_{50} values less than field concentration but falling in cautious group are also placed in Group I, i.e., dangerous group.

IX Category: <50% inhibition of spore germination and <50% inhibition of radial growth.

X Category: No inhibition of spore germination and no inhibition of radial growth

The above two categories were placed in "safe" group.

Further, all the chemicals with EC_{50} values more than field concentration were also considered "safe."

Among 18 fungicides evaluated, 15 fungicides, viz., Bordeaux mixture, mancozeb, thiram, captan, carbendazim, benomyl, carboxin, metalaxyl, propiconazole, hexaconazole, tricyclazole, tridemorph, chlorothalonil, fosetyl-Al, and azoxystrobin, were found toxic and grouped as dangerous, and the remaining three fungicides, viz., copper oxychloride, wettable sulfur, and dinocap, were placed in "cautious group" (Table 1.1). None of the test fungicide was found safe to the test *Trichoderma* isolates.

Table 1.1 Categorization of agrochemicals into various groups based on percent inhibition of radial growth and spore germination

Sl. no.	Percent inhibition: Spore germination	Percent inhibition: Radial growth	Category	Group	Agrochemicals
1.	100	100	I	*Dangerous*	*Fungicides*: Bordeaux mixture, mancozeb, thiram, captan, carbendazim, benomyl, carboxin, metalaxyl, propiconazole, hexaconazole, tricyclazole, tridemorph, fosetyl-Al, chlorothalonil, azoxystrobin
2.	100	>50	II		***Insecticides***: chlorpyriphos, quinalphos, dimethoate
3.	100	<50	III		*Fertilizers*: zinc sulfate
4.	>50	100	IV		*Herbicides*: pendimethalin, glyphosate, alachlor
5.	<50	100	V		
6.	>50	>50	VI	*Cautious*	*Fungicides*: copper oxychloride, dinocap, wettable sulfur
7.	>50	<50	VII		*Insecticides*: endosulfan, carbofuran, thiamethoxam, emamectin benzoate, fipronil, spinosad
8.	<50	>50	VIII		*Fertilizers*: urea, DAP
					Herbicides: 2,4-D sodium salt
9.	< 50	< 50	IX	*Safe*	*Insecticides*: imidacloprid, indoxacarb
10.	0	0	X		*Fertilizers*: MOP, SSP, ammonium sulfate, potassium nitrate

Two insecticides, chlorpyriphos and quinalphos, were placed in dangerous group. Though dimethoate showed differential sensitivity on test *Trichoderma* isolates, as the EC_{50} value was less than field concentration (indicating toxic nature of the dimethoate), it was placed in dangerous group. Endosulfan, carbofuran, thiamethoxam, emamectin benzoate, fipronil, and spinosad were placed in cautious group. Only two insecticides, viz., imidacloprid and indoxacarb, were placed in safe group. Fipronil, spinosad, and carbofuran showed differential sensitivity to test *Trichoderma* isolates. Further, as the EC_{50} values of these chemicals were more than field concentrations, they were placed in safe group (Table 1.1).

Among seven fertilizers evaluated, zinc sulfate was placed in dangerous group, urea and DAP in cautious group, and muriate of potash, single superphosphate, ammonium sulfate, and potassium nitrate in safe group (Table 1.1).

Three herbicides, viz., alachlor, glyphosate, and pendimethalin, out of four evaluated herbicides were placed in dangerous group, while the fourth herbicide 2,4-D showed differential sensitivity toward test *Trichoderma* isolates with least inhibition and thereby was placed in cautious group (Table 1.1).

1.5 Measures to Overcome Negative Effect of Agrochemicals

Ideally, agricultural systems should be designed in a way that pests, diseases, and weeds do not build up to a level that they cause significant damage to the crop. Suitable agronomic practices, the use of resistant varieties, and integrated pest management are key preventive measures. Biocontrol and the use of natural substances can complement these efforts. The safe application of minimal toxic synthetic pesticides should be used as a last resort.

1.5.1 List of Strategies

- Avoid the prophylactic sprays of pesticides.
- The use of optimum dose, at the right time and on the right pest, is the key measure to reduce the indiscriminate use of pesticides and thereby development of resistant strains.
- Appropriate plant nutrition and soil fertility management based on organic matter form the basis for healthy crops that are less susceptible to pests, diseases, and weeds.
- Following crop rotation that prevents the carryover of pest, pathogen, and weed populations.
- Appropriate timing of sowing or planting and of intercultural operations reduces pest pressure.
- Precision farming like spraying of hotspots and weeding with optical detectors.
- Crops and crop varieties differ in their susceptibility to pests and diseases and in their ability to compete with weeds. The use of resistant varieties together with rotations of non-susceptible crops can substantially limit pest buildup within a field, thereby limiting the use of pesticides.
- Use of genetically modified crops. For example, Bt cotton against bollworms in cotton substantially reduced the pest incidence and thereby pesticides.
- Use of less hazardous pesticides: Phasing out the use of highly hazardous pesticides and replacing them with less hazardous ones is therefore the most obvious way to reduce the negative side effects of pesticides.
- Various plant extracts and other natural materials are used that repel pests, reduce their feeding or reproductive activities, reduce proliferation of diseases, and act as biopesticides.
- Following integrated approach for pest management.
- Policies to reduce pesticide use and risks: International codes, treaties, conventions, commissions, and advisory bodies play an important role in plant protection and pesticide management. Each country needs to follow the strict regulations to reduce the risk of pesticide effect on human health and environment.

References

Adams PB (1990) The potential of mycoparasites for biological control of plant diseases. Annu Rev Phytopathol 28:59–72

Agrios GN (2005) Plant pathology. Elsevier Academic Press, Burlington

Akbari LF, Parakhia AM (2001) Effect of fungicides on fungal bioagents. J Mycol Plant Path 31(1):101–102

All India Network Project on Pesticide Residues (2013) Annual report. AINP ICAR, India, pp 1–62

Arias RN, Fabra PA (1993) Effects of 2,4-dichlorophenoxyacetic acid on *Rhizobium* sp. growth and characterization of its transport. Toxicol Lett 68:267–273

Barcelo D, Hennion MC (1997) Trace determination of pesticides and their degradation products in water. Elsevier, Amsterdam

Bhattiprolu SL (2007) Compatibility of *Trichoderma viride* with fungicides. Ind J Plant Prot 35(2):357–358

Bohra B, Vyas BN, Mistry KB (2006) Biocontrol agents ansd neem formulations for management of damping-off in brinjal and chilli. Ind Phytopathol 59:223–226

Bortleson G, Davis D (1987–1995) U.S. geological survey & Washington State Department of Ecology. Pesticides in selected small streams in the Puget Sound Basin, pp 1–4

Chakravarty P, Sidhu SS (1987) Effects of glyphosate, hexazinone and triclopyr on *in vitro* growth of five species of ectomycorrhizal fungi. Eur J For Pathol 17:204–210

Cramer HH (1967) Plant protection and crop production (trans: from German by Edwards JH). Pflanzenschutz-Nachr. 20. Farbenfabriken Bayer AG, Leverkusen

Das BC, Das BK, Dutta P, Sarmah DK (2006) Bioformulation of *Trichoderma harzianum* Rifai for management of soybean stem rot caused by *Rhizoctonia solani* Kuhn. J Biol Control 20(1):57–64

Desai SA, Kulkarni S (2004) Effect of fungicides, insecticides and weedicides on the growth and sporulation of native *Trichoderma harzianum* Rifai. K J Agric Sci 17(1):57–62

Desai SA, Kulkarni S (2005) A note on fungistatic effect of chlorpyriphos on the growth of *Trichoderma viride* Pers. K J Agric Sci 18(1):157–157

Desai SA, Thammaiah N, Kulkarni S (2002) Fungicidal tolerance by *Trichoderma harzianum* Rifai. K J Agric Sci 15(2):397–398

Elad Y, Chet I, Katan I (1980) *Trichoderma harzianum* a biocontrol agent effective against *Sclerotium rolfsii* and *Rhizoctonia solani*. Phytopathology 70:119–121

Estok D, Freedman B, Boyle D (1989) Effects of the herbicides 2,4-D, glyphosate, hexazinone and triclopyr on the growth of three species of ectomycorrhizal fungi. Bull Environ Contam Toxicol 42:835–839

Fabra A, Duffard R, Evangelista DDA (1997) Toxicity of 2,4-dichlorophenoxyacetic acid in pure culture. Bull Environ Contam Toxicol 59:645–652

Federation of Indian Chambers of Commerce and Industry (FICCI) and TATA Strategic Management Group (2016) A report on next generation Indian agriculture: role of crop protection solutions, pp 1–56

Forget G (1993) Balancing the need for pesticides with the risk to human health. In: Forget G, Goodman T, de Villiers A (eds) Impact of pesticide use on health in developing countries. IDRC, Ottawa, p 2

Gade RM, Sarika VA, Wardha S (2009) Effect of soil types and nutritional factors on growth and sporulation of *Trichoderma* spp. Ann Plant Prot Sci 17(1):111–113

Gaur RB, Sharma RN, Sharma PR, Gautham VS (2005) Efficacy of *Trichoderma* for *Rhizoctonia* root rot control in chickpea. J Myco Plant Pathol 35(1):144–150

Glotfelty DE, Schomburg CJ (1989) Volatilization of pesticides from soil in reactions and movements of organic chemicals in soil. In: Sawhney BL, Brown K (eds) . Soil Science Society of America Special Publications, Madison

Gupta V (2004) Compatibility of biocontrol agent *Trichoderma harzianum* with pesticides. J Myco Plant Pathol 34(2):504–505

Gupta S, Katre RK, Chhonkar PK, Gupta MK, Adil ML (2000) Significant research and development with reference to biofertilisers in Chhattisgarh. Fertilizer News 45(11):35–40
Harman GE (2000) Myths and Dogmas of biocontrol – changes in perception derived from research on *Trichoderma harzianum* T_{22}. Plant Dis 84:377–393
Harman GE, Howell CR, Viterbo A, Chet I, Lorito M (2004) *Trichoderma* species-opportunistic, avirulent plant symbionts. Nat Rev Microbiol 2:43–56
Hazarika DK, Sharma R, Paramanick T, Hazarika K, Phookan AK (2000) Biological management of tomato damping-off caused by *Pythium aphanidermatum*. Indian J Plant Pathol 18:36–39
Howell CR (2003) Mechanism employed by *Trichoderma* species in the biological control of plant diseases. The history and evolution of current concepts. Plant Dis 87:4–10
Igbedioh SO (1991) Effects of agricultural pesticides on humans, animals and higher plants in developing countries. Arch Environ Health 46:218
Indian Labour Statistics (1997) Employment information. Labour Bureau, Ministry of Labour, Chandigarh
Indu S, Mukhopadhyay AN (1990) Integration of metalaxyl with *Trichoderma harzianum* for the control of *Pythium* damping-off in sugarbeet. Ind Phytopathol 43(4):535–541
Jang SS, Han JK, Park CS, Kim HK (1993) Plant growth enhancement induced by strain of biocontrol agents *Trichoderma* Spp. and *Gliocladium* Spp. K J Plant Pathol 9:149–155
Jayaraj J, Ramabadran R (1996) Effect of certain soil-insecticides on in vitro growth, sporulation and cellulase production by *Trichoderma harzianum* Rifai. J Biol Control 10:111–115
Jayaraj J, Radhakrishnan NV (2000) Survival of biocontrol agent *Trichoderma harzianum* in soil as influenced by certain herbicides. Ind J Weed Sci 32(1&2):116–117
Jayaraj J, Ramabadran R (1998) Effect of nitrogenous fertilizers on the survival and competitive saprophytic ability of *Trichoderma harzianum* in soil. J Mycol Plant Pathol 28(1):21–23
Jeyaratnam J (1981) Health problems of pesticide usage in the third world. *BMJ* 42:505
Karpagavalli S (1997) Effect of different fungicides on the growth of *Trichoderma*. Ind J Plant Prot 25(1):82–83
Karunakaran (1958) The Kerala food poisoning. J Indian Med Assoc 31:204
Kelley WD, South DB (1978) *In vitro* effects of selected herbicides on growth and mycorrhizal fungi. Weed Science Society of America meeting. Auburn University, Auburn, Alabama, p 38
Khan AA, Sinha AP (2007) Screening of *Trichoderma* spp. against *Rhizoctonia solani* the causal agent of rice sheath blight. Ind Phytopathol 60(4):450–456
Kole RK, Banerjee H, Bhattacharyya A (2001) Monitoring of market fish samples for Endosulfan and Hexachlorocyclohexane residues in and around Calcutta. Bull Envirron Contam Toxicol 67:554–559
Lal B, Maharshi RP (2007) Compatibility of biocontrol agents *Trichoderma* spp. with pesticides. J Mycol Plant Pathol 37(2):369–370
Leo Daniel AE, Praveen Kumar G, Desai S, Mir Hassan ASK (2011) *In vitro* characterization of *Trichoderma viride* for abiotic stress tolerance and field evaluation against root rot disease in *Vigna mungo* L. J Biofertil Biopestici 2:111
Madhavi M, Pramod Chandra Kumar C, Reddy DR, Singh TVK (2008) Compatibility of mutant isolates of *Trichoderma* spp. with agrochemicals. J Biol Cont 22(1):51–55
Majewski M, Capel P (1995) Pesticides in the atmosphere: distribution, trends, and governing factors. Ann Arbor Press Inc, Chelsea, p 118
Manoranjitham SK, Prakasam V, Rajappan K, Amutha G (2000) Effect of two antagonists on damping-off disease of tomato. Ind Phytopathol 53(4):441–443
Mathur SC (1999) Future of Indian pesticides industry in next millennium. Pestic Inf 24(4):9–23
Melander AL (1914) Can insects become resistant to sprays? J Econ Entomol 7:167
Mohanan C (2007) Biological control of seeding diseases in forest nurseries in Kerala. J Biol Cont 21(2):189–195
Mondal G, Srivastava KD, Aggarwal R (1995) Antagonistic effect of *Trichoderma* spp. on *Ustilago segetum* Var. *tritici* and their compatibility with fungicides and biocides. Ind Phytopathol 48(4):466–470

Monga D (2001) Effect of carbon and nitrogen sources on spore germination, biomass production and antifungal metabolites by species of *Trichoderma* and *Gliocladium*. Ind Phytopathol 54(4):435–437

Mukherjee PK, Raghu K (1997) Effect of temperature on antagonistic and biocontrol potential of shape *Trichoderma* sp. on *Sclerotium rolfsii*. Mycopathologia 139(3):151–155

Mukhopadhyay AN (1987) Biological control of soil borne plant pathogens by *Trichoderma* spp. Ind J Mycol Plant Pathol 17:1–10

Muthamilan M, Jeyarajan R (1992) Effect of antagonistic fungi on *Sclerotium rolfsii* causing root rot of groundnut. J Biol Cont 6(2):88–92

Pan S, Bhagat S (2007) Antagonistic potential of *Trichoderma* and *Gliocladium spp.* from West Bengal. J Mycol Plant Pathol 37(2):235–239

Pandey KK, Upadhyay JP (1998) Sensitivity of different fungicides to *Fusarium udum, Trichoderma harzianum* and *Trichoderma viride* for integrated approach of disease management. Veg Sci 25(1):89–92

Pandey KK, Pandey PK, Upadhyay JP (2005) Mycoparasitism of *Trichoderma* spp. on *Fusarium and Rhizoctonia*. J Mycol Plant Pathol 35(1):174–176

Pandey KK, Pandey PK, Mishra KK (2006) Bio-efficacy of fungicides against different fungal bioagents for tolerance level and fungistatic behaviour. Ind Phytopathol 59(1):68–71

Papavizas GC (1985) *Trichoderma* and *Gliocladium*: Biology, ecology and potential for biocontrol. Annu Rev Phytopathol 23:23–54

Parakhia AM, Akbari LF (2001) Effect of weedicides on fungal bioagents. J Mycol Plant Pathol 31(1):106

Pradhan S (1960) Investigations on problems of insect resistance to insecticides carried out at the Indian Agricultural Research Institute, New Delhi; Proceedings of the Haffkine Institute Diamond Jubilee Symposia, pp 115–127

Prasanna AR, Nargund VB, Bheemanna M, Patil BV (2002) Compatibility of thiamethoxam with *Trichoderma harzianum*. J Biol Cont 16(2):149–152

Rai Ajay Kumar, Srivastava M, Narayanan U (2005) Assessment of compatibility of *Trichoderma harzianum* with agrochemicals. Ind J Plant Pathol 23(1&2):104–106

Rakesh Kumar, Indra Hooda (2007) Evaluation to antagonistic properties of *Trichoderma* spp. against *Pythium aphanidermatum* causing damping-off of tomato. J Mycol Plant Pathol 37(2):240–243

Ranganathswamy (2009) Sensitivity of *Trichoderma* isolates to selected agrochemicals *in vitro*. M.Sc. Thesis, ANGRAU, Hyderabad, p 260

Ranganathswamy M, Patibanda AK, Krishnamurthy KVM, Prasuna Rani P (2011) Sensitivity of *Trichoderma* isolates to selected insecticides *in vitro*. Andhra Agric J 58(4):490–495

Ranganathswamy M, Patibanda AK, Krishnamurthy KV, Prasuna Rani P (2012a) *In vitro* sensitivity of *Trichoderma* isolates to selected herbicides. Andhra Agric J. 59(4):680–684

Ranganathswamy M, Patibanda AK, Krishnamurthy KVM, Prasuna Rani P (2012b) Differential sensitivity of *Trichoderma* to selected fertilizers. Andhra Agric J. 59(3):432–436

Ranganathswamy M, Patibanda AK, Krishnamurthy KVM, Prasuna Rani P (2012c) Sensitivity of *Trichoderma* isolates to selected fungicides *in vitro*. Andhra Agric J 59(1):77–82

Ranganathswamy M, Patibanda AK, Nageswarrao G (2013) Evaluation of toxicity of agrochemicals on *Trichoderma* isolates *in vitro*. J Mycopathol Res 51(2):289–293

Rao DLN (1996–2000) All India coordinated research project on biological nitrogen fixation, Indian Institute of Soil Science, Bhopal.

Reetha S, Bhuvaneswari G, Selvakumar G, Thamizhiniyan P, Pathmavathi M (2014) Effect of temperature and pH on growth of fungi *Trichoderma harzianum*. J Chem Bio Phy Sci 4(4):3287–3292

Report of the Special Committee of ICAR (1972) Wadhwani AM, Lall IJ (eds) Harmful effects of pesticides. Indian Council of Agricultural Research, New Delhi, p 44

Roberts TR (1998) Metabolic pathway of agrochemicals. Part I. In: Herbicides and plant growth regulators. The Royal Society of Chemistry, Cambridge

Roberts TR, Hutson DH (1999) Metabolic pathway of agrochemicals Part II. In: Insecticides and fungicides. The Royal Society of Chemistry Cambridge

Sankaranarayanan C, Rajeswari S (2001) Influence of moisture and pH on the efficiency of VA-mycorrhiza, *Glomus mosseae* (Nicol & Gerd.) Gerd. & Trappe against *Meloidogyne incognita* (Kofoid and White) Chitw. on blackgram (*Vigna mungo* L.) Hepper. J Biol Control 15:69–72

Santos A, Flores M (1995) Effects of glyphosate on nitrogen fixation of free-living heterotrophic bacteria. Lett Appl Microbiol 20:349–352

Savonen C (1997) Soil microorganisms object of new OSU service. Good Fruit Grower. http://www.goodfruit.com/archive/1995/6other.html

Shanmugam V, Varma AS (1999) Effect of native antagonists against *Pythium aphanidermatum* the causal organisms of rhizome rot of ginger. J Mycol Plant Pathol 29:375–379

Sharma S, Mathur K (2008) Inorganic amendments for enhancement of biocontrol efficiency of *Trichoderma harzianum* against *Phythium* damping-off of tomato. Ind J Plant Prot 36(1):89–93

Sharma SD, Mishra A (1995) Tolerance of *Trichoderma harzianum* to agrochemicals. Ind J Mycol Plant Pathol 25(1&2):129–130

Singh JB, Singh S (1989) Effect of 2,4-dichlorophenoxyacetic acid and maleic hydrazide on growth of blue green algae (cyanobacteria) *Anabaena doliolum* and *Anacystis nidulans. Sci Cult* 55:459–460

Sinha KK, Bhatnagar D (1998) Mycotoxins in agriculture and food safety. Dekker, New York

Sivan A, Elad Y, Chet T (1984) Biological control effects of new isolate of *Trichoderma harzianum* on *Pythium aphanidermatum*. Phytopathology 74:498–501

Srinivasulu B, Doriswamy S, Aruna K, Rao DVR, Rabindran R (2002) Efficacy of bio-control agent, chemicals on *Ganoderma* sp. the coconut basal stem rot pathogen. J Plantation Crops 30(3):57–59

Subhalakshmi T, Mishra NK, Pan S (2006) Effect of some herbicides on the growth and sporulation of two fungal biocontrol agents. J Mycopathol Res 44(2):275–278

Surajit K, Subhendu J, Shaonli B, Manisha R, Sitansu P (2006) Evaluation of tolerance in *Macrophomina phaseolina, Trichoderma harzianum, Trichoderma viride* and *Gliocladium virins* to fungicides. J Mycophathol Res 44(1):109–111

Sushir AM, Pandey RN (2001) Tolerance of *Trichoderma harzianum* Rifai to insecticides and weedicides. J Mycol Plant Pathol 31(1):106

Tiwari RKS, Singh A (2004) Efficacy of *Rhizoctonia solani* and *Sclerotium rolfsii* and their effect on *Trichoderma harzianum* and *Rhizobium leguminosarum*. J Mycol Plant Pathol 34(2):482–484

Tiwari RKS, Rajput ML, Singh A, Thakur BS (2004) Non target effect of insecticides on the mycelial growth of *Trichoderma harzianum*. Ind J Plant Prot 32(2):140–141

U.S. Geological Survey (1998) National Water-Quality Assessment. Pesticide National Synthesis Project. Pesti-cides in surface and ground water of the United States; Summary of results of the National Water Quality Assessment Program. http://water.wr.usgs.gov/pnsp/allsum/fig02.gif

Upadhyay JP, Mukhopadhyay AN (1986) Biological control of *Sclerotium rolfsii* by *Trichoderma harzianum* in sugar beet. Trop Pest Manag 32:215–220

Upadhyay JP, Roy S, Lal HC (2000) Growth and sporulation of *Gliocladium virens* as influenced by fungicides, cakes and plant byproducts. J Res 10(1):41–46

Upadhyay JP, Lal HC, Roy S (2004) Effect of fungicides, cakes and plant byproducts on *Trichoderma viride*. J Mycol Plant Pathol 34(2):313–315

Vijayaraghavan R, Abraham K (2004) Compatibility of biocontrol agents with pesticides and fertilizers used in black pepper gardens. J Mycol Plant Pathol 34(2):506–510

Webster JPG, Bowles RG, Williams NT (1999) Estimating the economic benefits of alternative pesticide usage scenarios: wheat production in the United Kingdom. Crop Prod 18:83

Whipps JM (2001) Microbial interactions and biocontrol in the rhizosphere. J Exp Bot 52:487–511

Widden P, Hsu D (1987) Competition between *Trichoderma* species: effects of temperature and litter type. Soil Biol Biochem 19(1):89–93

World Health Organization (1990) Public health impact of pesticides used in agriculture. World Health Organization, Geneva, p 88

World Resources Institute (1994) World resources: a guide to the global environment. Oxford University Press, New York

Azotobacter salinestris: A Novel Pesticide-Degrading and Prominent Biocontrol PGPR Bacteria

2

G. Chennappa, M. Y. Sreenivasa, and H. Nagaraja

Abstract

The species *Azotobacter* belongs to plant growth-promoting rhizobacteria (PGPR) group and are ubiquitous, aerobic, free-living and nitrogen (N_2)-fixing bacteria commonly living in soil, water and sediments. Being the major group of soilborne bacteria, *Azotobacter* plays different beneficial roles by producing different types of secondary metabolites in the soil such as vitamins, amino acids, plant growth hormones, antifungal substances, hydrogen cyanide and siderophores. These secondary metabolites have direct influence on growth of shoot, root and seed germination of many agriculture crops. Among different species, *Azotobacter salinestris* is considered as efficient in N_2 fixation (29.21 μg Nm/L/day), production of indole acetic acid (24.50 μg/mL), gibberellic acid (GA) (15.2 μg/25 mL) and phosphate-solubilizing activity (13.4 mm). Molecular and biochemical studies confirmed the identity of the isolates (*A. salinestris* KF470807). *A. salinestris* found tolerant to a highest NaCl concentration (6–8%), to maximum temperature (45 °C) and also to varied pH ranges (8–9). The isolate was tested for pesticide resistance and biodegradation studies and showed 100% biodegradation of pendimethalin and did not recorded pendimethalin residues. It was also studied for antifungal efficacy against phytopathogen *Fusarium* species and influence on growth parameters of cereals. *A. salinestris* helps to replace

G. Chennappa (✉)
Department of Studies in Microbiology, Manasagangothri, University of Mysore, Mysuru, Karnataka, India

Department of Processing and Food Engineering, College of Agricultural Engineering, University of Agricultural Sciences, Raichur, Raichur, Karnataka 584104, India
e-mail: chinnagurikar@gmail.com

M. Y. Sreenivasa · H. Nagaraja
Department of Studies in Microbiology, Manasagangothri, University of Mysore, Mysuru, Karnataka, India

D. G. Panpatte et al. (eds.), *Microorganisms for Green Revolution*, Microorganisms for Sustainability 7,
https://doi.org/10.1007/978-981-10-7146-1_2

chemical fertilizer and restore the soil fertility and crop productivity for the sustainable agriculture.

Keywords
Biodegradation · PGPR · Abiotic stress · Formulations · Biocontrol

2.1 Introduction

All agriculture crops need continuous and stringent management practices. All cultivating crops need up to 40% of total cost of cultivation (Kadam and Gangawane 2005). Demand of pesticides is more for the control of pests and diseases of major agriculture crops. Among pathogens, bacteria and fungi are responsible for more than 60% of infection leading to serious economic losses. In countries like India, for the control of pests and diseases, farmers are applying excess dosages of pesticides and chemical fertilizers. For the cultivation of agriculture crops, several pesticides are being widely used, viz. pendimethalin, phorate, glyphosate, simazine, malathion, endosulfan, carbofuran, monocrotophos, diazinon, fenthion, phosphamidon, methyl parathion, chlorpyrifos, carbendazim, mancozeb and azinphos methyl (Moneke et al. 2010), although wide-scale application of pesticides is an essential part for improving crop yield.

Indiscriminate and unscientific use of pesticides may decline the population of beneficial microorganisms present in the soil. Agriculturally used pesticides and chemical fertilizer reaching the soil ecosystem in significant quantities have direct effect on soil microbiological aspects, environmental pollution and health hazards (Martin et al. 2011).

An ideal pesticide should have the ability to destroy only target pest and should be able to undergo degradation to non-toxic substances as quickly as possible. However, pesticides cause everlasting changes in the soil microflora, adverse effect on soil fertility and crop productivity (Aleem et al. 2003; Sachin 2009; Reinhardt et al. 2008). Pesticide concentration contaminates soil biodiversity, soil microbe interaction, organic matter decomposition and biogeochemical cycles. At present, different types of formulated pesticides are used worldwide to control widespread of diseases and to minimize the economic losses. Hence, degradation of such pollutant plays an important role in agricultural practice to reduce the load of pesticides from the soil and to serve safe food to the world and clean environment for future generation.

The number of soilborne bacteria and fungi has the ability to break down pesticides into simpler non-toxic compounds. To overcome pesticide effects, several biodegradation works have been carried out in order to minimize the pesticide residues in food and feed. For biodegradation of toxic pesticides, several types of soilborne bacteria and fungi are widely used (Castillo et al. 2011; Kadam and Gangawane 2005; Moneke et al. 2010). Among the bacterial genera, *Azotobacter* species are one of the most dominant bacteria in rhizosphere soils. *Azotobacter* species can break down different types of pesticide compounds including phenols, substituted phenolics and hazardous compounds (Castillo et al. 2011; Moneke et al. 2010). Limited data is available on biodegradation of pesticides by *Azotobacter* species which are commonly used for cultivation of many crops. It is confirmed that the *Azotobacter* isolates, viz. *A. vinelandii, A. tropicalis, A. armeniacus, A. salinestris* and *A. chroococcum*, proved their efficiency of PGPR activity. In this chapter, all the PGPR activities, abiotic tolerance, application of *Azotobacter* as bio-inoculants for seed treatments and biodegradation of prominent herbicide (pendimethalin) have been discussed.

2.2 PGPR Activities

Azotobacter are known to produce different types of secondary metabolites such as vitamins, amino acids, plant growth hormones, antifungal compounds and siderophore (Myresiotis et al. 2012). These growth-promoting substances such as IAA, GA, nicotinic acid, etc. have direct influence on plant growth and development of several agricultural crops (Ahmad et al. 2005). *Azotobacter* produces siderophore, polyhydroxybutyrate (PHB) and HCN (hydrogen cyanide), and siderophores are iron chelating agents and provide iron to the plants. PHB is used in large-scale production of alginic acid. Antifungal compounds and HCN can inhibit root-colonizing pathogens in plant rhizosphere.

2.2.1 Vitamins

Vitamins are essential compounds for the physiological functions of cells which are produced by several groups of bacteria. *Azotobacter* species are known to produce vitamins under favourable conditions. *A. vinelandii* (ATCC 12837) and *A. chroococcum* (CECT 4435) strains produced B-group vitamins which consist of niacin, pantothenic acid, riboflavin and biotin. Vitamins are used to maintain metabolic processes of living beings, but the production of vitamins is controlled by several factors such as growth conditions, pH, incubation temperatures, and availability of nitrogen and carbon sources (Revillas et al. 2000). Riboflavin is a vitamin B2 required for a wide variety of cellular processes, and it plays a key role in metabolism of fats, ketone bodies, carbohydrates and proteins, respectively. Genetically engineered *Bacillus subtilis* and *Corynebacterium ammoniagenes* are used for mass production of riboflavin by which they overexpress genes of the enzymes involved in riboflavin biosynthesis (Almon 1958; Revillas et al. 2000).

2.2.2 Amino Acids

Azotobacter vinelandii and *A. chroococcum* are known to produce different types of amino acids such as aspartic acid, serine, glutamic acid, glycine, histidine, threonine, arginine, alanine, proline, cysteine, tyrosine, valine, methionine, lysine, isoleucine, leucine and phenylalanine when media amended with glucose as sole carbon source under diazotrophic conditions (Revillas et al. 2000; Lopez et al. 1981).

2.2.3 Polyhydroxybutyrate (PHB)

Azotobacter species produces intracellular polyester, i.e. poly-b-hydroxybutyrate (PHB), and extracellular polysaccharide alginate and catechol compounds under nutritional and favourable environmental conditions (Barrera and Soto 2010). PHB are used in large-scale production of alginic acid which is biodegradable and biocompatible thermoplastic. Alginic acid is a substitute for plastics such as polyethylene and polypropylene used in food industry, for thickening soups and jellies.

2.2.4 HCN Production

Many of the rhizobacterial species such as *Azotobacter*, *Alcaligenes*, *Aeromonas*, *Bacillus*, *Pseudomonas* and *Rhizobium* are capable of producing HCN as a volatile, secondary metabolite that suppresses the growth and development of pathogenic microorganisms which in turn influences the growth of plants (Ahmad et al. 2008). Different species of PGPR bacteria are known to produce different types of HCN as by-product of bacterial metabolism. Among PGPR groups, *Azotobacter* species, viz. *A. salinestris*, has showed HCN production. *A. salinestris* releases high intensity light orange to golden yellow colour HCN compounds and which can be detected based on the Lorck (1948) method. Kannapiran and Ramkumar (2011) reported the ammonia- and HCN-producing activity of *Azotobacter chroococcum* and *A. beijerinckii* and found positive results.

2.2.5 Siderophore Production

Siderophores are low molecular weight iron (Fe) chelating compounds that are produced and utilized by many bacterial and fungi groups. These compounds are produced by various types of bacteria in response to iron deficiency which normally occurs in neutral to alkaline pH soils, due to low iron solubility at elevated pH (Johri et al. 2003). *Azotobacter* excretes siderophore under limited iron conditions and found that *A. vinelandii* produces at least five different siderophores which are antibiotic in nature. These include siderophore 2,3-dihydroxybenzoic acid, aminochelin, azotochelin, protochelin and azotobactin. The main biotechnological applications of siderophore are they are used as drug delivery agents and antimicrobial agents and used for soil remediation (Mollmann et al. 2009; Kraepiel et al. 2009; Barrera and Soto 2010; Page and Von Tigerstrom 1988). Siderophore-producing PGPR can prevent the proliferation of pathogenic microorganisms by sequestering Fe^{3+} in the vicinity of the root. Many plants can use various bacterial siderophore as iron sources.

Siderophore production was detected by chrome azurol S (CAS) assay method. *A. salinestris* produced orange yellow to golden yellow colour around the colonies, and the colour intensity can be visualized under UV light. *Azotobacte*r species can produce varied ranges of siderophore, and *Azotobacter salinestris* produced siderophore in the range of 2–4 mm diameter zone. In support of the siderophore production, Bhosale et al. (2013) described the siderophores and HCN-producing isolates of *Azotobacter armeniacus* and *Azotobacter vinelandii* which are having antifungal activities by producing 2,3-dihydroxy benzoic acid, azotobactin and aminochelin. Similarly Mali and Bodhankar (2009) reported that siderophore production by *A. chroococcum* is known to suppress plant pathogens by solubilizing Fe^{3+} ions in the soil.

2.2.6 Antifungal Compounds

The production of antibiotics by microbes is considered one of the most studied biocontrol mechanisms for the disease management and combating phytopathogens. Different types of antibiotics produced by PGPR have shown to be effective against different phytopathogens which are known to cause pre- and postharvest diseases (Bowen and Rovira 1999). *Azotobacter* is one of the important biocontrol agents and produces antifungal compounds which inhibit the growth of soilborne fungi such as *Aspergillus*, *Fusarium*, *Curvularia*, *Alternaria* and *Helminthosporium* in the rhizosphere soil (Khan et al. 2008; Mali and Bodhankar 2009). The species of *Azotobacter* produces different types of antibiotics which include 2,3-dihydroxybenzoic acid, aminochelin, azotochelin, protochelin and azotobactin. It can also provide protection for the crops against drought and stress conditions (Kraepiel et al. 2009).

Some of the antibiotics produced by PGPR group include 2-acetamidophenol, phenazine-l-carboxamide (PCN), HCN, tesin, butyrolactones, eumycin, pyocyanin, hernipyocyanin, tyrothricin, zwittermicin A, kanosamine, oligomycin A, oomycin A, phenazine-I-carboxylic acid, pyoluteorin, pyrrolnitrin, viscosinamide, xanthobaccin and 2,4-diacetylphloroglucinol (2,4-DAPG) (Whipps 2001). The 2,4-DAPG is one of the most efficient antibiotics in the control of plant pathogens which are produced by various strains of *Pseudomonas*. The 2,4-DAPG has a wide spectrum of activities such as antifungal, antibacterial and antihelminthic properties (Naik et al. 2013).

2.3 *Azotobacter* as a Potent Biocontrol Agent

The terms "biological control" or "biocontrol" have been used in different fields of life science notably entomology and plant pathology. In the field of plant pathology, it is the application of direct or indirect use of microbial source to suppress the disease either by antagonizing the pathogen or by complete inhibition of the pathogen, and such organisms which suppress the pathogens are scientifically referred as biological control agents.

Cereal crops are essential source of nutrition worldwide. They are having good economic importance both as food and feed (Bottalico 1998). Colonization of cereals by trichothecene-producing *Fusarium* spp. leads to economical and food quality loses (Foroud and Eudes 2009). Their colonization also implicates potential human and animal health risks (Yazar and Omurtag 2008). Several *Fusarium* species are widespread pathogens affecting small grain cereals (maize, sorghum, rice, wheat, barley, oats, rye and millets, etc.), causing root, stem and ear rot and results in reduction in crop yield to the extent of 10–40% worldwide. (Nelson et al. 1983).

Different fungicides and chemical fertilizers are available in the market to control *Fusarium* infection in cereals. But in agronomic system, the use of chemical fertilizer for higher yield of crops is inevitable and results in considerable economic losses in the longer run. Due to this, problems encountered, such as gradual aggravation of soil fertility, disrupt soil ecology and environment, development of pathogen resistance to fungicides, inability to reach the roots of mature plants, rapid degradation in soil and need for repeated applications which leads to harmful effects on human and animal health (Ayala and Prakasa Rao 2002) along with contaminating groundwater resources (Joshi et al. 2006). Among the different microorganisms used as biocontrol agents, the plant growth-promoting rhizobacteria (PGPR) such as *Pseudomonas*, *Bacillus*, *Rhizobium*, *Burkholderia* and *Azotobacter* spp. have been used to reduce fungal colonization (Cavaglieri et al. 2005; Nayaka et al. 2009; Aiyaz et al. 2015). Different PGPR strains such as species of Azotobacter and Pseudomonashave shown promising ability to control Fusarium spp. however control of trichothecene producing Fusarium spp. has not been reported widely so far.

In a recent study, Nagaraja et al. (2016) have revealed the potential anti-fusarial efficacy of *Azotobacter* species such as *A. nigricans* and *A. salinestris* against trichothecene-producing *Fusarium* species which are common contaminant of cereals. Preliminary antifungal assay showed that all the tested *Fusarium* spp. was effectively inhibited by *A. salinestris*. The efficacy of *A. salinestris* in controlling the trichothecene-producing *Fusarium* spp. was studied by dual culture technique against most of the *Fusarium* species tested.

Challenge inoculation of *Fusarium* spp. to maize, sorghum and wheat cereals showed 100% infection incidence. Reduction in *Fusarium* infection incidence was recorded upto 50% in treated cereals. Total mass of the maize seedlings increased twofold by *A. nigricans* treatment, but only a slight increase was observed in sorghum and wheat seedlings. The highest vigour index recorded in maize was 1229.48 against *F. poae*, 1616.71 against *F. sporotrichioides* in sorghum and 1582.16 against *F. poae* wheat treated with *A. nigricans* (Nagaraja et al. 2016). Highest germination incidence of 64% was in maize, 66% in sorghum and 56.1% in wheat treated with *A. nigricans* (Fig. 2.1).

The effect of *A. salinestris* on total mass of seedlings was also evaluated. Significant increase in total mass of seedlings was observed in maize, in sorghum and in wheat seedlings, respectively. The effect of *A. salinestris* against trichothecene-producing *Fusarium* spp. on germination incidence of maize, sorghum and wheat proved that *Azotobacter* had enhanced germination of seedlings.

Significant decrease in the infection incidence of *Fusarium* spp. treated with *A. salinestris* was observed. In maize *A. salinestris* treatment was more effective on *F. crookwellense*, and infection incidence had drastically reduced in all treated cereal samples. The efficacy of *A. salinestris* treatment under in situ conditions was evaluated. The germination index and the vigour index (VI) of maize, sorghum and wheat were significantly increased. Maize recorded highest vigour index followed by sorghum and wheat. The results clearly explain the importance of *Azotobacter*

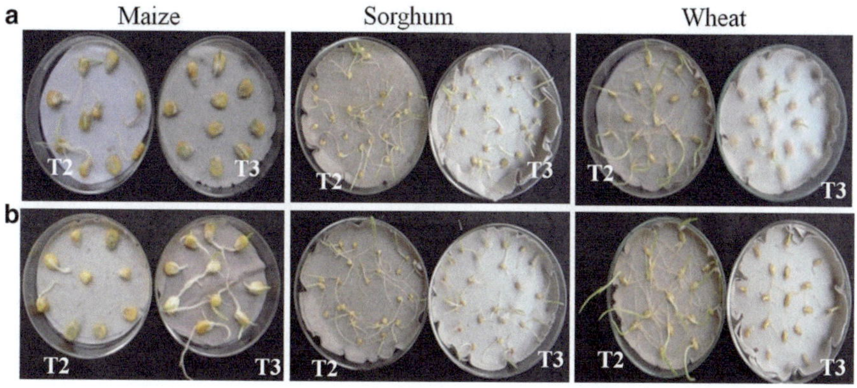

Fig. 2.1 Efficacy of *Azotobacter* on infection and germination incidence. (**a**) To reduce the infection incidence of *Fusarium* on maize, sorghum and wheat. (**b**) To influence on seed germination of maize, sorghum and wheat. *Note*:**T2**- treatment of cereals witn *Fusarium* sp. and **T3**- treatment of cereals with combination of *Fusarium* sp + *Azotobacter*

with a prominent attribute of PGPR properties leading to increase in root and shoot lengths, vigour index and total mass.

The concept of introducing PGPR bacterial strains into the rhizospheric region of maize, sorghum and wheat was to provide resistance against trichothecene-producing *Fusarium* spp. Rhizoplane colonization efficiency showed successful isolation of *A. salinestris* with high CFU/mL rate and also significant inhibition in colonization of *Fusarium* species when artificially inoculated to maize, sorghum and wheat. The greenhouse studies also enhanced the total plant growth and its root system. The bioformulation assay results revealed the viability after 6 months of storage of both *A. nigricans* and *A. salinestris* on N-free media, a selective media for isolation of *Azotobacter*. The study further revealed the promising PGPR and biocontrol nature of *Azotobacter* strains, namely, *A. nigricans* and *A. salinestris* for preventing the trichothecene-producing *Fusarium* species.

2.3.1 Nitrogen Fixation

Nitrogen fixation is the biological reaction where atmospheric N_2 gas is converted into NH_3. Ammonia is a form of nitrogen that can be easily utilized for biosynthetic pathways; nitrogen fixation is a critical process in the completion of the nitrogen cycle (Murcia et al. 1997; Barrera and Soto 2010). Nitrogen fixation can be considered as one of the most interesting microbial activity as it makes the recycling of nitrogen on earth (Aquilanti et al. 2004).

PGPR are root-colonizing microorganisms which are known to fix atmospheric molecular nitrogen through symbiotic, asymbiotic and associative nitrogen-fixing process. These bacteria belong to the genera *Azotobacter*, *Azospirillum*, *Bacillus*, *Arthrobacter*, *Enterobacter*, *Pseudomonas*, *Alcaligenes*, *Klebsiella*, *Rhizobium*,

Frankia, *blue-green algae*, *Azomonas* and *Serratia*. Among all the species, *A. chroococcum* and *A. vinelandii* are the most studied diazotrophic non-symbiotic nitrogen-fixing bacterial species and aerobic soil bacteria with a wide variety of metabolic capabilities (Mirzakhani et al. 2009; Khan et al. 2008). These bacteria are capable of synthesizing various plant growth-promoting substances, which may augment its performance as efficient inoculants for crop plants.

The species of *Azotobacter* are known to fix on an average 10 mg of N/g of carbohydrate under in vitro conditions. *A. chroococcum* happens to be the dominant inhabitant in arable soils capable of fixing N_2 (2–15 mgN_2 fixed /g of carbon source) in culture medium. Most efficient strains of *Azotobacter* would need to oxidize about 1000 Kg of organic matter for fixing 30 Kg of N/ha. Besides, soil is inhabited by a large variety of other microbes, all of which compete for the active carbon. Plant needs nitrogen for its growth and *Azotobacter* fixes atmospheric nitrogen non-symbiotically. Therefore, plants get benefited especially cereals, vegetables, fruits, etc. that are known to get additional nitrogen requirements from *Azotobacter* (Tilak et al. 2005; Khan et al. 2008; Mirzakhani et al. 2009; Tejera et al. 2005). The ability of the *Azotobacteraceae* family to grow in a medium free from nitrogen helps in their selective isolation. An organic carbon source and phosphate are usually the essential nutrients required for the development of *Azotobacter* under natural conditions (Becking 1981).

Among different species of *Azotobacter*, *A. salinestris* isolates showed the ability to fix N_2 at varied range from 22.70 to 34.50 μgNmL^{-1}day^{-1}. However, *A. salinestris* isolate recorded significant N_2 fixation of 34.50 μgNmL^{-1}day^{-1} as compared to the other isolates (Fig. 2.2). *A. salinestris* showed nitrogen fixation found equal to known

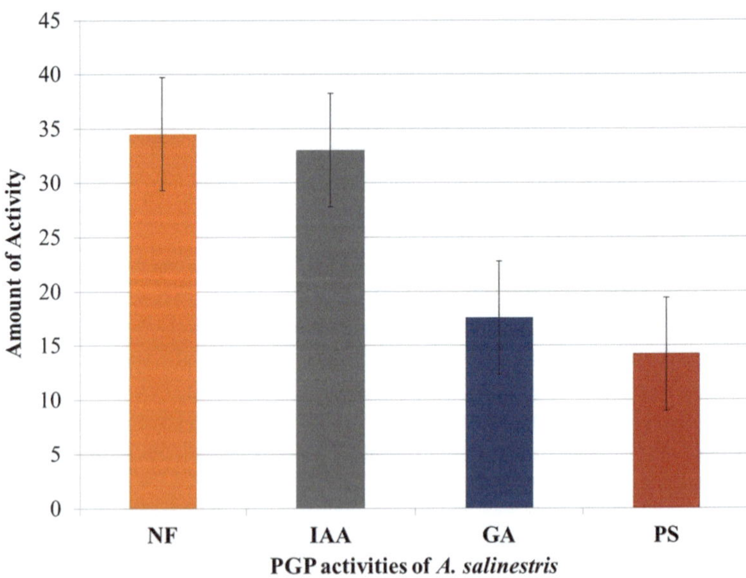

Fig. 2.2 Nitrogen fixation, NF (μg Nm/L/day); indole acetic acid, IAA (μg/mL); gibberellic acid, GA (μg/25 mL); and P-solubilization, PS (mm) activity of *A. salinestris* under in vitro conditions

species of *A. chroococcum* and *A. vinelandii*, respectively. Similarly, Page and Shivprasad (1991) identified a novel *A. salinestris* species isolated from saline soils of Canada and nitrogen fixation found almost similar to the present investigation. *A. salinestris* has proved to be one of the best nitrogen-fixing bacteria among the species.

2.3.2 IAA Production

The IAA-synthesizing ability has been detected in many rhizobacteria as well as in pathogenic, symbiotic and free-living bacterial species. Auxin-synthesizing rhizobacteria are the most studied phytohormone producers (Spaepen et al. 2007). These rhizobacteria synthesize IAA from tryptophan by tryptophan-dependent and tryptophan-independent pathways. Phytopathogenic bacteria mainly use the indole acetamide pathway to synthesize IAA, which has been implicated in tumour induction in plants. In contrast, the indole pyruvic pathway appears to be the main pathway present in plant growth-promoting beneficial bacteria (Patten and Glick 2002). Among PGPR species, *Azospirillum* is one of the best studied IAA producers, and other bacteria belonging to genera of *Aeromonas*, *Burkholderia*, *Azotobacter*, *Bacillus*, *Enterobacter*, *Pseudomonas* and *Rhizobium* species have been isolated from different rhizosphere soils (Ghosh et al. 2010; Ahmad et al. 2008).

Among different *Azotobacter* species, highest IAA was produced by *A. salinestris* (33.0 μgmL^{-1}) strain supplemented with tryptophan (Fig. 2.2). The production of IAA may vary from species to species and with the soil conditions. *Azotobacter* isolates proved the PGP potential with respect to *A. chroococcum*, and addition of tryptophan to the growth media has increased the metabolic activity of the *Azotobacter* species. Sachin (2009) studied IAA-producing activities of *A .chroococcum* and concluded that *A. chroococcum* have capacity to increase the seed germination and root and shoot length, respectively. IAA production of *Azotobacter* isolate found 1.47–32.80 μg/mL supplemented with 1–5 mg of tryptophan. The range of IAA production varies and depends on the efficiency of bacterial species. The addition of tryptophan has influenced the rate of a reaction as well as bacterium respiration rate (Ahmad et al. 2005).

Further Patil (2011) reported the highest amount (32.80 μgmL^{-1}) of IAA production by *Azotobacter* supplemented with 5 mg of tryptophan. Very low amount of IAA production was observed between the two experiments, but it was supplemented with 2 and 5 mg of tryptophan. As compared to all the previous results, *A. salinestris* found on par but produced highest amount of IAA at 1 mg of tryptophan as compared to 5 mg. This report clearly indicates that the diazotrophic bacteria have the same type of metabolic activities in the bacterial diversity.

2.3.3 Gibberellic Acid Production

Another important activity of *Azotobacter* is the production of gibberellins. Like nitrogen fixation, IAA and GA (gibberellic acid) are also one of the important plant growth-promoting substances produced by the different species of PGPR including *Azotobacter* species. GA includes a wide range of chemicals that are produced naturally within plant rhizospheric bacteria and fungi. GA production was first discovered by Japanese scientist Eiichi Kurosaw, which was produced by a fungus called *Gibberella fujikuroi* under abnormal growth stage in rice plants. Gibberellins are important in seed germination and enzyme production that mobilizes growth of new cells. In grain (rice, wheat, corn, etc.) seeds, a layer of cells called the aleurone layer wraps around the endosperm tissue, and absorption of water by the seed causes production of GA. GA promote flowering, cellular division and seed growth after germination (Upadhyay et al. 2009).

The *Azotobacter* isolates were tested for the GA production, and the results revealed that all the isolates produced GA in the range of 10.81–17.55 μg25 mL^{-1}. *A. salinestris* produced the highest amount of GA (17.55 μg 25 mL^{-1}) as compared to the other species (Fig. 2.2). GA is also one of the important plant growth-promoting substances produced by the different species of PGPR. Similarly, Upadhyay et al. (2009) reported the production of gibberellins by PGPR strain of *Arthrobacter* (130.74 μgmg^{-1}) and *Bacillus* species (7.67 μgmg^{-1}) isolated from wheat rhizosphere under 1–8% of NaCl saline conditions. The GA production variation is due to NaCl concentration, and it has influenced bacterial growth rate, and all PGPRs will not produce equal quantity of GA, and it varies from species to species and geographical conditions.

2.3.4 Phosphate Solubilization

Microbes play a significant role in the transformation of phosphorous and referred as phosphobacteria. Phosphate-solubilizing bacteria are a group of beneficial bacteria capable of hydrolysing organic and inorganic phosphorus from insoluble compounds. P-solubilization ability of the microorganisms is considered to be one of the most important traits associated with plant phosphate nutrition. Phosphate-solubilizing bacteria such as *A. chroococcum*, *B. subtilis*, *B. cereus*, *B. megaterium*, *Arthrobacter ilicis*, *E. coli*, *P. aeruginosa*, *E. aerogenes* and *Micrococcus luteus* were identified (Garg et al. 2001; Kumar et al. 2000). Currently, different strains of bacteria have been identified as biofertilizer, and three new strains *Pantoea agglomerans* strain, *Microbacterium laevaniformans* strain and *Pseudomonas putida* strain have been recently identified as the highly efficient phosphate solubilizer (Garg et al. 2001; Upadhyay et al. 2009).

The formation of clear halo zone indicates that the isolate was able to solubilize the available tricalcium phosphate to phosphate. The diameter of the clear zone from the edge of the colony was 7.7–14.2 mm. *A. salinestris* isolate formed larger clear zone (14.2 mm) compared to other isolates (Fig. 2.2). From the previous reports of Garg et al. (2001), phosphate-solubilizing activity of *Azotobacter* was reported under in vivo conditions and which confirms the present study. Kumar et al. (2000) reported the P-solubilizing activity of *A. chroococcum* under greenhouse condition of wheat, and P-solubilization has increased the total vigour of the plant. *A. salinestris* showed similar P-solubilization values as compared to the *A. chroococcum*, and this is also good competitor for PGPR group.

Upadhyay et al. (2009) isolated P-solubilizing PGPR isolates at 1–8% of NaCl conditions and all the isolates able to solubilize P at the rate of 4.5–8.0 mm. As compared to the present study, *A. salinestris* solubilized almost double area of zone formation. All P-solubilizing species will not have similar efficiency rate, and *A. salinestris* isolate found more efficient than the previous isolates of *A. chroococcum*. These isolates can be used as alternate P solubilizers compared to the existing isolates.

2.3.5 Abiotic Stress Tolerance

In the soil ecosystem, all the living creatures has to struggle for nutrients produced by plants and other sources available in the soil by competing with each other for their survival. In the last few decades, agriculture soil is receiving enormous amount of pesticide chemicals and other toxic materials, and these chemical materials include a variety of pesticides and fertilizers. The excessive use of chemical fertilizers has generated several environmental problems including the greenhouse effect, ozone layer depletion and acidification of water and soil rhizosphere (Saadatnia and Riahi 2009). Apart from these effects, all the microbes have to survive under different physiological conditions which include abiotic and biotic stresses that occurred in the soil such as alternations in salt, temperature and pH values. Change in pH and NaCl concentration may reduce the physiological activity of *Azotobacter* species.

2.3.5.1 Salt Tolerance

Mrkovacki et al. (2002) reported that some of the pesticides such as glyphosate not only inhibited the N_2 fixation but also reduced *A. chroococcum* respiration rate by 60%. *A. salinestris* recorded maximum growth up to 8% of NaCl concentration, but *A. chroococcum* was grown up to 6% concentration. The highest growth rate was observed at 6% NaCl concentration, but after 6% concentration the rate declined and it continues up to 10%. The highest CFU was recorded by *A. chroococcum* (1.75×10^{-4}) and *A. salinestris* (1.4×10^{-4}) at 6% NaCl concentration (Chennappa et al. 2016). *A. chroococcum* and *A. salinestris* were tolerant to 8% NaCl concentration (OD 0.9), but the total CFU values were reduced at 8% concentration at 600 nm. All the five *Azotobacter* isolates recorded growth up to 6% of NaCl concentration among them. *A. chroococcum*, *A. vinelandii* and *A. salinestris* were grown up to 8% NaCl concentration (Fig. 2.3). Upadhyay et al. (2009) reported the NaCl-tolerant *Arthrobacter* sp. isolated from wheat rhizosphere and documented that the isolate was tolerant to 8% NaCl concentration having same PGP activity.

Similarly Akhter et al. (2012) reported salt-tolerant *Azotobacter* spp. isolated from saline soils of Bangladesh and found that the *Azotobacter* species tolerated NaCl concentration (10%) and few of them were sensitive even at 6% NaCl. Salt tolerance study shows that the isolated *Azotobacter* can grow and survive at higher salt concentration and salinity of the paddy soils was increased due to the application of excess quantity of pesticides.

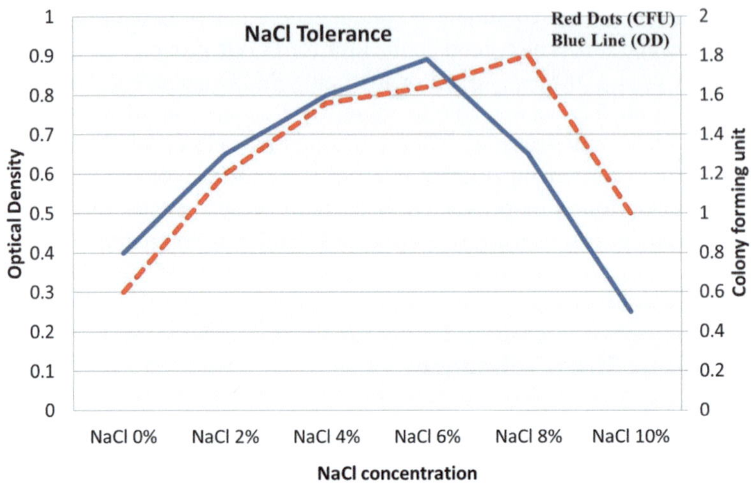

Fig. 2.3 Salt tolerance activity of *A. salinestris* at different concentrations of NaCl and their colony-forming units (CFU) and optical density (OD)

2.3.5.2 Temperature Tolerance

Azotobacter strains, viz. *A. salinestris* and *A. chroococcum*, were grown at 15–35 °C for 7 days at pH 7, but *A. salinestris* survived up to 45 °C. All the *Azotobacter* isolates grew at different temperatures, but the optimum temperature was found to be 35 °C, and the rate of a reaction was decreased from 35 °C onwards (Chennappa et al. 2016). The peak of the growth curve was log phase for up to 35 °C and which was incubated at 45 °C but *Azotobacter* strains documented reduced with increasing temperature (Fig. 2.4).

Moreno et al. (1986) reported the survival of *Azotobacter* in dry soils for a period of 24 years and concluded that the isolated species was *A. chroococcum* and it could be able to characterize the species based on cultural features which are grown on Burk's N-free media. All the *Azotobacter* species have been isolated from major paddy cultivation area of northern parts of Karnataka, and the temperature of this region is around 40–45 °C in summer, and *Azotobacter* are known to survive at higher temperatures. Temperature-tolerant *Azotobacter* population will help to improve the soil fertility of this region which is highly sprayed with pesticides and fertilizers.

2.3.5.3 pH Tolerance

Among two *Azotobacter* strains, *A. salinestris* survived at a pH of 9.0 and did not observe any inhibition of growth at higher pH range. *A. chroococcum* was sensitive to pH of above 8.0 and no growth was observed above pH 10. The highest CFU was recorded at a pH of 8.0 and 9.0 by *A. salinestris*. The soil pH has increased because of the maximum use of pesticides and chemical fertilizer in order to control the pests and diseases of crops. Earlier reports indicate that *Azotobacter* can survive at extreme temperature and varied pH ranges.

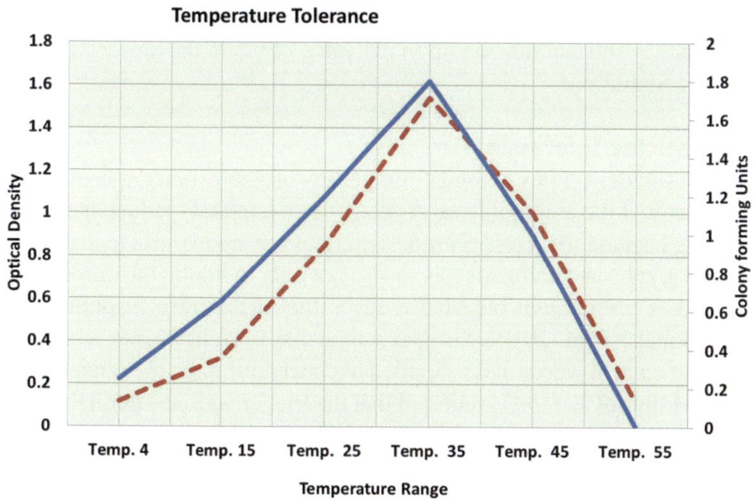

Fig. 2.4 Temperature tolerance activity of *A. salinestris* at different temperatures and their colony-forming units (CFU) and optical density (OD)

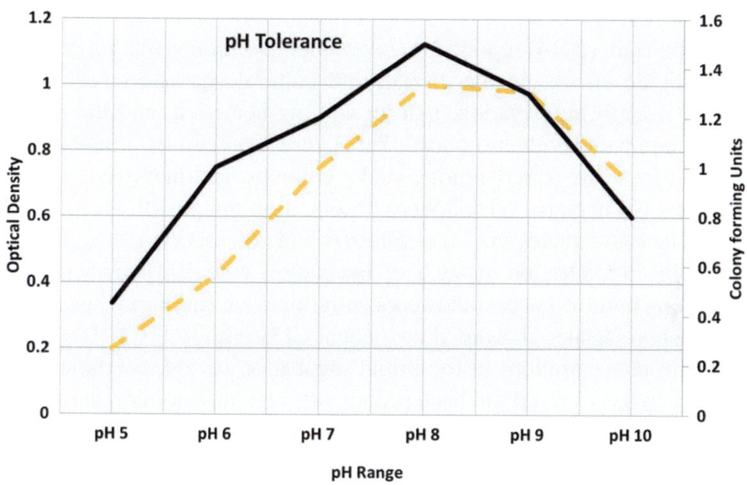

Fig. 2.5 pH tolerance activity of *A. salinestris* at different pH ranges and their colony-forming units (CFU) and optical density (OD)

Azotobacter chroococcum survived at a pH of 9.0, and decline in the growth at higher pH range was recorded. *A. salinestris* was sensitive to the pH of above 9.0 and no growth was observed above pH 10. The highest CFU was recorded at a pH of 8.0 and 9.0 by *A. chroococcum* and *A. salinestris*, respectively (Fig. 2.5). The genus *Azotobacter* is ubiquitous in nature, and they can grow with pH ranging from 6.0 to 9.0 of different climatic temperatures (Jimenez et al. 2011; Chennappa et al.

2016). This result clearly indicates that the species of *Azotobacter* are resistant to higher pH values and can tolerate up to pH value of 9.5 in the presence of different pesticide concentrations.

2.3.5.4 Pesticide Tolerance

Effects of pesticides on physiological properties of *Azotobacter* species have been reported. Many of the commonly used pesticides completely reduce the population of free-living bacteria. Herbicides used for agriculture are harmful to many bacterial species, and glyphosate herbicide has shown not only to inhibit the nitrogen fixation process in *A. chroococcum* but also reduces the bacterium's respiration rate by 40–60% (Moneke et al. 2010). Similarly simazine is an herbicide widely used in agriculture to control weeds and is applied at concentration in the range from 1.0 to 4.0 Kg/ha. Murcia et al. (1997) reported that the triazine-derived herbicide simazine affects nitrogenase activity and ATP content of *Azotobacter*. Niewiadomska (2004) reported the effect of carbendazim, thiram and imazetapir on nitrogenase activity of bacteria in soil. Among them, glyphosate has negative effect on N_2 fixation, and pendimethalin concentration did not affect the metabolic activity but was lethal to bacterial respiration and reduced the N_2 fixation ability. Similarly, effect of endosulfan was also reported on IAA production and nitrogen fixation activity of *A. chroococcum* (Castillo et al. 2011).

Chennappa et al. (2014) reported that *Azotobacter* cultures did not recorded any kind of pesticide effects on the growth and cultural characters such as colony appearance, size, form, elevation, margin and pigmentation, and the consistency was same as compared with the control. The *Azotobacter* culture produced brown to black and milky white water-soluble, sticky colonies, and they were similar with control plates. No negative effects were observed on the growth and survivability (Fig. 2.6). The two isolates viz. *A. salinestris* and *A. chroococcum* grown at 1% pendimethalin concentration at 48 h of incubation with 100% growth rate. The colony-forming units at 1% pesticide concentration were similar to the control standards. These two isolates showed almost equal CFU/mL (12.0–13.30 × 10^3) at 3% pendimethalin concentration. In the initial incubation period, pendimethalin concentration (5%) has reduced the bacterial growth rate but not inhibited the growth, and after 3–4 days of incubation period, complete growth was observed on the plates containing 5% pendimethalin.

Castillo et al. (2011) reported that the *Azotobacter* can grow in the presence of endosulfan as carbon or sulphur source, and the cell growth was slow in the presence of endosulfan as carbon source when compared with glucose and magnesium sulphate. *Azotobacter* species was able to grow up to 10 mgL^{-1} of endosulfan concentrations as sulphur source. Moneke et al. (2010) reported that the species of *Azotobacter*, *Pseudomonas*, *Escherichia*, *Acetobacter*, etc. can tolerate and survive glyphosate herbicide-contaminated soils. Inhibition of growth was not observed at higher concentrations (100–250 mg/mL), indicating that the isolated bacteria can tolerate up to 250 mg/mL of glyphosate. However, there was subsequent decrease in growth of isolates as the concentration of glyphosate increased above 250 mg/mL. Present report clearly indicates that all the isolates were resistant to 1, 3, and

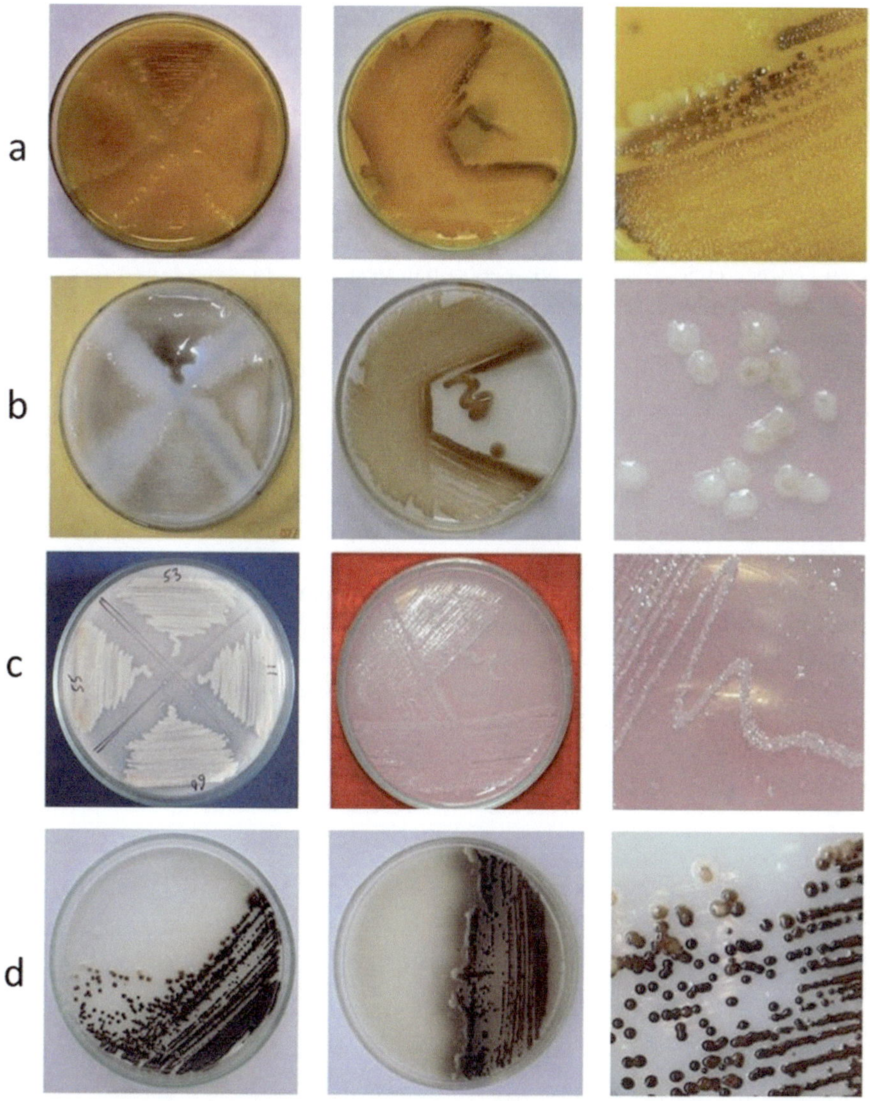

Fig. 2.6 Pesticide tolerance level of *A. salinestris* at different pesticide concentrations (1, 3, 5%), viz. pendimethalin (**a**), chlorpyrifos (**b**), phorate (**c**) and control (**d**); their colony-forming units (CFU)

5% of pesticides but slightly reduced the CFU/ml at higher concentration as compared to the control.

The *Azotobacter* isolates showed tolerance to the 5% glyphosate, but the previous reports of *Pseudomonas* and *Acetobacter* did not show tolerance to higher pesticide concentration, and the highest tolerance was exhibited by the species of *Azotobacter* isolates (Chennappa et al. 2013). In contrary to the present study, research conducted by Shafiani and Malik (2003) reported that *Pseudomonas*, *Azotobacter* and *Rhizobium* species can tolerate the pesticides such as endosulfan, carbofuran and malathion isolated from agricultural soils. Among all, *Azotobacter* isolate was able to tolerate 1600 μg/mL concentration of all the three pesticides, but these *Azotobacter* species isolated from pesticide-applied paddy soils. *Azotobacter* was the only bacteria that showed maximum tolerance towards the growth against the pesticide as compared to the *Pseudomonas* and *Rhizobium* species.

However, Castillo et al. (2011) reported that the considerable pleomorphism can be seen between endosulfan and non-endosulfan treatments. In control, cyst formation was observed after 5 days, while in the endosulfan treatment, microscopic analysis revealed that a more rapid development of cyst cells measuring 2.7 μm was observed after 2 days and 3.67 μm after 3 days of incubation. As in case of *A. salinestris* and *A. chroococcum* study, no such considerable pleomorphic changes were observed.

Similarly *A. salinestris* and *A. chroococcum* isolates were resistant up to 5% of glyphosate and showed similar CFU/ml values (13.00 and 13.00 × 10^4) as compared to the control. With all the concentration, pendimethalin and glyphosate did not inhibit the growth of *Azotobacter* isolates as compared to the control. Two isolates were resistant up to 5% chlorpyrifos concentration with a higher growth rate (100%). *A. chroococcum* and *A. salinestris* showed maximum growth rate at 48 h of incubation with CFU/ml value ranging from 13.33 to 12.66 × 10^4 as compared to the control (Fig. 2.7). As in case of phorate, the isolates showed maximum growth up to 3% concentration, and more than 3% concentration has slightly reduced the growth rate of *Azotobacter* (Chennappa et al. 2013).

Two isolates were grown in media containing 1–5% carbendazim, but initially the isolates showed slower growth rate above 1% concentration. Carbendazim

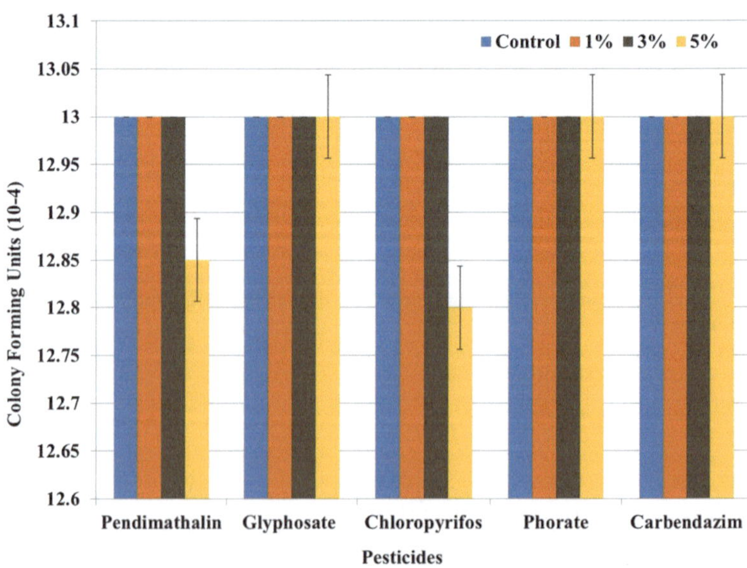

Fig. 2.7 Pesticide tolerance level of A. salinestris at different pesticide concentrations (1, 3, 5%) and their colony-forming units (CFU)

influenced the growth of *Azotobacter* isolates at longer incubation period and produced higher CFU/ml values over 3% concentration. The CFU/ml of the isolates *A. salinestris* (13.00×10^4) and *A. chroococcum* (12.80×10^4) were in the same range as compared to the control.

The species of *Azotobacter* are known to grow under extreme environmental conditions (Garg et al. 2001; Jimenez et al. 2011), and these two isolates showed fast growth with all the concentration of herbicides, insecticides and fungicides during 2–3 days of incubation. This study clearly indicates that two *Azotobacter* isolates tolerate up to 5% concentration of pesticides and can survive in soils containing heavy amount of pesticides that can be utilized as carbon and sulphur source for the metabolic activities. *Azotobacter* species are able to grow under pesticide-applied paddy soils, and isolates showed their ability to survive and tolerate higher pesticides concentration.

2.3.6 Biodegradation of Pendimethalin

Azotobacter chroococcum and *A. salinestris* were used for biodegradation of pesticides such as pendimethalin at 1% concentration. Assay showed that all the *Azotobacter* strains degraded pesticide compounds efficiently. Pendimethalin biodegradation assay showed that *A. salinestris* strain showed 100% degradation rate. The GC detector did not record even single ppm of pendimethalin over the standard sample (1 ppm/ 429104.25area). The retention time of pendimethalin in gas chromatography was 9.78 min in the sample area (Fig. 2.8).

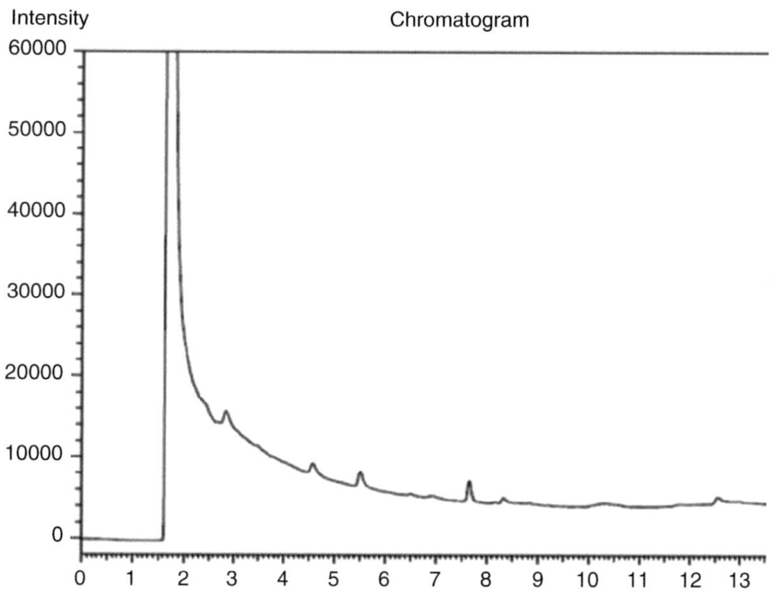

Fig. 2.8 Gas chromatographic analyses of biodegradation of pendimethalin by *Azotobacter salinestris*

Azotobacter chroococcum was degraded pendimethalin up to 98.86%, and very minimum quantity of the residue has been recovered from the analyses. Two isolates degraded the pendimethalin in a range of 30.50–61.50% under in vivo conditions, and *A. salinestris* showed maximum degradation (61.50%) of pendimethalin. In the present study, *A. tropicalis*, *A. vinelandii* and *A. salinestris* isolates degraded the pendimethalin by 100% over the control, and *A. armeniacus* showed least biodegradation of pendimethalin (62.16%) under in vitro conditions within 2 weeks of incubation period. As in case of in vivo studies, all the three isolates showed biodegradation rate up to 61.50%, and this is because of the availability of growth supplements which is less as compared to the in vitro conditions.

Megadi et al. (2010) reported the biodegradation of pendimethalin by *Bacillus circulans* by growing on media containing pendimethalin (1 g/L). Two different metabolites were recorded and accumulated, viz. 6-aminopendimethalin and 3,4-dimethyl-2,6-dinitroaniline in the culture medium. Similarly in the present study, from the biodegradation of pendimethalin, two residues were detected such as 3,4-dimethyl-2,6-dinitroaniline and 6-aminopendimethalin under in vitro.

Similarly Moneke et al. (2010) reported another important herbicide used for the paddy cultivation, i.e. glyphosate for the control of grasses and herbaceous plants. The biodegradation abilities of bacterial species, viz. *Escherichia* sp., *Azotobacter* sp., *Alcaligenes* sp., *Acetobacter* sp. and *P. fluorescens*, were tested for glyphosate and found that all species were degraded glyphosate up to a maximum level. Similarly Elsyaed and Nady (2013) reported that the optimum pH (7) and temperature (30 °C) for the growth of *P. putida* and degraded pendimethalin from mineral liquid medium with half-life of 5.46 days. Very limited work was done on the biodegradation of herbicides by *Azotobacter* species from paddy soils, and *Azotobacter* proved the efficiency of biodegradation of pendimethalin under in vitro and in vivo conditions which help to restore the soil fertility.

2.4 Significance of Study

The present data investigation reveals that *Azotobacter* isolate is a better PGPR, and they can fix maximum amount of N_2 and produce IAA, as GA also have P-solubilization activity. *Azotobacter* isolates can withstand higher pesticide load and break down these toxic materials into non-toxic forms efficiently and survive for longer periods. Among the five isolates, *A. salinestris* recorded better performance for sustainable agriculture and environmental clean-up processes.

References

Ahmad F, Ahmad I, Khan MS (2005) Indole acetic acid production by the indigenous isolated of *Azotobacter* and fluorescent *Pseudomonas* in the presence and absence of tryptophan. Turk J Biol 29:29–34

Ahmad F, Ahmad I, Khan MS (2008) Screening of free living rhizospheric bacteria for their multiple growth promoting activities. Microbiol Res 163(2):173–181

Aiyaz M, Divakara ST, Nayaka SC (2015) Application of beneficial rhizospheric microbes for the mitigation of seed borne mycotoxigenic fungal infection and mycotoxins in maize. Biocontrol Sci Tech 25:1105–1119

Akhter MS, Hossain SJ, Hossain SKA, Datta RKD (2012) Isolation and characterization of salinity tolerant *Azotobacter* sp. Greener J Biol Sci 2(3):43–51

Aleem A, Isar J, Malik A (2003) Impact of long term application of industrial wastewater on the emergence of resistance traits of *Azotobacter vinelandii* isolated from rhizosphere soil. Bioresource Tech 86:7–13

Almon L (1958) The vitamin B12 content of *Azotobacter vinelandii*. J Nutr 65(4):643–648

Aquilanti L, Favilli F, Clementi F (2004) Comparison of different strategies for isolation and preliminary identification of *Azotobacter* from soil samples. Soil Biol Biochem 36:1475–1483

Ayala S, Prakasa Rao EVS (2002) Perspectives of soil fertility management with a focus on fertilizer use for crop productivity. Curr Sci 82:797–807

Barrera DA, Soto E (2010) Biotechnological uses of *Azotobacter vinelandii* current state limits and prospects. Afr J Biotechnol 9:5240–5250

Becking JH (1981) The family *Azotobacteraceae*. In: Ballows A, Truper HG, Dworkin M, Harder W, Schleifer KH (eds) The procaryotes: a handbook on habitats, isolation and identification of bacteria. Springer, Heidelberg, pp 795–817

Bhosale HJ, Kadam TA, Bobade AR (2013) Identification and production of *Azotobacter vinelandii* and its antifungal activity against *Fusarium oxysporum*. J Environ Biol 34:177–182

Bottalico A (1998) *Fusarium* diseases of cereals: species complex and related mycotoxin profiles in Europe. J Plant Pathol 80(2):85–103

Bowen GD, Rovira AD (1999) The rhizosphere and its management to improve plant growth. Adv Agron 66:101–102

Castillo JM, Casas J, Romero E (2011) Isolation of an endosulfan- degrading bacterium from a coffee farm soil: persistence and inhibitory effect on its biological functions. Sci Total Environ 412–413:20–27

Cavaglieri LR, Andres L, Ibanez M, Etcheverry MG (2005) Rhizobacteria and their potential to control *Fusarium verticillioidies*, effect of maize bacterisation and inoculums density. Antonie Van Leeuwenhoek 87:179–187

Chennappa G, Adkar-Purushothama CR, Suraj U, Tamilvendan K, Sreenivasa MY (2013) Pesticide tolerant *Azotobacter* isolates from paddy growing areas of northern Karnataka, India. World J Microbiol Biotechnol 30:1–7

Chennappa G, Purushothama ACR, Naik MK, Sreenivasa MY (2014) Impact of pesticides on PGPR activity of *Azotobacter* Sp. isolated from pesticide flooded paddy soils. Greener J Biol Sci 4(4):117–129

Chennappa G, Naik MK, Adkar-Purushothama CR, Amaresh YS, Sreenivasa MY (2016) PGPR, abiotic stress tolerant and antifungal activity of Azotobacter sp. Isolated from paddy soils. Indian J Exp Biol 54:322–331

Elsyaed BB, Nady MF (2013) bioremediation of pendimethalin contaminated soil. Afr J Microbiol Res 7(21):2574–2588

Foroud NA, Eudes F (2009) Trichothecenes in cereal grains. Int J Mol Sci 10:147–173

Garg SK, Bhatnagar A, Kalla A, Narula N (2001) *In vitro* nitrogen fixation, phosphate solubilization, survival and nutrient release by *Azotobacter* strains in an aquatic system. Bioresource Tech 80:101–109

Ghosh PG, Sawant NA, Patil SN, Aglave BA (2010) Microbial biodegradation of organophosphate pesticides. Int J Biotech Biochem 6:871–876

Jimenez DJ, Montana JS, Martinez MM (2011) Characterization of free nitrogen fixing bacteria of the genus Azotobacter in organic vegetable grown Colombian soils. Braz J Microbiol 42:846–858

Johri BN, Sharma A, Virdi JS (2003) Rhizobacterial diversity in India and its influence on soil and plant health. Adv Biochem Eng Biotechnol 84:49–89

Joshi KK, Kumar V, Dubey RC, Maheshwari DK (2006) Effect of chemical fertilizer adaptive variants, *Pseudomonas aeruginosa* GRC2 and *Azotobacter chroococcum* AC1 on *Macrophomena phaseolina* causing charcoal rot of *Brassica juncea*. Korean J Environ Agric 25:228–235
Kadam TA, Gangawane LV (2005) Degradation of phorate by *Azotobacter* isolates. Indian J Biotechnol 4:153–155
Kannapiran E, Ramkumar SV (2011) Inoculation effect of nitrogen-fixing and phosphate-solubilizing bacteria to promote growth of black gram (*Phaseolusmungo* Roxb; Eng). Annals Biol Res 2(5):615–621
Khan HR, Mohiuddin M, Rahman M (2008) Enumeration, isolation and identification of nitrogen-fixing bacterial strains at seedling stage in rhizosphere of rice grown in non-calcareous grey flood plain soil of Bangladesh. J Fac Environ Sci Tech 13:97–101
Kraepiel A, Bellenger J, Wichard T, Morel F (2009) Multiple roles of siderophores in free living nitrogen-fixing bacteria. Biometals 22:573–581
Kumar V, Behl RK, Narula N (2000) Establishment of phosphate solubilizing strains of *Azotobacter chroococcum* in the rhizosphere and their effect on wheat cultivars under greenhouse conditions. Microbiol Res 156:87–93
Lopez JG, Toledo MV, Reina S, Salmeron V (1981) Root exudates of maize on production of auxins, gibberellins, cytokinins, amino acids and vitamins by *Azotobacter chroococcum* chemically defined media and dialysed soil media. Toxicol Environ Chem 33:69–78
Lorck H (1948) Production of hydrocyanic acid by bacteria. Physiolo Plant 1(2):142–146
Mali GV, Bodhankar MG (2009) Antifungal and phytohormone potential of *Azotobacter chroococcum* isolates from ground nut (*Arachis hypogea* L.) rhizosphere. Asian J Exp Sci 23:293–297
Martin XM, Sumathi CS, Kannan VR (2011) Influence of agrochemical and *Azotobacter* spp. application on soil fertility in relation to maize growth under nursery conditions. Eurasia J Biosci 5:19–28
Megadi VB, Tallur PN, Hoskeri RS, Mulla SI, Ninnekar HZ (2010) Biodegradation of pendimethalin by *Bacillus circulans*. Indian J Biotechnol 9:173–177
Mirzakhani M, Ardakani MR, Band AA, Rejali F, Rad SAH (2009) Response of spring safflower to co-inoculation with *Azotobacter chroococcum* and *Glomus intraradices* under different levels of nitrogen and phosphorus. Am J Agri Biol Sci 4:255–261
Mollmann U, Heinisch L, Bauernfeind A, Kohler T, Ankel-Fuchs D (2009) Siderophores as drug delivery agents: application of the Trojan horse strategy. Biometals 22:615–624
Moneke AN, Okpala GN, Anyanwu CU (2010) Biodegradation of glyphosate herbicide *in vitro* using bacterial isolates from four rice fields. Afr J Biotechnol 9:4067–4074
Moreno J, Lopez JG, Vela GR (1986) Survival of *Azotobacter* spp in dry soils. Appl Environ Microbiol 51:123–125
Mrkovacki NB, Nikola A, Cacic, Milic VM (2002) Effects of Pesticides on *Azotobacter chroococcum*. Proc Nat Sci 102:23–28
Murcia R, Rodelas B, Salmeron V, Toledo MVM, Lopez GJ (1997) Effects of herbicide simazine on vitamin production by *Azotobacter chroococcum* and *Azotobacter vinelandii*. Appl Soil Ecol 6:187–193
Myresiotis CK, Vryzas Z, Mourkidou EP (2012) Biodegradation of soil applied pesticides by selected strains of plant growth promoting rhizobacteria (PGPR) and their effects on bacterial growth. Biodegradation 23:297–310
Nagaraja H, Chennappa G, Rakesh S, Naik MK, Amaresh YS, Sreenivasa MY (2016) Anti fusarial activity of *Azotobacter nigricans* against trichothecene producing *Fusarium* spp., associated with cereals. Food Sci Biotechnol 25(4):1197–1204
Naik MK, Rajalaxmi K, Amaresh YS et al (2013) Search for 2, 4 DAPG positive genes in fluorescent *Pseudomonas* and their exploitation for sustainable disease management. Recent Advances in biofertilizer and bio fungicides (PGPR) for sustainable agriculture. In: Proceedings of the 3rd Asian PGPR Conference on Plant growth promoting rhizobacteria (PGPR) and other microbials, pp 21–24

Nayaka SC, Uday Shankar AC, Reddy MS, Niranjana SR, Prakesh HS, Shetty HS, Mortensen CN (2009) Control of *Fusarium verticillioides*, cause of ear rot of maize, by *Pseudomonas fluorescens*. Pest Manag Sci 65:769–775

Nelson PE, Toussoun TA, Marasas WFO (1983) *Fusarium* species: an illustrated manual for identification. Pennsylvania State University Press, University Park

Niewiadomska A (2004) Effect of carbendazim, imazetapir and thiram on nitrogenase activity, the number of microorganisms in soil and yield of red clover (*Trifolium pratense* L.) Pol J Environ Stud 13:403–410

Page WJ, Shivprasad S (1991) *Azotobacter salinestris* spp. nov, a sodium dependent, micro aerophilic and aero adaptive nitrogen fixing bacteria. Int J Syst Bacteriol 41:369–376

Page W, Von Tigerstrom M (1988) Aminochelin, a catecholamine siderophore produced by *Azotobacter vinelandii*. J Gen Microbiol 134:453–460

Patil V (2011) Production of indole acetic acid by *Azotobacter* sp. Recent Res Sci Technol 3(12):14–16

Patten CL, Glick BR (2002) Role of *Pseudomonas putida* indole acetic acid in development of the host plant root system. Appl Environ Microbiol 68:3795–3801

Reinhardt EL, Ramos PL, Manfio GP, Barbosa HR, Pavan C, Filho CAM (2008) Molecular characterization of nitrogen fixing bacteria isolated from Brazilian agricultural plants at also Paulo state. Braz J Microbiol 39:414–422

Revillas JJ, Rodelas B, Pozo C, Toledo MV, Gonzalez-Lopez J (2000) Production of B-group vitamins by two *Azotobacter* strains with phenolic compounds as sole carbon source under diazotrophic and a diazotrophic conditions. J Appl Microbiol 89(3):486–493

Saadatnia H, Riahi H (2009) Cyanobacteria from paddy fields in Iran as a biofertilizer in rice plants. Plant Soil Environ 55(5):207.212

Sachin DN (2009) Effect of *Azotobacter chroococcum* (PGPR) on the growth of bamboo (*Bambusa bamboo*) and maize (*Zea mays*) plants. Biofrontiers 1:24–31

Shafiani S, Malik A (2003) Tolerance of pesticides and antibiotic resistance in bacteria isolated from wastewater irrigated soil. World J Microbiol Biotechnol 19:897–901

Spaepen S, Vanderleyden J, Remans R (2007) Indole 3 acetic acid in microbial and microorganism plant signaling. FEMS Microbiol Rev 31:425–448

Tejera NC, Lluch MV, Martinez T, Gonzalez JL (2005) Isolation and characterization of *Azotobacter* and *Azospirillum* strains from the sugarcane rhizosphere. Plant Soil 270:223–232

Tilak KVBR, Ranganayaki N, Pal KK, De R, Saxena AK, Nautiyal SC, Mittal S, Tripathi AK, Johri BN (2005) Diversity of plant growth and soil health supporting bacteria. Curr Sci 89:136–150

Upadhyay SK, Singh DP, Saikia R (2009) Genetic diversity of plant growth promoting rhizobacteria isolated from rhizospheric soil of wheat under saline condition. Curr Microbiol. https://doi.org/10.1007/s00284-009-9464-1

Whipps JM (2001) Microbial interactions and biocontrol in the rhizosphere. J Exp Bot 52:487–512

Yazar S, Omurtag GZ (2008) Fumonisins, trichothecenes and zearalenone in cereals. Int J MolSci 9:2062–2090

Microbial Management of Organic Waste in Agroecosystem

3

M. Patchaye, B. Sundarkrishnan, S. Tamilselvan, and N. Sakthivel

Abstract

Organic waste materials occur universally as surplus agricultural waste, municipal solid waste, industrial waste and poultry waste. Microbial management of organic waste is a promising, environment-friendly approach to manage the large quantities of organic waste due to its recycling potential. Composting is a biotechnological process by which different microbial communities play a key role in biotransformation of complex organic substrates into useful decomposed residues. In this biological oxidative decomposition process, different types of diverse compost microbiota are involved. Due to the synergistic activities and functional diversity of microbiota, relatively low-cost, stable biomass, humus products are produced. Microbial compost materials are used as bioorganic manures to increase soil fertility that facilitates sustainable agriculture. In this chapter, we describe the history and science of composting, types of organic waste, microbial parameters, different types of composting, diversity and function of composting microbiota, mechanism of decomposition, microbiological methods for the analysis of compost maturity and socio-economic significance of microbial management of organic waste.

Keywords

Organic waste · Biotransformation · Microbiota · Organic manures · Composting · Soil fertility

M. Patchaye · B. Sundarkrishnan · S. Tamilselvan · N. Sakthivel (✉)
Department of Biotechnology, School of Life Sciences, Pondicherry University, Kalapet, Puducherry, India
e-mail: puns2005@gmail.com

D. G. Panpatte et al. (eds.), *Microorganisms for Green Revolution*, Microorganisms for Sustainability 7,
https://doi.org/10.1007/978-981-10-7146-1_3

3.1 Introduction

Due to urbanization, large quantity of organic wastes is produced. The organic wastes include food industrial wastes, farmyard wastes and agro-industrial wastes (Hoornweg and Bhada-Tata 2012; Hoornweg et al. 2013; Lohri et al. 2017). Large quantities of cellulosic wastes occur universally as surplus and waste residues of agricultural practices. They occur in the form of crop residues, usually the straws of cereals of grains such as rice, barley, wheat, corn, rye and sugarcane bagasse. The substrates contain by-products such as hydrocarbons from refineries, effluents from the cellulose and paper industries, distiller grains from distilleries and sugarcane molasses, etc. (Kuhad et al. 1997; Zino et al. 1999). Plant production from agriculture represents a considerable amount, 123×10^6 tons per year, and approximately half of the agricultural biomass produced is not used as food or feed representing lignocellulosic residues (Smith et al. 1988). The routine production of solid organic wastes worldwide will reach over 11 million tons by the end of the twenty-first century (Hoornweg et al. 2013).

Composting is a biotechnological process by which different microbial communities convert organic waste into a stabilized form. Composting converts organic wastes to CO_2, humus and heat by compost microorganisms. It is found to be the most promising environment-friendly technology to manage organic wastes due to its potential to eliminate the plant and human pathogenic microbes, ectoparasites and weeds during manure applications (Hao et al. 2004). It is further considered as an efficient way of recycling organic wastes to restore organic nutrients and improves soil fertility. The major components of organic matter are the complex substrates of carbohydrates, protein, lipids and lignin. Therefore, the synergistic activities of microbial populations are required to degrade the complex organic materials into smaller decomposed residues (Golueke 1992). In this process of biodegradation, the solid organic substrates are converted to stable, humidified and suitable product for agroecosystem under aerobic conditions through the action of different microbial populations (Insam and de Bertoldi 2007) (Fig. 3.1). In addition, composting strategy is used as a management tool for organic wastes.

The organic wastes go through different phases such as a mesophilic phase which is characterized by the proliferation of the microbiota; a thermophilic phase where a high rate of biodegradation, the growth of thermophilic organisms and the inhibition of non-thermotolerant organisms occur; and the final phase that includes a period of cooling, stabilization and maturation, characterized by the growth of mesophilic organisms and the humification of the compost (Ryckeboer et al. 2003a). Composting is a spontaneous biological oxidative decomposition of solid or semisolid organic waste materials of any natural source under optimum temperature, moisture, aeration, porosity, pH and C/N ratio (Dees and Ghiorse 2001; Golueke 1973; Lim et al. 2016) using mesophilic and thermophilic microbial populations. Composting yields relatively low-cost, stable biomass, thermal energy and humus product (lignoproteins) for soil amendments (Leonard and Ramer 1993; Miller 1993; Tuomela et al. 2000; Ryckeboer et al. 2003a; Insam and de Bertoldi 2007; Bernal et al. 2009; Lim et al. 2016; Onwosi et al. 2017). Microbial composts are used as manure in agroecosystem. Microbial compost-based-humidified final products are termed as

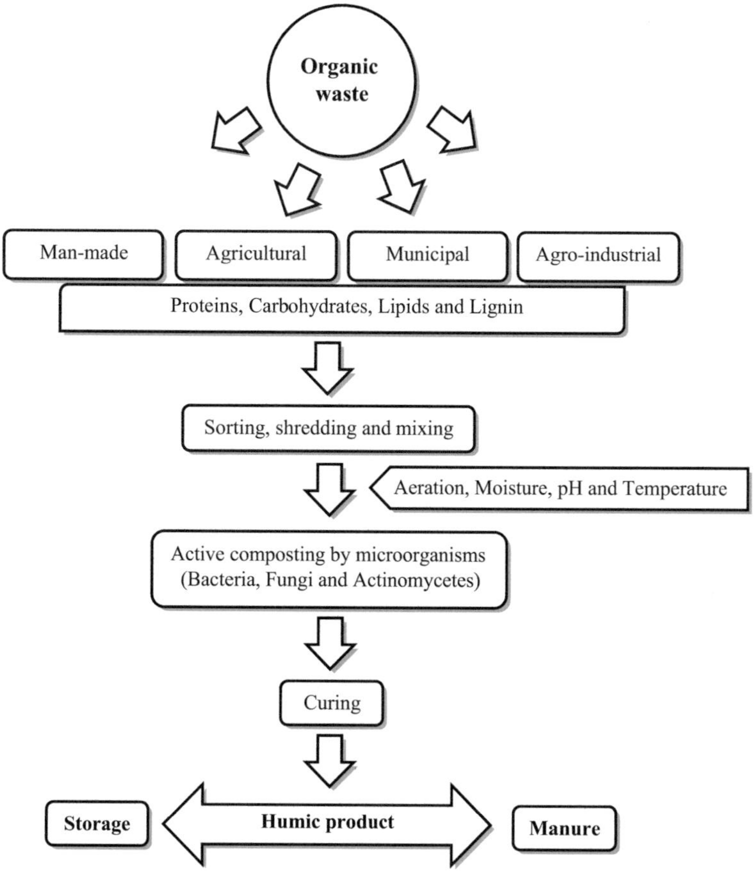

Fig. 3.1 Composting process of organic waste (Schaub and Leonard 1996)

"bioorganic manure or biofertilizers". These biofertilizers are capable of improving soil fertility and subsequently, increasing crop production.

3.2 Types of Organic Waste

Due to human, farming, industrial and municipal activities, a variety of organic waste has been accumulated. All the wastes have the potential to be recycled back into the natural system.

3.2.1 Municipal Solid Waste

Sustainable management of municipal solid waste is a critical issue for municipal authorities all around the world. Landfilling is least cheap and a favourable option in the management of municipal solid waste in many countries (Jeswani and

Azapagic 2016). The municipal solid waste is normally assumed to include all non-industrial community waste. Regulatory measure has been drawn up by the Environmental Protection Agency (EPA) and the Council of Agriculture (COA) to promote the agricultural development towards being more organic and sustainable in the future. Composting of municipal solid wastes is a potential and beneficial recycling tool in urban situation (Braber 1995; Tsai 2008).

3.2.1.1 Food Waste

The management of food wastes, especially generated from kitchen and restaurant, is of great importance because of their potential environmental hazards and public health risks. The best way of managing food wastes is to prevent and minimize its generation at the source level. However, the preventive approach is somewhat limited and sometimes impractical because of the ready biodegradability and the increasing ratio of eat-outs. The recycling of food wastes via composting and reusing it in agroecosystem not only ease the environmental and economic burdens of society but also improve the quality of agricultural soils and upgrade land productivity. The management of food waste can be divided into three major approaches such as on-site recycling, collection and transfer and off-site treatment (Tsai 2008).

3.2.1.1.1. On-Site Recycling

Food waste is generally placed in separate plastic bags or small collection bins with covers. The sorted food garbages may also be directly reused as an organic fertilizer at the place (on-site) of generation source (Tsai 2008).

3.2.1.1.2. Collection and Transfer

The generators carry the sorted food wastes and deposit it into a collection bin, which is the characteristic of closed-compact equipment. The collection was generally made every day at a fixed route, location and specified time. Then, the food wastes were transported by clearance vehicle to the recycling facility for further treatment and utilization (Tsai 2008).

3.2.1.1.3. Off-Site Treatment

At present, an array of food waste recycling methods such as compost, swine-raising and anaerobic digestion has been followed. Major portion of food wastes is recycled through composting for manufacturing organic fertilizer. Because of their organic and biodegradable nature, food wastes are also treated by anaerobic digestion to produce organic fertilizer, raw material for soil modification and plant cultivation soil. The compost produced from the food waste has been listed as "Green Mark" item to facilitate its green procurement and promotion in the market (Tsai 2008).

3.2.1.2 Sewage Sludge and Waste Water

Sewage sludge and waste water are defined as the combination of domestic effluent (urine, faecal sludge, kitchen and bathing water) and water from commercial institutions, restaurants, hospitals, schools, etc. It is characterized by offensive odour, poor settling, foam of agitation and high organic and solid content. Spreading sewage sludge water from municipal wastewater treatment on open land is employed as a common practice. However, it is well known that sewage sludge without special treatment contains various pollutants and contaminants, which are (re)introduced into the environment by sludge land spreading and in turn have harmful effects on the environment and human health. The wastewater sludge is commonly reused in many parts of the world (Polprasert and Koottatep 2017).

3.2.1.3 Human Waste

The human excreta contain faeces and urine normally contains daily average of 30 g of carbon, 90 g of organic matter, 10–12 g of nitrogen and 2–3 g of phosphorous and potassium. Decomposed excreta can be used as fertilizer, soil conditioner, to increase soil water holding capacity and neutralize soil toxins and heavy metals. A proper management and adequate sanitation are needed to overcome a range of pathogenic organisms like bacterial, viral, protozoans and helminths present in excreta (Polprasert and Koottatep 2017).

3.2.2 Agricultural Wastes

Various agricultural activities such as horticulture, dairy farming, seed growing, nursery and market gardens and livestock breeding produce huge amount of agricultural wastes. Due to poor management and limited access to disposal facilities, most of the agricultural waste are burned or incinerated (Polprasert and Koottatep 2017).

3.2.2.1 Crop Waste and Residues

Crop residues are the most abundant, cheapest and most readily available organic waste with high lignocellulose content to be biodegraded naturally. It includes rice straw, wheat straw, stover of sorghum, maize, pearl millet, finger millet and sugarcane, leaves, yard trimmings and cut woods. In India, the major portion of crop residues is used as animal feed, as fuel in homes and as animal bedding. The lignocellulosic agro-wastes enrich the N-microbial biomass of soil upon decomposition (Polprasert and Koottatep 2017).

3.2.3 Agro-Industrial Waste

The agro-industries are the manufacturers of food, beverages, rice products, coffee, tea, coir products, tobacco, textiles, furniture, printing and paper products, rubber, etc. The major waste generated from these industries are husks, sawdust, flour wastes, dye wastes and rubber wastes which are plant products that offer potential recycling. The utilization of agro-industrial waste products as soil amendments poses problems due to their resistance to biodegradation and extremely low nutrient content (Polprasert and Koottatep 2017).

3.2.3.1 Oil Cakes

Oil mill effluents and oil cakes are generated in large quantities mainly from extraction of oil, washing and cleaning processes in the mill. These usually contain cellulosic material, fat, oil and grease, and if discharged, untreated effluent into water streams may cause considerable environmental problems. Since these wastes have a high content of organic matter and mineral elements, composting can reduce the mixture volume by 40–50% and, therefore, be potentially used as soil amendment or fertilizer to improve soil structure (Singh et al. 2010).

3.2.3.2 Sugarcane Waste

India is the largest producer of sugarcane and consumer of sugar in the world. The press mud or filter cake, bagasse, spent wash and molasses are generated as by-products of sugarcane industries and distilleries. In India, the annual by-product production by these industries is more than 8 Mt of press mud, 7.5 Mt of molasses, 43.8 Mt of bagasse and 7.4 Mt of bagasse ash. These wastes as can be used as soil manure to supply macro- and micronutrients (Dotaniya et al. 2016).

3.2.3.3 Coir Waste

Coir waste is a biomass residue generated during the extraction of coir fibre from coconut husk. Well-decomposed coir waste of 12.5 t/ha with recommended fertilizer is recommended to increase the agricultural yield.

3.2.4 Animal and Poultry Waste

The animal wastes are characterized into urinary waste, i.e. slurry or liquid manure from livestock and poultry waste, solid manure or farmyard manure and waste water produced from farms, feedlot, bedding, silage juices, etc. The livestock involved in production of large amount of organic wastes are dairy cattle, beef cattle, chickens, broilers, swine, sheep and lambs, ducks and horses. The composition of wastes differs variably and depends on the livestock, i.e. weight, species, size and age of the animal, feed and water intake, climatic conditions, etc. (Polprasert and Koottatep 2017). Good quality organic fertilizer from animal waste provides an opportunity for the agricultural sector to reduce their reliance on chemical fertilizer which improves the soil fertility and sustainability. Chicken litter has the mixture of faeces, wasted feeds, bedding materials and feathers (Wilkinson et al. 2011). Over several million tons of chicken litter is produced every year in the world, most of which is usually recycled and spread on arable land as a low-cost organic fertilizer (Enticknap et al. 2006). Poultry manure contains significant amounts of nitrogen because of the presence of high levels of protein and amino acids. Due to its high nutrient content, chicken litter has been used as the most valuable organic fertilizer (Chen and Jiang 2014).

3.3 History and Science of Composting

Composting is a very old process as old as the civilization of the mankind. The process of composting has been followed from about 10,000 years ago; in the ancient Akkadian empire in the Mesopotamian valley, people have started to use organic manure in agriculture. These evidenced for small-scale evolution in the Neolithic Bronze Age and Iron Age in Scotland about 6000 years ago. They used waste pits for the management of organic waste applied on soil field. Marcus Porcius Cato (234–149 BC) a Roman general described composting in his book named *De Agri Cultura*. During civilization (1800–1200 BC), he used large pits for making manure. It is found at Knossos, Crete. In the thirteenth century, Templars developed descriptions of composting which helps to understood soil biology and parameters of composting. There are many references to composting in the Bible, Talmud, ancient Arab writing, mediaeval texts and Renaissance literature. It was in India that the modern composting got its big start. Sir Albert Howard, a government agronomist, developed the so-called Indore method, named after a city in southern India. Sir Albert Howard, the father of organic method, published his ideas on organic gardening in the 1940 book named *An Agricultural Testament*. Composting remained a basic activity of farming until the twentieth century. Later on, due to the steadily increasing population of the world, in order to increase productivity, the costly synthetic fertilizers made in factories from non-renewable resources were introduced in farming. However, it was realized that the regular use of synthetic fertilizers does not improve the structure of the soil, and gradually harm the soil, and water due to phosphate, nitrosamine and heavy metal contamination.

3.3.1 Science of Composting

Composts are the stabilized and sanitized products that improve soil amendments and agro-industrial production (Zucconi and de Bertoldi 1987; Haug 1993; Jakobsen 1995; Gajalakshmi and Abbasi 2008). These humidified end products are mainly formed from composting of lignin, polysaccharides and nitrogenous compounds (Varadachari and Ghosh 1984; Fustec et al. 1989; Inbar et al. 1989). Oxygen uptake rates and carbon dioxide production are directly related to decomposition (Wang et al. 2017). The humic substances are classified into (i) humins which are insoluble in water at any pH, (ii) humic acids that are insoluble in water under acidic conditions and (iii) fulvic acids that are soluble in water under all pH conditions. In general, immature composts are obtained as intermediate products of thermophilic stage. During this period, fulvic acid content is very high when compared to humic acids. When decomposition starts, there is an increase in humic acid content without decreasing fulvic acids in compost (Aiken et al. 1985). The ratio between humic acids and fulvic acids as a percentage of the total organic matter is regarded as humification index. This also helps in evaluating maturity of composts (Riffaldi et al. 1986; Inbar et al. 1990; Chen et al. 1996). Compost acts as a slow release manure to improve soil quality and further to improve microbial population (Diaz et al. 2007a, b; D'Hose et al. 2016). On the basis of preparation of composts, they are distinguished as stable, mature and cured (Khan et al. 2014). On the basis of nature of the organic matter being composed, temperature and the degree of aeration, and agitation, composting process can be grouped into four main phases such as (i) mesophilic phase (moderate temperature phase), (ii) thermophilic phase, (iii) second mesophilic phase (cooling) and (iv) maturation phase (curing period). Several kinds of microbes are involved in all phases of composting (Ryckeboer et al. 2003b; Awasthi et al. 2017).

Mesophilic phase is an early stage and moderate temperature phase of composting. Mesophilic bacteria with an ambient temperature in the range of 20–50 °C are responsible for decomposition of mass. This phase usually retains the acidic nature. Easily degradable organic matters such as carbon sources which include monosaccharides, starch and lipids are readily utilized by mesophilic microorganisms. Organic acids are formed from these compounds which decrease pH of the reactions (Crawford 1983). The mesophilic phase lasts for 1–3 days (Keener et al. 2000).

In thermophilic phase, the exothermic process produces heat energy that results into increased biomass heat, and therefore, mesophilic bacteria are decreased but thermophilic bacteria are increased (Rynk et al. 1992). The increasing population of thermophilic bacteria is capable of degrading proteins, fats, cellulose, hemicellulose and lignin of organic matters and resulting in the liberation of ammonium and also increasing pH (Crawford 1983). Thus, maximum degradation was achieved along with the elimination of pathogenic bacteria, weed seeds and other non-thermotolerant organisms (Zucconi and de bertoldi 1987; Haug 1993; Jakobsen 1995; Cournoyer 1996; Ryckeboer et al. 2003b; Gajalakshmi and Abbasi 2008).

Cooling is the second mesophilic phase. The maturation phase of the composting is the curing phase. Due to mesophilic microorganisms, there is a decrease of the temperature, slower decomposition rate and depletion of degradable organic substrates (Gajalakshmi and Abbasi 2008). Mesophilic microorganisms are able to degrade the remaining sugar, cellulose and lignin (Miller 1993; de Bertoldi 1995; Zucconi and Bertoldi 1987). This phase usually lasts few months to years (Cournoyer 1996). Maturation, the curing phase, has been studied to determine when compost is mature enough to be added to the soil by establishing maturity and stability of the final product (Bernal et al. 2009; Insam and de Bertoldi 2007; Paradelo et al. 2010). CO_2, water, minerals, ammonium and lignin-humus complexes (Mathur et al. 1993) are produced during maturation phase. When compared to mineral fertilizers, the composts continuously release nutrients for the susceptibility of microbial biomass population for prolonged period (Murphy et al. 2007).

3.3.2 Microbial and Other Associated Parameters for Composting

Different critical parameters such as microbes, temperature, pH, C/N ratio, moisture, oxygen supply, porosity, etc. affect composting process (Wakchaure et al. 2010; Bernal et al. 2009; Diaz et al. 2007a, b; Viaene et al. 2017). Therefore, it is important to optimize the parameters for minimizing the time and costs and for improving the quality of end products (Michel et al. 2004; Solano et al. 2001).

3.3.2.1 Microbiological Parameters

Microbial biodiversity and activity are varying in composting process (Pathma and Sakthivel 2012). Different microbial parameters, including microbial population, rate of oxygen consumption, ATP content, dehydrogenase activity, microbial biomass and humic acid along with other compost parameters such as temperature, determine the composting process and compost quality. Positive correlations were found between temperature and microbial properties, including oxygen consumption rate, ATP content, dehydrogenase activity and microbial biomass. The humification parameters also showed significant correlations with microbial properties of the manure compost. Humic acid content in manure was positively correlated with the population of aerobic heterotrophic microbes and microbial biomass and negatively correlated with oxygen consumption rate, ATP content and dehydrogenase activity. Among these microbial parameters, dehydrogenase activity was the most important factor affecting compost temperature and humification parameters (Tiquia 2005).

3.3.2.2 Temperature

Temperature plays an important role in mesophilic and thermophilic stages of composting. Optimum temperature (65–70 °C) increases the biotransformation of organic materials into humified products and further eliminates the harmful pathogenic microbes (Hassen et al. 2001; Czekała et al. 2016). Biochar such as bamboo, pine chips, wood and rice straw are added as an additive in sewage sludge and poultry wastes to ameliorative compost quality in terms of enhanced decomposition and reduced greenhouse gases (Onwosi et al. 2017; Khan et al. 2014; López-Cano et al. 2016; Vandecasteele et al. 2016; Zhang et al. 2016; Chen et al. 2017).

3.3.2.3 pH

The pH range of 6.5–7.00 is required for active degradation (Zhang et al. 2016). The pH range of more than 8 results in impaired microbial activity, volatilization of ammonium and low stability of metals. In order to maintain the optimum pH for active degradation, elemental sulphur is used in the process of composting (Mari et al. 2005).

3.3.2.4 C/N Ratio

The ratio of C/N is required to be below 50 at the beginning and below 15 at the end of the process (Khan et al. 2014). Wet green materials such as leaf and meat contain more nitrogen, while, dry brown materials which include straw, sawdust and paper are rich in carbon than nitrogen (Rynk et al. 1992). At the beginning of the reaction, carbon source is high when composting progresses lead to gradual reduction in available carbon sources (Hao et al. 2004). The optimum C/N ratio at the start of the process should be below 30:1 and 20:1 at the end the process (Kavitha and Subramanian 2007). Reduction of carbon leads to loss of nitrogen. If the carbon ratio is more, the nitrogen source is used up before completion of degradation.

3.3.2.5 Moisture

In order to transfer of nutrients in the degrading organic matter, the moisture is important (Onwosi et al. 2017). Generally, the mixture should contain an optimum moisture between 50 and 60% (Gajalakshmi and Abbasi 2008). Large quantity of water can evaporate to control temperature, but reduced level diminishes microbial activity as well as decreases decomposition (Das and Keener 1997).

3.3.2.6 Oxygen Supply

As aeration is an important factor for composting, it should be provided at the beginning of composting process (Gao et al. 2010). Proper aeration is required to control temperature, moisture level and proper oxygen supply for condensation of organic matter. The ambient aeration is achieved between 15 and 20% (Miller 1992). Proper aeration requires the temperature range below 60–65 °C (Finstein et al. 1985).

3.3.2.7 Porosity

Substrate porosity is important in composting process. Porosity depends on particle size, and the maintenance of adequate porosity ensures proper aeration. Smaller particles in mixtures give balanced surface area for growth of microorganism. Bulking agents such as wood were added to increase the porosity of the sludge and enhance the composting process (Schaub and Leonard 1996; Bernal et al. 2009).

3.4 Types of Composting

In controlled composting, there are five major areas to be optimized. The balance between green and brown organic materials, homogeneous nature of compost mixture to control particle size, moisture content that transports substances within compost, oxygen flow and temperature are the required areas to be controlled for proper compost management. Different types of composting systems are most widely used to manage organic waste (Polprasert and Koottatep 2017).

3.4.1 Turned Windrow Composting

The organic waste materials are placed into long, narrow piles approximately six feet high by twelve feet wide and increase subject to convenience. This is the least sophisticated system which involves periodic turning of the windrow with a front-end loader about daily or once a week. Turning the pile will increase aeration and porosity of the mixture, which helps to improve decomposition rate (Ahmad et al. 2007; Polprasert and Koottatep 2017).

3.4.2 Aerated Pile Composting

In this method, an aeration piping system is constructed and the organic mixture is placed in piles over the base. The aerated static piles are constructed individually to allow the composting process in batches. The piles can be constructed as extended piles. A blower is connected to draw air through the pile or to blow into the pile by temperature-controlled sensor placed within the pile (Ahmad et al. 2007, Polprasert and Koottatep 2017).

3.4.3 In-Vessel Composting

In-vessel composting systems are highly controlled aeration system which includes provisions for aeration, stirring, turning, temperature and moisture control for a better composting. The composting process takes place in a variety of containers or vessels or reactors and can be run in continuous or batch modes. In-vessel systems are generally the costliest and proprietary and, hence, require greater operation, higher skill level and maintenance costs (Ahmad et al. 2007; Polprasert and Koottatep 2017).

3.4.4 Vermicomposting

Earthworms are often noticed in composting heaps and act as pulverizers of organic materials to promote microbial activity and facilitate aeration during composting (Edwards et al. 1985). Earthworms feed on organic wastes, ingest more than their body weight and use only 5–10% of the feedstock for their growth. Worm casts are rich in macro- and micronutrients and, also, rich in vitamins, enzymes, antibiotics, growth hormones and immobilized microflora (Bhawalkar and Bhawalkar 1993; Kale et al. 1982). Therefore, vermicomposting is an appropriate technique for the disposal of non-toxic solid and liquid organic wastes. It helps in cost-effective and efficient recycling of animal wastes, agricultural residues and industrial wastes using low energy (Jambhekar 1992).

3.5 Diversity and Function of Composting Microbiota

Biocycle of nutrients is mediated by microorganisms. The biotransformation is a biological modification that alters the chemical structure of the compounds and includes both syntheses of atoms or simple molecules into complex molecules, i.e. biosynthesis or breakdown of complex structure into simpler elements by biodegradation or mineralization.

Biodegradation of organic compounds usually occurs in soil and sediments on the soil surfaces. The decomposing substrates and the degraded materials have physical contact; thus, exchange of nutrients of the degrading material and the matrix is possible. This dispersal of degrading material takes place at ambient temperature. Understanding the role of microbiota in composting is depending on the efficient approaches for isolation and identification of microbes (Miller 1996). Composting induces high metabolic activities of microorganisms. The change in conditions such as temperature, pH, aeration, moisture and availability of substrate results in stages of exponential growth and stationary phases of various organisms. The microbial consortium of the compost at any point of time is replaced by other in short intervals. The problem is only a minor fraction of the microbes are cultivable. In microbiological view, biodegradation in nature is commonly comparable to continuous culture, and the most important determinant factors are external such as substrate quality, temperature and moisture. Continuous composting processes may be regarded as a sequence of continuous culture, each of them with their own physical, chemical and biological properties. These changes make it difficult to study the process which is virtually impossible to study in the laboratory conditions.

In contrast, composting resembles a batch culture with steady changes in substrate concentration and biochemical conditions, and this microbiological process of composting is divided into major types based on the characteristic features of the microbiota involved in the process of decomposition of organic waste.

3.5.1 Function of Microbiota in Different Phases of Composting

3.5.1.1 Mesophilic Phase

This phase is also called as starting phase, in which energy-rich, easily degradable compounds such as sugars and proteins are abundant and degraded by fungi, actinobacteria and bacteria, which act as the primary decomposers along with compost worms, mites, millipedes and other mesofauna, which act as catalysts. The number of mesophilic organisms in the original substrate is three orders of magnitude higher than the number of thermophilic organisms, but the activity of primary decomposers induces a temperature rise (Insam and de Bertoldi 2007).

3.5.1.2 Thermophilic Phase

The organisms that adapt to high temperature get a competitive advantage and replace the mesophilic flora. The decomposition continues to be fast until the temperature of about 62 °C. Thermophilic fungi have maximum growth between 35 and 55 °C, while higher temperature inhibits fungal growth. Thermotolerant and thermophilic bacteria including actinobacteria are known to remain active at high temperatures. Even at high temperatures of about 65 °C, the temperature-stable enzymes produced by actinobacteria are involved in exothermic reactions. There are four major zones identified within a compost pile. The outer zone is the coolest and well supplied with oxygen, the inner zone is poorly supplied with oxygen, the lower zone is the hot zone and well supplied with oxygen and the upper zone is the hottest zone and fairly supplied with oxygen (Eliot Epstein 1996; Insam and de Bertoldi 2007). The thermophilic phase is important for hygienization where the human and plant pathogens are destroyed. It is not only because of the temperature but also the specific flora dominated by actinobacteria is important for hygienization through the production of antibiotics (Insam and de Bertoldi 2007; Polprasert and Koottatep 2017).

3.5.1.3 Cooling Phase

When the activity of the thermophilic organisms ceases due to the exhaustion of substrates, the temperature starts to decrease. Mesophilic organisms recolonize the substrates, either originating from surviving spores, through spread from protected micro-niches, or from the external inoculation. While in the starting phase, organisms with the ability to degrade sugars, oligosaccharides and proteins dominate; the second mesophilic phase is characterized by an increasing number of organisms that degrade starch or cellulose; among them are both bacteria and fungi (Insam and de Bertoldi 2007).

3.5.1.4 Maturation Phase

In maturation phase, the quality of the substrate declines, and in several successive steps, the composition of the microbial community is essentially altered. Usually, the proportion of fungi increases, while bacterial population decreases. Compounds that are not further degradable such as lignin-humus complexes are formed and found predominantly. The dry weight loss of materials during composting increases the concentration mineral. The total N concentration increases during composting due to concentration effect, and the evolution of N forms shows the mineralization of the organic compounds with formation of ammonia (Bernal et al. 2009; Polprasert and Koottatep 2017).

3.5.2 Microbial Decomposition of Organic Waste

It was first proved that self-heating of composts is due to the biological activity (Browne 1933). The microbial population dynamics of the compost was first studied by Waksman (1932). Since appropriate isolation and culturing procedures are considered as basis for studying the microbial communities of compost. It is known that from soil, only a minor fraction of organisms, i.e. less than 1%, is cultivable. The DNA- and RNA-based approaches have shown that many unknown species of microorganisms are yet to be identified from compost (Beffa et al. 1996). The development of next-generation sequencing analysis has inspired numerous recent studies in environmental microbiology. Blanc et al. (1999) investigated a clone library of bacterial 16S rRNA genes in order to characterize thermophilic communities in hot compost. Kowalchuk et al. (1999) characterized ammonia-oxidizing bacteria by analysing genes encoding ribosomal and ammonia monooxygenase. Peters et al. (2000) investigated microbial community successions during composting by means of single-strand conformation polymorphism (SSCP) profiles. It is shown that the microbial community analysis of the compost using different methods such as denaturing gradient gel electrophoresis (DGGE) and SSCP resulted in different community compositions (Alfreider et al. 2002).

3.5.3 Diversity of Composting Microbiota

3.5.3.1 Bacteria

Bacteria are responsible for most of the decomposition and heat generation in compost. They are the most nutritionally diverse group of compost organisms, using a broad range of enzymes to chemically break down a variety of organic materials (Eliot Epstein 1996). The importance of non-mycelial bacteria during the composting process was long neglected, probably because of the better visibility of fungi and actinobacteria. In the composting of sewage sludge, bacteria are more important than fungi. More than 40% of the solids are degraded by bacterial activity (Strom 1985). The temperature range of 50–65 °C is of advantage for bacteria, such as the genus *Bacillus*. If the temperature exceeds 65 °C, *B. stearothermophilus* dominates the process. Obligate anaerobic bacteria are also common in composts. Sulphate reduction under thermophilic conditions during the preparations of *Agaricus* substrates is also reported (Eicker 1981).

Anaerobic bacteria have the potential to degrade materials in the absence of oxygen through fermentation process. However, this anaerobic composting is a time-consuming process which requires over a period of 4–12 months. Also, this process produces lesser heat production which leads to partial eradication of pathogenic microbes. Anaerobic microbes during composting process release foul odour-producing gases such as methane (CH_4) and hydrogen sulphide (H_2S). Further, the nutrients are not retained in composts after decomposition (Sait et al. 2002).

Aerobic bacteria perform decomposition in the presence of oxygen. This process is usually a rapid process that takes only 8–10 weeks to complete decomposition of organic matter. Active composting produces reduced unpleasant odour gases but higher exothermic heat which ultimately leads to rapid destruction of pathogenic organism (Larney et al. 2006; Bernal et al. 2009). This process results composts with rich bioavailability of multiple nutrients (Golabi et al. 2003; Larney et al. 2006; Li et al. 2012; Zhang et al. 2016; Awasthi et al. 2017).

The source of the organic materials plays an important key role in determining nature and diversity of microbial population involved in composting process (Klammer et al. 2008; Ryckeboer et al. 2003b; Vargas-Garcia et al. 2010). The nature of starting materials and parameters decides the microbial behaviour, structure and microbial biodiversity (López-González et al. 2015).

3.5.3.2 Actinobacteria

These are class of bacteria that prefer neutral or slightly alkaline pH and are able to degrade relatively complex compounds. Species of actinobacteria are thermotolerant with a temperature range from 50 to 60 °C. They require moisture and good oxygen supply for growth. These conditions are naturally provided when the easily degradable substrates are digested by primary decomposers and the temperature rises beyond 45 °C. Actinobacteria are visible by the naked eye in thick mats (Insam and de Bertoldi 2007). *Thermus* or *Deinococcus* group of organisms grow on organic substrates at temperatures from 40 to 80 °C, with optimum growth between 65 and 75 °C. Beffa et al. (1996) reported *Thermus* species, as high as 10^7–10^{10} g^{-1} dry weight of compost. Autotrophic bacteria are also present in the compost.

3.5.3.3 Archaea

Many archaea are known to be thermophilic or hyperthermophilic. They have primarily isolated from hypothermal vents. A few cases have also been isolated from composts; considerable methanogenesis in compost piles has been reported (Cabanas-Vargas and Stentiford 2006). The reason for the relatively low abundance of archaea is that they are usually oligotrophic, and their generation times are much higher compared to bacteria, which make them unsuited for the rapidly changing conditions.

3.5.3.4 Fungi

Fungi are responsible for the decomposition of many complex plant polymers in soil and compost. In compost, fungi are important because they break down tough debris including cellulose. They can attack organic residues that are too dry, acidic or low in nitrogen for bacterial decomposition (Eliot Epstein 1996). From the starting phase of the composting, fungi compete with bacteria for easily available substrates. Since the maximum growth rate of bacteria exceeds that of fungi by one order magnitude (Griffin 1985), fungi are soon outcompeted. A good supply of oxygen is more important for fungal growth than for bacteria. Also, fungi are less thermotolerant and, therefore, play a very little role in the thermophilic phase. In composting of substrates rich in cellulose and lignin, fungi remain the most important throughout the entire process. The water potential decrease in the later part of the composting process favours fungi mostly (Insam and de Bertoldi 2007).

3.6 Mechanism That Mediates Bioconversion of Organic Substrates

Transformation of organic materials into useful products or energy source by biological processes or agents is known as bioconversion. In this process, microbial spores or cells are used to convert an organic compound to a structurally related compound through one or a few enzymatic reactions. The decomposition of organic materials follows common biochemical pathways of any other degradation process such as mineralization (aerobic respiration), humification (anaerobic respiration) and partial digestion (fermentation) (Insam and de Bertoldi 2007). The biomass of the world is synthesized through photosynthetic activity or consumer biomass. Essentially all the available biomass originates as a result of plant, animal or microbial activity. The lignin, cellulose, hemicellulose, xylan, pectin, starch and chitin are the natural substrates available in compost.

3.6.1 Lignin

Lignin degraders are mainly of white-rot fungi that contain thousands of species. In addition to white-rot fungi, brown-rot fungi, soft-rot fungi and also bacteria such as actinomycetes and myxobacteria are involved in lignin degradation (Buswell and Odier 1987; Rayner and Boddy 1988; Eriksson et al. 1990; Blanchette 1995). Lignin degrading eubacteria can be divided into erosion, cavitation and tunnelling bacteria (Eriksson et al. 1990; Blanchette 1995). Bacteria that belong to genera such as *Pseudomonas*, *Alcaligenes*, *Archaebacteria*, *Nocardia* and *Streptomyces* can degrade significantly aromatic compounds (Vicuna 1988; Ruttimann et al. 1991). Aerobic microbes are the primary lignin degrader; anaerobic microbes are also involved in the degradation of plant fibres (Kuhed et al. 2006). Complete organic compounds such as lignin are degraded by thermophilic microorganisms (Tuomela et al. 2000). Composting is a general treatment method for organic wastes. Organic materials are converted to CO_2, humus and heat by group of microorganisms. Humus is formed mainly from lignin. Lignin is not totally mineralized during composting. Elevated temperature in thermophilic phase is essential for rapid degradation of lignocellulose.

Lignin is a class of organic complex polymer that forms the major structural component of the plants. It is important for the formation of cell walls especially wood and bark; the lignin content of wood varies from 18 to 30% and is hard to degrade (Insam and de Bertoldi 2007). The extraordinary variety of bonding among the basic monomer compounds makes its degradation very complex. Yet, nature has found ways to degrade lignin through a dedicated ligninolytic enzyme system (Gonzalo et al. 2016). The degradation of lignin has been thoroughly studied in fungi, which are often pathogens such as *Trametes versicolor* and *Stereum hirsutum*. Bacteria also degrade lignin due to their enzymatic potential. Heme-containing peroxidases have received attention due to their ability to degrade lignin and other compounds (Sugano 2009; Colpa et al. 2014; Singh and Eltis 2015). Peroxidases, from *Bjerkandera adusta*, were isolated and characterized in 1999 (Kim and Shoda 1999). A large number of bacterial enzymes have been reported (Lambertz et al. 2016). Most studied bacterial laccases in lignin degradation are from actinomycetes, particularly from *Streptomyces* species (Fernandes et al. 2014). Glutathione-dependent etherases are capable of catalysing the reductive cleavage of ether bonds in lignin-related compounds (Masai et al. 1989). Two bacterial manganese-dependent superoxide dismutases (MnSODs) are reported to have lignin-degrading activity (Rashid et al. 2015). A peculiar dioxygenase and lignin bacterial enzyme with dioxygenase domain and a lignin-binding domain have been reported (Bianchetti et al. 2013).

3.6.2 Cellulose

Cellulose is the most abundant component found in plants. It is also the major component in every type of organic waste and is most prominent in wastes dominated by plant remains with a high percentage of structural elements. Cellulose molecules are chain of β-D-glucose with a polymerization degree of 40,000. The glucose molecules are combined by β-1,4-glycosidic bonding. Enzymatic cleavage results from the activity of three enzymes such as (i) endo-β-1,4-glucanases cleave the β-1,4 bonds within the molecules, which results in long-chain free ends, (ii) exo-β-1,4-glucanases separate the disaccharide cellobiose from the free ends and (iii) β-glucosidases hydrolyse the cellobiose and the resulting glucose is taken up by microorganisms.

Under aerobic conditions, fungi, bacteria and myxomycetes are involved in the degradation of cellulose. The catalytic action of the enzyme or the mechanical destruction of large structural elements by the microfauna is considered as important. Fungi are particularly more important in cellulose degradation compared to bacteria. Since cellulose is rich in carbon and does not contain nitrogen or other essential elements, the mycelia structure of fungi is a competitive advantage, for example, *Chaetomium*, *Fusarium* and *Aspergillus*. Among the bacteria, myxomycetes, *Cytophaga*, *Polyangium*, *Sorangium* and *Pseudomonas* are known to degrade cellulose, but only few actinobacteria are involved (Inzam and de Bertoldi 2007).

3.6.3 Hemicellulose

Hemicellulose, known as polyose, is any of the several forms of heteropolysaccharides (matrix polysaccharides) present along with cellulose. While cellulose is crystalline, strong and resistant to hydrolysis, hemicellulose has a random, amorphous structure with little strength. Hemicelluloses are polysaccharides present almost in all plant cell walls that have beta-(1-4)-linked backbones with an equatorial configuration. The most important biological role of hemicelluloses is their contribution to strengthening the cell wall by interaction with cellulose and, in some cases, with lignin (Scheller and Ulvskov 2010).

3.6.4 Xylan

Among the hemicelluloses, xylan is the most important and major component present in straw, bagasse (up to 30%) and wood (2–25%). Xylan is made up of pentoses such as xylose, arabinose and hexoses such as glucose, mannose and galactose. The degree of polymerization is 30–100. The main degrading enzymes are xylanases, produced by many bacteria and fungi (Insam and de Bertoldi 2007).

3.6.5 Pectin

Pectin is also known as polygalacturonides and is made up of unbranched chain of polygalacturonic acid, degraded by pectinases that are found in a wide range of fungi and bacteria. The enzymes that degrade pectin substances are of two types:

those that attack polyuronic acid chain (depolymerases) and those that remove the methyl group from pectins to produce pectic acid and methanol (de bertoldi 2013).

3.6.6 Starch

Starch is composed of amylase and amylopectin. Amyloses are unbranched chains of D-glucose, due to 1,4-β-glycosidic bonding, in contrast to cellulose. Amylopectin is branched at 1,6 position and contains phosphate residues with Ca and Mg ions. The two important enzymes involved in the degradation of starch are phosphorolysis by phosphorylases, acting on the non-reducing end of amylase releasing glucose-1-phosphate molecules and the other enzyme α-amylase that cleaves the α-1,4 bonds within the molecule (Insam and de Bertoldi 2007).

3.6.7 Chitin

Chitin, a polymer of N-acetyl glucosamine, is the most abundant biological molecule in nature and interacts with both carbon and nitrogen cycles (Beier and Bertilsson 2013). Chemically cellulose and chitin are very similar; also murein in bacterial cell wall is considered as chitin structural homologue. Chitin is the most important and structural element in many organisms such as fungi, crustaceans, insects or algae. Chitin is degraded through exoenzymes to N-acetyl glucosamine and transformed to fructose-6-phosphate and incorporated in carbohydrate metabolism (Insam and de Bertoldi 2007). The process of degradation of chitin is called chitinoclastic; if the process involves initial hydrolysis of 1-4-β-glycoside bond through chitinase-catalysed chitin degradation, then it is called chitinolytic degradation. Fungi such as *Aspergillus* and bacteria such as *Flavobacterium*, *Cytophaga* and *Pseudomonas* are capable of utilizing chitin as carbon and nitrogen source (Beier and Bertilsson 2013).

3.7 Microbiological and Associated Methods for Analyses of Compost Maturity

The stability and maturity of the compost of a particular sample are very difficult to determine by using any single analytical parameter. The compost which is not undergoing rapid decomposition is called “stable compost”. In contrast, the compost which may release nitrogen into soil or tie up nitrogen from soil is called “unstable compost”. The term mature refers to the degree of phytotoxicity of compost. The immature compost will contain more growth-inhibiting compounds than in mature compost. Conventional compost analysis has focused on NPK and micronutrient concentration as done for fertilizer analysis, by analysing few chemical parameters such as pH, ammonia and C/N. However, compost is much more complex than fertilizer, and its significant value for plants may be far more than its mineral contribution to soil. The microbiological component determines how the compost toils as a soil inoculant and plant disease suppressant. Concise compost maturity and quality criteria are still lacking, and usually do not include any microbiological test, despite the fact that microbial activity is the major process for compost production. Rather than chemical parameter analysis, microbiological test can provide much more information on the maturity and quality of the compost.

3.7.1 Prerequisites

Composting is the process of biological activities; continuous process of degradation takes place even after sampling. Hence, for this reason, the storage of samples is very important for biological assays than chemical tests. The samples should be stored in sterile insulated container that allows exchange of oxygen, provided that the samples have to be analysed within 24 h. For samples stored more than 24 h, a temperature of <4 °C is recommended. The biosolid compost samples were found to be unaffected by refrigeration up to 11 weeks (Butler et al. 1999). While any microbiological method that employs living cells depends on recovery of the cells after storage, molecular biological methods using drying at room temperature also appear to be appropriate.

3.7.2 Estimation of Microbiota

The standard microbiological content analysis of the compost is determined by the estimation of six functional groups of microorganism, namely, aerobic bacteria, anaerobic bacteria, fungi, actinobacteria, pseudomonads and nitrogen-fixing bacteria (Bess 1999).

3.7.3 Heat Evolution

A simple and rapid means to evaluate compost maturity is to determine the temperature. If the temperature of the compost is more than 8 °C than the ambient air, the compost is fairly unstable. The self-heating test in Dewar flasks is a frequently used test to assess the quality of the compost. It is based on the self-heating potential of the biodegrading organic material through the activities of microbes. A slow or no rise in temperature indicates a high degree of maturation of the compost. The Dewar flask test is considered a very robust test, and it is the most sensitive indicator of compost maturity (Insam and de Bertoldi 2007).

3.7.4 Respiration

Bioconversion of organic materials continues throughout the composting process. Microbial degradation of organic materials into water, CO_2, inorganic compound and humic matter under aerobic condition is called respiration. The respiratory activity is directly related to the endogenous supply of degradable material. The respiration activity may be estimated either by estimation of O_2 consumption or CO_2 production (Insam and de Bertoldi 2007).

3.7.5 Community-Level Physiological Profiling

This method is based on the utilization of different carbon substrates. The compost extracts are inoculated onto microtiter plates containing 31 different carbon substrates. The utilization pattern of these substrates indicates if the compost is mature or not (Lulu et al. 2001). Screening for various enzymes such as reductase, endocellulase, glucosidase, lipase, phosphatase, dehydrogenase and arginine ammonification has been proposed for analysing the maturity of the compost.

3.7.6 Community-Based Maturity Test

Nucleic acid-based methods for the detection of the microbial ecology of the compost have been recently intensified. Steger et al. (2007) have evidenced that compost maturity is linked to the microbial community structure of the compost. The combination of signature lipid and nucleic acid-based analyses greatly expands the specificity and the scope for assessing the microbial community composition in composts. Many compost microorganisms have been detected with molecular-based methods that had not been detected with classical isolation procedure. DNA chip technology offers the possibility to screen for the presence or absence of hundreds of thousands of microbial species in the environmental samples (Ogram 2000). Specific compost-targeted microarrays are suitable to investigate bacterial (Franke-Whittle et al. 2005, 2009) and fungal community (Hultman 2009) including general decomposers as well as plant growth-promoting microorganisms and pathogens.

3.7.7 Risk Management of Pathogens in Composting

The composting of human or animal waste involves the risk of pathogens. The primary pathogens such as bacteria, viruses, protozoa and helminths can initiate an infection in healthy individuals, while secondary pathogens usually infect people with debilitated immune system, weakened by primary infections. Microorganisms such as *Salmonella enteritidis*, *Entamoeba histolytica*, *Ascaris lumbricoides*, *Hepatitis virus*, *Aspergillus fumigatus* and *Micromonospora* spp. are found during composting of human or animal waste. The contact or inhalation of air containing high density of spores of secondary pathogens can cause hazards to the compost workers and users (Polprasert and Koottatep 2017).

The two major parameters that control or responsible for pathogen die-offs during composting are time and temperature. The complete inactivation of pathogens in a compost pile is hard to achieve because of reasons such as the heterogeneity of the compost materials, uneven distribution of temperature in the compost pile and partial inactivation of pathogens. Many pathogens such as spore-forming bacteria, cysts and helminth ova are partially inactivated during composting (Polprasert and Koottatep 2017). Several researches (Kawata et al. 1977; Copper and Golueke 1979) reported several orders of magnitude reduction of bacteria and viruses during composting. Better inactivation rate is observed with the Beltsville Aerated Rapid Compost (BARC) system in which bacteria including *Salmonella* were reduced after 10 days of composting and F_2 bacteriophages in about 15–20 days. The secondary pathogen such as *Aspergillus fumigatus* may be controlled through moisture control management practice, as fungi tend to grow in dry conditions (Millner et al. 1977; Finstein et al. 1980).

3.8 Significance and Conclusions

Organic waste composts have several benefits to application in agricultural crops as well as on soil structural fertility and plant growth and development (Esse et al. 2001). The utilization of composts in agriculture is related to the sustainability of both agriculture and waste disposal. Composting, as sustainable transformation of potential wastes into organic fertilizers, tunes up with sustainable agriculture and must be optimized and encouraged. The importance of the use of compost in agriculture is a commonplace and does not correspond to reality, as if rules and technical parameters for its application are not well defined. The use of compost in agriculture has much more than the importance of the application of an amendment to the soil. The use of compost is shown to enhance sustainability, not only of the agricultural activity but in a more general context. Sustainable agriculture and the use of compost in agriculture can be considered as essential activities for a sustainable society (Sequi 1996).

The mature compost differs in its quality and stability. It depends upon the composition of raw materials used for the compost production (Gaur and Singh 1995; Ranalli et al. 2001). Microbial community plays a key role in composting process, and appearance of some microorganisms reflects the quality of maturing compost (Ishii et al. 2000; Ryckeboer et al. 2003a). Lack of adequate nutrient supply and poor soil structure are the principal constraints to crop productivity under low inputs. Application of organic residue material or composted organic residue influences the soil structure and fertility with special addition of nutritional content (Courtney and Mullen 2008). Intensified agricultural practices by using organic and chemical fertilizer affect the biodiversity of soil microorganisms and adversely affect soil quality. According to D'Hose et al. (2014), long-term farm compost amendments resulted in strengthening of the soil quality, while effectiveness of plant hormones blended with recycled organic waste was observed to improve growth and yield of wheat (Zahir et al. 2007). Moss et al. (2002) examined that the nonrecyclable petroleum-based fertilizers associated with the "green revolution" displaced the use of manures and human waste as compost. Martnez-Blanco et al. (2014) reviewed that compost through life cycle assessment ascertained nutrient supply, carbon sequestration and pest and fungal disease suppression, increased crop yield, decreased soil erosion and retention of soil moisture, increased soil work ability, enhanced soil biological properties and biodiversity and enhanced crop nutritional quality. Compost application could contribute to increase food availability, and therefore, efforts should be made to alleviate the socio-economic constraints to the adoption of compost technology (Ouédraogo et al. 2001). The compost application increases soil organic carbon content, hot water extractable carbon, earthworm's numbers and microbial biomass and reduces soil bulk density and crop yield (D'Hose et al. 2012). The end use of the finished product decides the importance of the product in the market. In addition, the availability of the market that provides an important source of revenue defrays some of the cost of processing and contributes to the financial availability of the overall waste management strategy.

Compost has the potential to be used as soil amendment in various applications and, therefore, has innate potential in agriculture, landscaping, gardening, nurseries, top dressing and land reclamation (Diaz et al. 2007a, b). Screening, grinding or combination of similar processes should be done to remove plastics, glass and other materials from the compost, otherwise objectionable to use. Immature compost is used for landfilling or land reclamation. For agricultural purpose, a coarse grind is satisfactory, whereas for horticulture and gardening, the compost must be finer. Composts are used to improve the overall soil fertility through the addition of organic waste by modifying the soil pH. The long-term use of composts in agricultural soil results in reduced erosion, increased water retention capacity, improved soil structure and rapid establishment of vegetation (Polprasert and Koottatep 2017). Microbe-fortified composts act as organic amendment for soil fertility, soil reclamation, greenhouse transplant media and pathogen and weed suppression (Dukare et al. 2011).

Organic matter plays a key role in preserving soil fertility and increases the nutrient value of the soil. Nutrients are originally present in an organic form that are biotransformed to an available form for plants through mineralization which takes place at different rates based on the degree of degradation of the organic matter in the compost, soil fertility and climate. A good quality of compost for soil amendment is expected to have high concentration of organic matter, high degree of organic matter stability and low content of nutrient in order to increase the amount of compost added to the soil (Sequi et al. 2000). Beneficial effect of compost on grapevines has been reported. The compost-treated vines are healthier with grow and greener foliage, show fewer nutrient deficiencies in the leaves and suffer less from drought. Compost-treated vines also resist disease and have a longer productive life.

Adding nutrients to the landscape is an important part of land stewardship. Landfill reclamation extends the life of the current facility by removing recoverable materials and reducing waste volume through composting. Compost is effectively used for soil reclamation. However, lower-quality compost is used for landfilling. The process of reclamation depends either on soil or the type of waste. In order to reduce the contamination and other health risks, environmental technologies are required to maintain the levels of concentration that do not create any harm to the human health and safety. There are provisions for the quality management of environmental projects such as characterization of environment system, characterization and quantification of hazardous waste in the environment and their biological effects (Sequi et al. 2000).

The nursery and greenhouse growers integrate the growth media with mineral fertilizers. The transplant industry for the production of plants relies on peat moss as a major ingredient in soil-less media; the economic competitiveness of compost is evident, as peat is expensive and non-renewable and has a low nutrient. Compost supports seed emergence and seedling growth. However, the use of compost in transplanting medium depends upon the characteristics such as pH, salinity, pore space and water retention capacity (Sequi et al. 2000).

Due to its physical surface cover and phytotoxic compounds, the organic material has the ability to suppress the weed. Chemical effects of phytotoxic compounds (volatile fatty acids and/or ammonia) in compost decrease the germination of weed seeds. Inhibition of germination or subsequent growth due to physical effect of the mulch and the presence of phytotoxic compounds (fatty acids) in the compost are considered important for weed control. Antagonistic bacteria amended in compost preparations also facilitate the suppression of diseases caused by plant pathogenic fungi such as *Fusarium oxysporum*, *Pythium debaryanum*, *P. aphanidermatum* and *Rhizoctonia solani*. Therefore, beneficial microbe-fortified compost act as amendment for soil reclamation, disease and weed suppression (Dukare et al. 2011).

Socio-economic factors relate to health regulations, state laws regarding the sale of fertilizers, soil conditions, public acceptance and marketing value. The distribution and utilization of compost produced from sewage has to be regulated for stabilization, pathogen reduction and chemical analysis. The extent of compost utilization depends upon the type of market and its proximity to the processing facility, since the transportation and distance to the markets will affect the product value and its potential use (Polprasert and Koottatep 2017). Economic dimension of the compost depends not only on the quantity of the product sold in the market but also upon its cost-effectiveness in the market. Studies on compost marketing hierarchy with market prices and volume, in which different market segments with relative size and price, were reported (Carlsbaek 2000). Further research on microbial management of organic waste in agroecosystem is required to improve quality and cost-effectiveness.

References

Ahmad R, Jilani G, Arshad M, Zahir ZA, Khalid A (2007) Bio-conversion of organic wastes for their recycling in agriculture: an overview of perspectives and prospects. Ann Microbiol 57(4):471–479

Aiken GR, McKnight DM, Wershaw RL, MacCarthy P (1985) An introduction to humic substances in soil, sediment, and water. In: Aiken GR et al (eds) Humic substances in soil, sediment and water: geochemistry, isolation and characterization. Wiley-Interscience, New York, pp 1–9

Alfreider A, Peters S, Tebbe CC et al (2002) Microbial community dynamics during composting of organic matter as determined by 16S ribosomal DNA analysis. Compost Sci Util 10:303–312

Awasthi M, Wang M, Chen H et al (2017) Heterogeneity of biochar amendment to improve the carbon and nitrogen sequestration through reduced greenhouse gases emissions during sewage sludge composting. Bioresour Technol 224:428–438
Beffa T, Blanc M, Lyon PF et al (1996) Isolation of Thermus strains from hot composts (60 to 80 degrees C). Appl Environ Microbiol 62:1723–1727
Beier S, Bertilsson S (2013) Bacterial chitin degradation—Mechanisms and ecophysiological strategies. Front Microbiol 4:149
Bernal MP, Alburquerque JA, Moral R (2009) Composting of animal manures and chemical criteria for compost maturity assessment. A review. Bioresour Technol 100(22):5444–5453
Bess V (1999) Evaluating microbiology of compost. Biocycle 40(5):62–64
Bhawalkar US, Bhawalkar UV (1993) Vermiculture biotechnology. In: Thampan PK (ed) Organic in soil health and crop production. Peekay Tree Crops Development Foundation, Cochin, pp 69–85
Bianchetti CM, Harmann CH, Takasuka TE (2013) Fusion of dioxygenase and lignin-binding domains in a novel secreted enzyme from cellulolytic Streptomyces sp. SirexAA-E. J Biol Chem 288:18574–18587
Blanc M, Marilley L, Beffa T, Aragno M (1999) Thermophilic bacterial communities in hot composts as revealed by most probable number counts and molecular (16S rDNA) analysis. FEMS Microbiol Ecol 28:141–149
Blanchette RA (1995) Degradation of lignocellulose complex in wood. Can J Bot 73:S999–S1010
Braber K (1995) Anaerobic digestion of municipal solid waste: a modern waste disposal option on the verge of breakthrough. Biomass Bioenergy 9:365–376
Browne CA (1933) The spontaneous heating and ignition of hay and other agricultural products. Science 77:223–229
Buswell JA, Odier E (1987) Lignin biodegradation. CRC Crit Rev Biotechnol 6:1–60
Butler TA, Sikora LJ, Steinhilber PM (1999) Compost age and sample storage effects on compost stability/maturity indicators, U.S. Dept. of Agriculture, Agricultural Research Service. http://www.nal.usda.gov/ttic/tektran/data/000010/59/0000105929.html
Cabanas-Vargas DD, Stentiford EI (2006) Oxygen and CO_2 profiles and methane formation during the maturation phase of composting. Compost Sci Util 14:86–89
Carlsbaek MS (2000) In Amlinger F. Composting in Europe: where do we go? Paper for the international forum on recycling, Madrid
Chen Z, Jiang X (2014) Microbiological safety of chicken litter or chicken litter-based organic fertilizers: a review. Agriculture 4(1):1–29. https://doi.org/10.3390/agriculture4010001
Chen Y, Chefetz B, Hadar Y (1996) Formation and properties of humic substance originating from composts. In: The science of composting. Springer, Dordrecht, pp 382–393
Chen Y, Liu Y, Li Y et al (2017) Influence of biochar on heavy metals and microbial community during composting of river sediment with agricultural wastes. Bioresour Technol 243:347–355
Colpa DI, Fraaije MW, Van Bloois E (2014) DyP-type peroxidases: a promising and versatile class of enzymes. J Ind Microbiol Biotechnol 41:1–7
Copper RC, Golueke CG (1979) Survival of enteric bacteria and viruses in compost and its leachate. Compost Sci Land Util 20:29–35
Cournoyer MS (1996) Sanitation and stabilization of slaughterhouse sludges through composting. In Proceedings of the Canadian meat research institute technology symposium, pp 1–7
Courtney RG, Mullen GJ (2008) Soil quality and barley growth as influenced by the land application of two compost types. Bioresour Technol 99:2913–2918
Crawford JH (1983) Composting of agricultural wastes – a review. Process Biochem 18:14–18
Czekała W, Malińska K, Cáceres R et al (2016) Co-composting of poultry manure mixtures amended with biochar–the effect of biochar on temperature and C-CO 2 emission. Bioresour Technol 200:921–927
D'Hose T, Cougnon M, De Vliegher A et al (2012) Influence of farm compost on soil quality and crop yields. Arch Agron Soil Sci 58:S71–S75

D'Hose T, Cougnon M, De Vliegher A (2014) The positive relationship between soil quality and crop production: a case study on the effect of farm compost application. Appl Soil Ecol 75:189–198

D'Hose T, Ruysschaert G, Viaene N (2016) Farm compost amendment and non-inversion tillage improve soil quality without increasing the risk for N and P leaching. Agric Ecosyst Environ 225:126–139

Das K, Keener HM (1997) Moisture effect on compaction and permeability in composts. J Environ Eng 123(3):275–281

De Bertoldi M (1995) Composting food processing waste in the European economic community. Compost Sci Util 3(2):87–92

de bertoldi M (2013) The science of composting. Springer, Dordrecht, p 1452

Dees PM, Ghiorse WC (2001) Microbial diversity in hot synthetic compost as revealed by PCR-amplified rRNA sequences from cultivated isolates and extracted DNA. FEMS Microbiol Ecol 35(2):207–216

Diaz LF, de Bertoldi M, Bidlingmaier W, Stentiford E (eds) (2007a) Compost science and technology – waste management series 8. Elsevier, Boston, p 357

Diaz LF, Savage GM, Eggerth LL, Chiumenti A (2007b) Systems used in composting. In: Diaz LF et al (eds) Compost science and technology – waste management series 8. Elsevier, Boston, pp 67–87

Dotaniya ML, Datta SC, Biswas DR, Dotaniya CK, Meena BL, Rajendiran S, Regar KL, Lata M (2016) Use of sugarcane industrial by-products for improving sugarcane productivity and soil health. Int J Recycl Org Waste Agric 5(3):185–194. https://doi.org/10.1007/s40093-016-0132-8

Dukare AS, Prasanna R, Chandra DS et al (2011) Evaluating novel microbe amended composts as biocontrol agents in tomato. Crop Prot 30(4):436–442

Edwards CA, Burrows I, Fletcher KE, Jones BA (1985) The use of earthworms for composting farm wastes. In: Crasser JKR (ed) Composting of agricultural and other wastes. Elsevier Applied Science, Oxford, pp 229–241

Eicker A (1981) The occurrence and nature of sulphur crystals in phase I mushroom compost. Mushroom Sci 11:27–34

Eliot Epstein (1996) The science of composting. CRC Press, USA, p 504

Enticknap JJ, Nonogaki H, Place AR, Hill RT (2006) Microbial diversity associated with odor modification for production of fertilizers from chicken litter. Appl Environ Microbiol 72:4105–4114

Eriksson KEL, Blanchette RA, Ander P (1990) Microbial and enzymatic degradation of wood and wood components. Springer, Berlin

Esse PC, Buerkert A, Hiernaux P, Assa A (2001) Decomposition of and nutrient release from ruminant manure on acid sandy soils in the Sahelian zone of Niger, West Africa. Agric Ecosyst Environ 83(1):55–63

Fernandes TAR, da Silveira WB, Lopes Passos FM, Domingues Zucchi T (2014) Laccases from actinobacteria—What we have and what to expect. Adv Microbiol 4:285–296

Finstein MS, Morris ML, Strom PF (1980) Microbial ecosystem responsible for anaerobic digestion and composting. J Water Poll Control Fed 52:2675–2685

Finstein MS, Miller FC, MacGregor ST, Psarianos KM (1985) The Rutgers strategy for composting: process design and control. National Technical Information Service, Springfield

Franke-Whittle IH, Klammer SH, Insam H (2005) Design and application of an oligonucleotide microarray for the investigation of compost microbial communities. J Microbiol Methods 62:37–56

Franke-Whittle IH, Knapp BA, Fuchs J et al (2009) Application of COMPOCHIP microarray to investigate the bacterial communities of different composts. Microb Ecol 57:510–521

Fustec E, Chauvet E, Gas G (1989) Lignin degradation and humus formation in alluvial soils and sediments. Appl Environ Microbiol 55(4):922–926

Gajalakshmi S, Abbasi SA (2008) Solid waste management by composting: state of the art. Crit Rev Environ Sci Technol 38(5):311–400

Gao M, Liang F, Yu A et al (2010) Evaluation of stability and maturity during forced-aeration composting of chicken manure and sawdust at different C/N ratios. Chemosphere 78(5):614–619

Gaur AC, Singh G (1995) Recycling of rural and urban wastes through conventional and vermi-composting. In: Tandon HLS (ed) Recycling of crop, animal, human and industrial wastes in agriculture. Fertiliser Development and Consultation Organisation, New Delhi, pp 31–49
Golabi SM, Nourmohammadi F, Saadnia A (2003) Electrosynthesis of organic compounds.: Part II: electrooxidative amination of 1, 4-dihydroxybenzene using some aliphatic amines. J Electroanal Chem 548:41–47
Golueke CG (1973) Composting: a study of the process and its principles. Rodale Press, Emmaus
Golueke CG (1992) Bacteriology of composting. BioCycle, USA
Gonzalo DG, Colpa DI, Habib MHM, Fraaije MW (2016) Bacterial enzymes involved in lignin degradation. J Biotechnol 236:110–119
Griffin DM (1985) A comparison of the roles of bacteria and fungi. In: Leadbetter ER, Poindexter JS (eds) Bacteria in nature. Plenum Publishing, London, pp 221–255
Hao X, Chang C, Larney FJ (2004) Carbon, nitrogen balances and greenhouse gas emission during cattle feedlot manure composting. J Environ Qual 33(1):37–44
Hassen A, Belguith K, Jedidi N et al (2001) Microbial characterization during composting of municipal solid waste. Bioresour Technol 80(3):217–225
Haug RT (1993) The practical handbook of compost engineering. CRC Press, Boca Raton
Hoornweg D, Bhada-Tata P (2012) What a waste: a global review of solid waste management. Urban Dev Se Knowl Pap 15:1–98
Hoornweg D, Bhada-Tata P, Kennedy C (2013) Environment: waste production must peak this century. Nature 502(7473):615
Hultman J (2009) Microbial diversity in the municipal composting process and development of detection methods. Ph. D. thesis Vikki Graduate School in Bioscience. University of Helsinki, Finland
Inbar Y, Chen Y, Hadar Y (1989) Solid-state carbon-13 nuclear magnetic resonance and infrared spectroscopy of composted organic matter. Soil Sci Soc Am J 53(6):1695–1701
Inbar Y, Chen Y, Hadar Y, Hoitink HAJ (1990) New approaches to compost maturity. BioCycle 31(12):64–69
Insam, de Bertoldi M (2007) Microbiology of the composting process. In: Compost science and technology – waste management series, vol 8. Elsevier, Boston, pp 26–45
Ishii K, Fukui M, Takii S (2000) Microbial succession during a composting process as evaluated by denaturing gradient gel electrophoresis analysis. J Appl Microbiol 89(5):768–777
Jakobsen ST (1995) Aerobic decomposition of organic wastes 2. Value of compost as a fertilizer. Resour Conserv Recycl 13(1):57–71
Jambhekar HK (1992) Use of earthworms as a potential source to decompose organic waste. In: Proceedings of national seminar on organic farming. MPKV, Pune, India, pp 52–53
Jeswani HK, Azapagic A (2016) Assessing the environmental sustainability of energy recovery from municipal solid waste in the UK. Waste Manag 50:346–363
Kale RD, Bano K, Krishnamoorthy RV (1982) Potential of *Perionyx excavatus* for utilization of organic wastes. Pedobiologia 23:419–425
Kavitha R, Subramanian P (2007) Bioactive compost-a value added compost with microbial inoculants and organic additives. J Appl Sci 7(17):2514–2518
Kawata K, Cramer WW, Burge WD (1977) Composting destroys pathogens in sewage solids. Water Sew Works 124:76–79
Keener HM, Dick WA, Hoitink HA (2000) Composting and beneficial utilization of composted by-product materials. Soil Sci Soc Am Book Series 6:315–342
Khan N, Clark I, Sánchez-Monedero MA et al (2014) Maturity indices in co-composting of chicken manure and sawdust with biochar. Bioresour Technol 168:245–251
Kim SJ, Shoda M (1999) Purification and characterization of a novel peroxidase from *Geotrichum candidum* Dec 1 involved in decolorization of dyes. Appl Environ Microbiol 65:1029–1035
Klammer S, Knap B, etal IH (2008) Bacterial community patterns and thermal analyses of composts of various origins. Waste Manag Res 26(2):173–187

Kowalchuk GA, Naoumenko ZS, Derikx PJ et al (1999) Analysis of ammonia-oxidizing bacteria of the β subdivision of the class Proteobacteria in compost and composted material. Appl Environ Microbiol 65:396–403
Kuhad RC, Singh A, Tripathi KK et al (1997) Microorganisms as an alternative source of protein. Nutr Rev 55:65–75
Kuhed RC, Sing A, Eriksson KL (2006) Microorganisms and enzymes involved in the degradation of plant fiber cell wall. Biotechnol Pulp Pap Ind 57:45–125
Lambertz C, Ece S, Fischer R, Commandeur U (2016) Progress and obstacles in the production and application of recombinant lignin-degrading peroxidases. Bioengineering 7:145–154
Larney FJ, Sullivan DM, Buckley K, Eghball B (2006) The role of composting in recycling manure nutrients. Can J Soil Sci 86(4):597–611
Leonard J' and Ramer S. (1993). Physical properties of compost – what do we know and why should we care? proceedings of the 4th Annual Meeting of the Composting Council Canada. Ottawa
Li R, Wang JJ, Zhang Z et al (2012) Nutrient transformations during composting of pig manure with bentonite. Bioresour Technol 121:362–368
Lim SL, Lee LH, TY W (2016) Sustainability of using composting and vermicomposting technologies for organic solid waste biotransformation: recent overview, greenhouse gases emissions and economic analysis. J Clean Prod 111:262–278
Lohri CR, Diener S, Zabaleta I et al (2017) Treatment technologies for urban solid biowaste to create value products: a review with focus on low-and middle-income settings. Rev Environ Sci Biotechnol 16:1–50
López-González JA, Suárez-Estrella F, Vargas-García MC et al (2015) Dynamics of bacterial microbiota during lignocellulosic waste composting: studies upon its structure, functionality and biodiversity. Bioresour Technol 175:406–416
López-Cano I, Roig A, Cayuela ML et al (2016) Biochar improves N cycling during composting of olive mill wastes and sheep manure. Waste Manag 49:553–559
Lulu B, Egger W, Neunhäuserer C et al (2001) Can community level physiological profiles be used for compost maturity testing? Compost Sci Util 9:6–18
Mari I, Ehaliotis C, Kotsou M et al (2005) Use of sulfur to control pH in composts derived from olive processing by-products. Compost Sci Util 13(4):281–287
Martnez-Blanco J, Lehmann A, Muñoz P et al (2014) Application challenges for the social LCA of fertilizers within life cycle sustainability assessment. J Clean Prod 69:34–48
Masai E, Katayama Y, Nishikawa S et al (1989) Detection and localization of a new enzyme catalyzing the beta-aryl ether cleavage in the soil bacterium (*Pseudomonas paucimobilis* SYK-6). FEBS Lett 249:348–352
Mathur SP, Owen G, Dinel H, Schnitzer M (1993) Determination of compost biomaturity. I. Literature review. Biol Agric Hortic 10(2):65–85
Michel Jr FC, Pecchia JA, Rigot J, Keener HM (2004) Mass and nutrient losses during the composting of dairy manure amended with sawdust or straw. Compost Sci Util 12(4):323–334
Miller FC (1992) Composting as a process based on the control of ecologically selective factors. In: Meeting FB Jr (ed) Soil microbial ecology. Applications in Agricultural and Environmental Management. Marcel Dekker, Inc., New York, pp 515–544
Miller FC (1993) Composting as a process based on the control of ecologically selective factors. Soil Microbiol Ecol 18:515–544
Miller FC (1996) Composting as a process based on the control of ecologically selective factors. In: Metting FB Jr (ed) Soil microbial ecology. Dekker, New York, pp 515–544
Millner PD, Marsh PB, Snowden RB, Parr JF (1977) Occurrence of Aspergillus fumigatus during composting of sludge. Appl Environ Microbiol 34:765–772
Moss LH, Epstein E, Logan T (2002) Comparing the characteristics, risks and benefits of soil amendments and fertilizers used in agriculture. Proc Water Environ Fed 2002(3):602–623
Murphy DV, Stockdale EA, Brookes PC, Goulding KW (2007) Impact of microorganisms on chemical transformations in soil. In: Soil biological fertility. Springer, Dordrecht, pp 37–59

Ogram A (2000) Soil molecular microbial ecology at age 20: methodological challenges for the future. Soil Biol Biochem 32:1499–1504

Onwosi CO, Igbokwe VC, Odimba JN et al (2017) Composting technology in waste stabilization: on the methods, challenges and future prospects. J Environ Manag 190:140–157

Ouédraogo E, Mando A, Zombré NP (2001) Use of compost to improve soil properties and crop productivity under low input agricultural system in West Africa. Agric Ecosyst Environ 84:259–266

Paradelo R, Moldes AB, Prieto B et al (2010) Can stability and maturity be evaluated in finished composts from different sources? Compost Sci Util 18(1):22–31

Pathma J, Sakthivel N (2012) Microbial diversity of vermicompost bacteria that exhibit useful agricultural traits and waste management potential. SpringerPlus 1(1):26

Peters S, Koschinsky S, Schwieger F, Tebbe CC (2000) Succession of microbial communities during hot composting as detected by PCR-single-strand-conformation polymorphism-based genetic profiles of small-subunit rRNA genes. Appl Environ Microbiol 66:930–936

Polprasert C, Koottatep T (2017) Organic waste recycling: technology, management and sustainability, 4th edn. IWA Publishing, London

Ranalli G, Bottura G, Taddei P et al (2001) Composting of solid and sludge residues from agricultural and food industries. Bioindicators of monitoring and compost maturity. J Environ Sci Health, Part A 36(4):415–436

Rashid GMM, Taylor CR, Liu Y et al (2015) Identification of manganese superoxide dismutase from *Sphingobacterium* sp: t2 as a novel bacterial enzyme for lignin oxidation. ACS Chem Biol 10:2286–2294

Rayner ADM, Boddy L (1988) Fungal decomposition of wood. Wiley, Chichester

Riffaldi R, Levi-Minzi R, Pera A, De Bertoldi M (1986) Evaluation of compost maturity by means of chemical and microbial analyses. Waste Manag Res 4(4):387–396

Ruttimann C, Vicuna R, Mozuch MD, Kirk TK (1991) Limited bacterial mineralization of fungal degradation intermediates from synthetic lignin. Appl Environ Microbiol 57:3652–3655

Ryckeboer J, Mergaert J, Coosemans J et al (2003a) Microbiological aspects of biowaste during composting in a monitored compost bin. J Appl Microbiol 94(1):127–137

Ryckeboer J, Mergaert J, Vaes K et al (2003b) A survey of bacteria and fungi occurring during composting and self-heating processes. Ann Microbiol 53(4):349–410

Rynk R, van de Kamp M, Willson GB et al (1992). On-farm composting handbook. Rynk, R. (Ed). Northeast engineering agricultural services, New York.

Sait M, Hugenholtz P, Janssen PH (2002) Cultivation of globally distributed soil bacteria from phylogenetic lineages previously only detected in cultivation-independent surveys. Environ Microbiol 4(11):654–666

Schaub SM, Leonard JJ (1996) Composting: an alternative waste management option for food processing industries. Trends Food Sci Technol 7(8):263–268

Scheller HV, Ulvskov P (2010) Hemicelluloses. Ann Rev Plant Biol 61:263–289

Sequi P (1996) The role of composting in sustainable agriculture. In: de Bertoldi M, Sequi P, Lemmes B, Papi T (eds) The science of composting. Springer, Dordrecht

Sequi P, Tittarelli F, Benedetti A (2000) The role of sludge in the reintegration of soil fertility. In: Langenkamp, M (ed) Workshop on problems around sludge. European Comission Research Centre, EUR 19657 EN

Singh R, Eltis LD (2015) The multihued palette of dye-decolorizing peroxidases. Arch Biochem Biophys 574:56–65

Singh RP, Ibrahim MH, Esa N, Iliyana MS (2010) Composting of waste from palm oil mill: a sustainable waste management practice. Rev Environ Sci Biotechnol 9(4):331–344. https://doi.org/10.1007/s11157-010-9199-2

Smith JF, Fermor TR, Zadražil F (1988) Pretreatment of lignocellulosics for edible fungi. In: Zadražil F, Reiniger P (eds) Treatment of lignocellulosics with white-rot fungi. Elsevier, Essex, pp 3–13

Solano ML, Iriarte F, Ciria P, Negro MJ (2001) SE—Structure and environment: performance characteristics of three aeration systems in the composting of sheep manure and straw. J Agri Engi Res 79(3):317–329
Steger K, gren S¨, AM JA et al (2007) Development of compost maturity and actinobacteria populations during full-scale composting of organic household waste. J Appl Microbiol 103:487–498
Strom PF (1985) Effect of temperature on bacterial diversity in thermophilic solid-waste composting. Appl Environ Microbiol 50:899–905
Sugano Y (2009) DyP-type peroxidases comprise a novel heme peroxidase family. Cell Mol Life Sci 66:1387–1403
Tiquia SM (2005) Microbiological parameters as indicators of compost maturity. J Appl Microbiol 99(4):816–828
Tsai WT (2008) Management considerations and environmental benefit analysis for turning food garbage into agricultural resources. Bioresour Technol 99:5309–5316
Tuomela M, Vikman M, Hatakka A, Itävaara M (2000) Biodegradation of lignin in a compost environment: a review. Bioresour Technol 72(2):169–183
Vandecasteele B, Sinicco T, D'Hose T et al (2016) Biochar amendment before or after composting affects compost quality and N losses, but not P plant uptake. J Environ Manag 168:200–209
Varadachari C, Ghosh K (1984) On humus formation. Plant Soil 77(2–3):305–313
Vargas-Garcia MC, Suárez-Estrella F, Lopez MJ, Moreno J (2010) Microbial population dynamics and enzyme activities in composting processes with different starting materials. Waste Manag 30(5):771–778
Viaene J, Agneessens L, Capito C et al (2017) Co-ensiling, co-composting and anaerobic co-digestion of vegetable crop residues: product stability and effect on soil carbon and nitrogen dynamics. Sci Hortic 220:214–225
Vicuna R (1988) Bacterial degradation of lignin. Enzym Microb Technol 10:646–655
Wakchaure VN, Zhou J, Hoffmann S, List B (2010) Catalytic asymmetric reductive amination of α-Branched Ketones. Angew Chem 122(27):4716–4718
Waksman SA (1932) Principles of soil microbiology. Williams & Wilkins, Baltimore, p 619
Wang X, Zhao Y, Wang H et al (2017) Reducing nitrogen loss and phytotoxicity during beer vinasse composting with biochar addition. Waste Manag 61:150–156
Wilkinson K, Tee E, Tomkins R, Hepworth G, Premier R (2011) Effect of heating and aging of poultry litter on the persistence of enteric bacteria. Poult Sci 90(1):10–18
Zahir ZA, Iqbal M, Arshad M et al (2007) Effectiveness of IAA, GA3 and kinetin blended with recycled organic waste for improving growth and yield of wheat (*Triticum aestivum* L.) Pak J Bot 39(3):761–768
Zhang J, Chen G, Sun H et al (2016) Straw biochar hastens organic matter degradation and produces nutrient-rich compost. Bioresour Technol 200:876–883
Zino M, Lo Curto RB, Salvo F et al (1999) Lipid composition of *Geotrichum candidum* in single-cell protein grown in continuous submerged culture. Bioresour Technol 67:7–11
Zucconi F, De Bertoldi M (1987) Compost specifications for the production and characterization of compost from municipal solid waste. In: De Bertoldi M, Ferranti MP, L'Hermite MP, Zucconi F (eds) Compost: production, quality and use. Elsevier, London, pp 276–295

Fungal Cellulases Production for Biodegradation of Agriculture Waste

4

Neha Srivastava, Manish Srivastava, Ambepu Manikanta, P. W. Ramteke, Ram Lakhan Singh, P. K. Mishra, and S. N. Upadhyay

Abstract

Cellulases have several existing and potential applications in various industries including biofuel, pulp and paper, detergent, juice, and textile. Different enzyme components found in the cellulase system can effectively depolymerize the cellulose into monomeric sugars. Fungi are regarded as an ideal candidate to produce a complete cellulase system. Cellulase production via solid-state fermentation (SSF) using fungi is one of the most desirable and cost-effective routes. The present chapter provides an outline on the major steps of fungal cellulase production using SSF. This chapter presents a concise account of fungal cellulases as well as their production and key role in bioconversion of lignocellulosic waste.

Various aspects of fungal cellulase production using the agriculture wastes and their components have been discussed. Additionally, potential fungal sources of cellulase along with effective bioconversion of agricultural biomass are also summarized.

Keywords

Agricultural waste · Cellulosic biomass · Cellulase enzymes · Solid state fermentation · Fungi

N. Srivastava · A. Manikanta · P. K. Mishra · S. N. Upadhyay (✉)
Department of Chemical Engineering and Technology, IIT, (BHU), Varanasi, India
e-mail: pkmishra.che@iitbhu.ac.in; snupadhyay.che@itbhu.ac.in

M. Srivastava
Department of Physics and Astrophysics, University of Delhi, Delhi, India

P. W. Ramteke
Department of Biological Sciences, Sam Higginbottom University of Agriculture Technology & Sciences, Allahabad 221007, Uttar Pradesh, India

R. Lakhan Singh
Department of Biochemistry, Dr. R. M. L. Avadh University, Faizabad 224001, India

D. G. Panpatte et al. (eds.), *Microorganisms for Green Revolution*, Microorganisms for Sustainability 7,
https://doi.org/10.1007/978-981-10-7146-1_4

4.1 Introduction

Cellulose is the most abundant and renewable component of plant biomass (Srivastava et al. 2015a, b, c; Srivastava and Jaiswal 2016). It is considered as the main product of photosynthesis, and ~100 billion dry tons/year of cellulosic biomass is produced in the biosphere (Wang et al. 2016). Besides plants, some animals and bacteria also produce cellulose. However, it is the major component of the plant cell walls and abundantly available in the environment. In the lignocellulosic biomass structure, cellulose is mainly associated with hemicelluloses and lignin and sometimes with silica (e.g., rice straw, rice husk). Further, it accounts ~35–50% of plant dry weight, whereas hemicelluloses as well as lignin cover ~ 20–35 and ~ 5–30% of plant dry weight, respectively (Zabeda et al. 2016). Cellulose is a linear polysaccharide, made up of combined units of glucose monomers bound by β-1,4-glycosidic linkage. To release these monomeric molecules, cellulase enzymes are required for carrying out enzymatic hydrolysis. Cellulases are the combination of enzymes capable of degrading the insoluble cellulose polymer present in the lignocellulosic biomass into fermentable sugars, predominantly small chain of cellobiose and glucose molecules (Taherzadeh and Karimi 2007). Endoglucanases, cellobiohydrolases, and β-glucosidases are the main subcomponents which make the cellulase enzyme system (Thota et al. 2017). Nevertheless, intensive researches are going on to achieve novel cellulase producing systems, not only to increase the yield and economic feasibility but also to expand the use of such systems by progressing toward more and more industrially flexible bacteria or fungi production systems. Additionally, many enzymatic hydrolysis researches have also been aimed on the production of enzymes with high enzymatic activity for improvement in the current technology to produce cellulase enzyme (Chandel et al. 2012; Garvey et al. 2013). Besides biofuel production, other major industrial applications of cellulases are biopolishing; bio-stoning; bio-finishing in textile industry; starch processing; grain alcohol fermentation; malting and brewing in beer and wine industry; extraction and processing of fruit and vegetable juices in food, pulp, and paper industry (Cavacopaulo 1998; Gao et al. 2008; Ibrahim et al. 2015); controlling plant pathogen and disease in agriculture as well as in household laundry detergents for improving the fabrics' softness and brightness, etc. (Wilson 2009).

This chapter deals with the production of fungal cellulases, their type, and efficiency to degrade lignocellulosic biomass. Additionally, recent advancements in cellulase production and future prospects have also been discussed.

4.2 Fungal Cellulases

Cellulase enzyme plays a key role in the hydrolysis of cellulosic substrate and converts it into monomeric sugars. For the effective hydrolysis of cellulosic substrate, three types of synergistically acting subcomponent enzymes are essential: endoglucanases (EG), exoglucanases (CBH), and beta-glucosidase (BGL). Cellobiohydrolases or exoglucanases which attack on the crystalline ends of

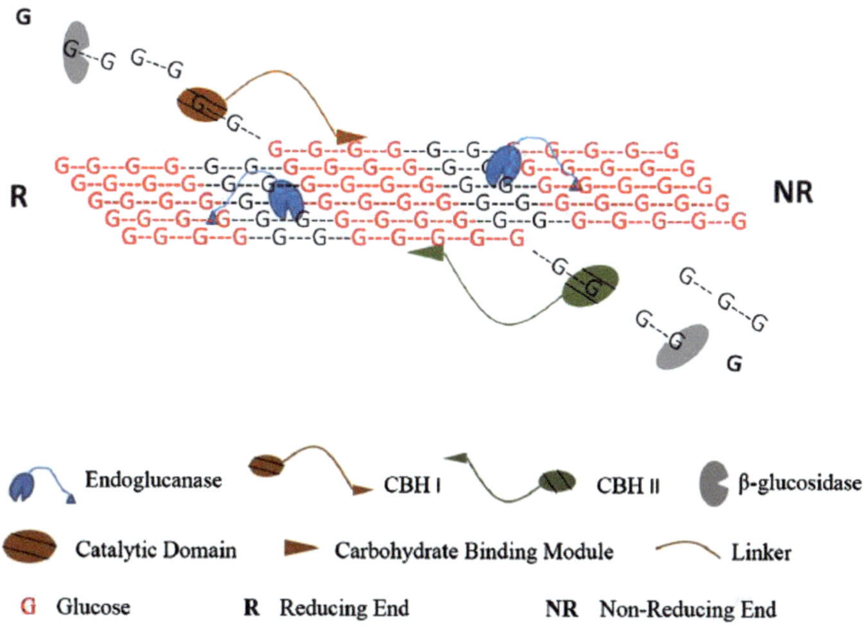

Fig. 4.1 Action of various cellulases enzymes on surface layer of cellulose. Figure illustrates the action of enzymes from three main classes of cellulose. Glucose molecules in red color represents crystalline region and glucose molecules in black color are in amorphous region (Adopted from Kumar and Murthy 2013)

cellulosic substrate produces cellobiose, while endoglucanases divide glycosidic bonds within the amorphous part of the cellulosic substrate (Zhang and Lynd 2004; Yoon et al. 2014). Further, the liberated cellobiose is sliced by β-glucosidases (BGL) and releases glucose molecules (Liu et al. 2012; Wang et al. 2013). Figure 4.1 shows all three components and functions of cellulase responsible for enzymatic hydrolysis of agriculture waste. Cellulase enzymes are widely distributed in nature, and fungi are known as the potential producer of cellulase. Additionally, cellulolytic fungi have the major advantages to utilize secretory pathways as well as the production of high yields of protein. Intense research on other fungi like *Penicillium*, *Acremonium*, and *Chrysosporium* are under way for the potential production of cellulase. Table 4.1 summarizes cellulase production from different fungal sp. using variety of wastes *via* SSF and SmF.

4.2.1 Cellulase Market Scenario

At present, cellulase enzymes which are commercially available from fungi are utilized for biofuel production (Mood et al. 2013; Menon and Rao 2012). *Aspergillus* and *Trichoderma* are the main fungal source of commercial cellulases (Conesa et al. 2016; Mosier et al. 2015). The estimated value for the enzymes in the global

Table 4.1 Cellulase production from different fungal sp. using various types of wastes *via* SSF and SmF

Fungi	Type of agricultural waste	Cellulase productivity	Fermentation process (SmF/SSF)	References
Aspergillus terreus	Rice straw	FPase: 10.96 IU/g	SSF	Narra et al. (2012)
Aspergillus fumigatus ABK9	Wheat bran: rice straw(1.1:1)[a]	CMCase: 826.2 IU/g FPase: 102.5 U/g β-glucosidase: 255.16 IU/g	SSF	Das et al. (2013)
Aspergillus protuberus	Rice husk	β-glucosidase: 26.06 IU/g	SSF	Yadav et al. (2016)
T. asperellum SR1-7	Wheat bran	FPase: 2.2 IU/gds CMCase: 13.2 IU/gds β-glucosidase 9.2 IU/gds	SSF	Raghuwanshi et al. (2014)
Aspergillus niger N402	Wheat straw	FPase: 24 IU/g	SSF	Pensupa et al. (2013)
Aspergillus fumigatus NITDGPKA3	Rice straw	CMCase: 64.18 IU/gds FPase: 3.1 IU/gds	SSF	Sarkar and Aikat (2012)
Myceliophthora thermophila JCP 1-4	Sugar cane bagasse: wheat bran (1: 1)[a]	CMCase: 357.51 IU/g β-glucosidase: 45.42 IU/g	SSF	Pereira et al. (2015)
Aspergillus niger KK2	Rice straw	FPase: 19.5 IU/g CMCase: 129 IU/g β-glucosidase :100 IU/g	SSF	Kang et al. (2004)
Aspergillus niger NCIM 548	Wheat bran : corn bran, kinnow peel (2:1:2)[a]	FPase : 5.54 IU/g	SmF	Kumar et al. (2011)
Aspergillus fumigatus P40M2	Soybean bran	CMCase: 160.1 IU/g	SSF	Delabona et al. (2012)
Aspergillus ellipticus	Wheat straw	FPase: 117.25IU/g CMCase: 725.11 IU/g β-glucosidase: 29.65 IU/g	SSF	Agrawal and Matkar (2016)
Penicillium echinulatum 9A02S1	Sugar cane bagasse	FPase: 12.5 IU/g	SmF	Camassola and Dillon (2014)
Trichoderma viride VKF3	Sugar cane bagasse	CMCase: 33IU/g FPase: 10.09 IU/g	SSF	Nathan et al. (2014)
Rhizopus oryzae CCT 7560	Rice husk and rice bran	CMCase: 5.1U/g FPase: 2.3 IU/g	SSF	Kupski et al. (2014)

[a]Ratio of different used biomass for cellulase production

industrial market was \$3.3 billion in 2010 and \$4.2 billion in 2014 and is expected to reach about \$6.2 billion by 2020 (Adrio and Demain 2014; Singh et al. 2016). Cellulases contribute ~20% in the commercial enzyme market (Srivastava et al. 2015a; Kapoor et al. 2016). A wide range of industrial applications is the main reason for the increased demand of cellulase enzymes in the global market (Juturu and Wu 2014; Srivastava et al. 2015b). Further, due to the crisis of fossil fuels, there is a growing demand of production of biofuels and chemicals from the renewable resources (like starch, lignocellulosic biomass, etc.) (Parmar and Rupasinghe 2012). Production of biofuels using lignocellulosic biomass as the substrate is becoming a hot area worldwide since enormous amount of biomasses are being generated and disposed as wastes (Behera et al. 2014, 2016; Behera and Ray 2016; Chakraborty et al. 2016).

Companies like "Genencor International Inc." and "Novozyme (Copenhagen, Denmark)" are playing significant role with innovative research to supply cocktails of cellulase enzyme for biomass conversion (Singhania et al. 2010; Srivastava et al. 2015b; Kuhad et al. 2016). Although both the companies claim a 20–30 fold reduction in the production cost of cellulase enzymes (~ 15–20 cents per gallon of ethanol produced), still the cost of cellulase enzymes is 5–10 fold higher than that of amylase which is used for starch-based biofuel production (Singhania et al. 2010; Kuhad et al. 2016; Liu et al. 2016). Therefore, the cost of cellulase enzyme should be decreased dramatically to produce bioethanol cost effectively. Other companies like Logen, Amano Enzyme Inc., Japan, and MAP's India, etc. are also involved in the production of cellulase enzymes (Srivastava et al. 2015b; Singh et al. 2016; Kuhad et al. 2016). Commercially, cellulase and hemicellulase enzymes for biomass conversion are produced by fungal strains of *Hypocrea jecorina (Trichoderma reesei) and Aspergillus niger* via submerged fermentation (SmF) (Phitsuwan et al. 2013; Adrio and Demain 2014; Kuhad et al. 2016). Cellic®CTec3 is the latest innovation of Novozymes which performs 1.5 times better than the previous market leading product. Cellic® CTec2 and biomass conversion needs only one-fifth of the enzyme dose compared to competing enzymes (http.// www.novozymes.com). This latest enzyme allows the cost of ethanol production from biomass to approach the level of corn ethanol and gasoline (http.// www.novozymes.com). In Cellic CTec3, the cellulase enzyme is based on *Trichoderma reesei*, and it contains at least two main CBHs, five different EGs, BGL, β-xylosidase and also contains other proprietary hydrolysis boosting proteins (Rasmussen et al. 2016, 2017). Accellerase® TRIO™, a product of Genecor, is an enzyme cocktail of cellulases, hemicellulase, and β-glucosidase to unlock 70–90% sugars from cellulosic biomass to produce ethanol (http:// www.genencor.com). Other Genecor products include Accellerase® 1500 and Accellerase® 1000 (http:// www.genencor.com).

4.2.2 Production of Fungal Cellulases

Fungal cellulase production can be carried out by two methods: (i) solid-state fermentation (SSF) and (ii) submerged fermentation (SmF) (Pandey A. 2003; Singhania et al. 2009; Bansal et al. 2012). In case of SSF, solid substrates are used such as agriculture waste of rice straw, wheat bran, sugarcane bagasse, etc. for the production of cellulases (Xia and Cen 1994; Subramaniyam and Vimala 2012; Cherian et al. 2016). SSF involves the fermentation process carried out in absence or nearly in absence of free water using solid substrate. On the other hand, SmF involves fermentation in presence of water. Moreover, SmF uses primarily free molecules soluble in water as liquid substrates, like molasses in broth (Subramaniyam and Vimala 2012). The main advantage of SSF technique is easy recycling of cheap waste material and less cost, whereas SmF offers ease of purification and product recovery (Pandey et al. 2000; Couto and Sanromán 2006). Further, SSF is mainly favorable for microorganisms which require less moisture content, while due to high water activity, SmF is suited to bacteria for cellulase production (Babu and Satyanarayana 1996). Additionally, SSF has been known for fermentation in Asian and Western countries since the ancient time (Ryu and Mandels 1980; Zhuang et al. 2007; Swain and Ray 2007). However, SSF has gained importance in Western countries after the discovery of penicillin via SmF technique in the 1940s. Though, in the last two decades, SSF has gained attention because of many biotechnological advantages like high fermentation ability, more stable end product, subordinate catabolic repression, as well as cost-effective technology (Sukumaran et al. 2009; Kasana et al. 2008; Liang et al. 2010; Dhillon et al. 2013). In the past 10 years, interest in SSF has been renewed because microorganisms including genetically modified organisms (GMO) may produce cellulase effectively through the SSF (Chahal 1983). Additionally, cellulase production via SSF is preferred over SmF because of two to three times higher enzyme production, high protein rate, and direct accessibility of dried fermentable solids as source of enzyme which easily eliminate the cost involved in downstream processing (Sanchez 2009; Hendriks and Zeeman 2009; Sadhu and Maiti 2013; El-Bakry et al. 2015).

4.2.3 Source of Fungal Cellulase Production

Degradation of cellulose via fungal cellulase is well documented, and several cellulose degrading fungi like *Aspergillus niger*, *Cladosporium cladosporioides*, *Cladosporium sphaerospermum*, *Penicillium chrysogenum*, *Scopulariopsis brevicaulis*, *Stachybotrys chartarum*, *Verticillium cycolsporum*, and *Chaetomium hamadae* have been investigated for cellulase production (El-Morsy 2000; Luo et al. 2005). Further, various fungi have also been screened for the cellulase production based on their habitat (El-Morsy 2000; Luo et al. 2005). In one of the study by Maria et al. (2005), twenty nine different fungal isolates were reported for cellulase production. Moreover, most of these fungal isolates produce endoglucanases where most of them belong to *Ascomycetes*. Kathiresan and Manivannan (2006)) isolated seven fungal species such as *Acremonium* sp., *Alternaria*, chlamydospore, *Aspergillus* sp., *Fusarium* sp., as well as *Pestalotiopsis* sp. from the southwest coast of India and used for cellulase production. Among all these fungi, *Aspergillus* sp. was found to be efficient cellulase producers.

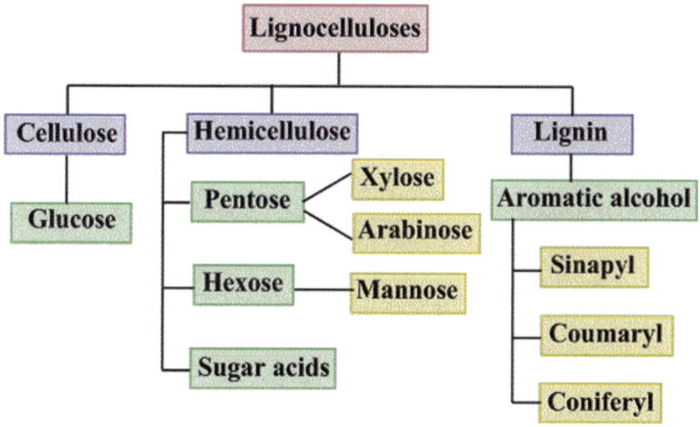

Fig. 4.2 Schematic representation showing the contents present in the lignocelluloses (Sajith et al. 2016)

Besides *Aspergillus* sp., *T. reesei* are also known as prominent cellulase producers that possess a complete cellulase system (Rasmussen et al. 2010). Though, action of fungi is different in terms of degradation of lignocellulosic biomass in decaying pattern and the structural changes found in the degraded substrates, category belonging to white-rot fungi (WRF) degrades all the components of biomass, namely, lignin, cellulose, and hemicellulose by colonizing themselves on the lignocellulosic substrate (Kuhad and Singh 2007). On the other hand, fungi belonging to category brown-rot fungi (BRF) favorably degrade the cellulosic component and hemicellulosic part with the modified lignin part (Schwarze et al. 2000; Schmidt 2006). Apart from the fungi type, production of cellulases might also be dependent on the initial amount of cellulose, hemicellulose, and lignin present in biomass (Liu et al. 2014). Figure 4.2 shows different subcomponents of lignocellulosic biomass and their obtained product as sugars after enzymatic hydrolysis. A separate group of fungi are responsible for degrading these biomass via bioconversion reaction of cellulases.

4.3 Biodegradation of Agricultural Biomass

Bioconversion of lignocellulosic biomass via cellulase enzyme is also known as enzymatic hydrolysis or saccharification. For effective conversion of cellulose into monomers a complete cellulase system is required (endoglucanase, exoglucanase, and β-glucosidase) to act synergistically. Biomass, which undergoes pretreatment before the enzymatic hydrolysis can be easily converted into sugars, effectively (Megan et al. 2013; Jose and Arnold 2014). Moreover, released sugars can undergo fermentation to produce biofuel. Although pretreatment process removes lignin effectively, the hemicellulose and cellulose part is converted by the synergistic

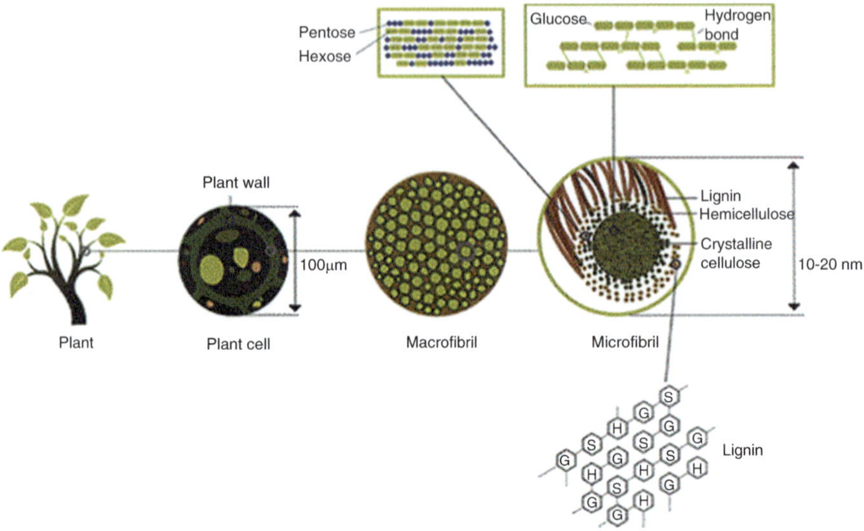

Fig. 4.3 Schmatic structure of lignocellulose. The hexagons denote the lignin subunits p-coumaryl alcohol (H), coniferyl alcohol (G) and sinapyl alcohol (S) [Adopted from Streffer F (2014)]

action of hemicellulases and cellulases (Dashtban et al. 2009) (Fig. 4.3). Enzymatic hydrolysis of lignocellulosic biomass is divided into primary hydrolysis and secondary hydrolysis. Primary hydrolysis generally takes place on the surface of substrate, and cellobiose is released due to the catalytic action of exo- and endoglucanases (Fig. 4.3). The cellobiose is further converted into glucose via β-glucosidase in secondary hydrolysis (Zhang et al. 2006).

4.4 Biofuel Production from Biomass Waste Degradation

Over the past decade, several research groups have focused on biofuel production using biomass-based process as cost-effective technology. Cellulosic biomass to sugar conversion is the key step for biofuel production, and efforts have been made to explore the technology related to biomass conversion via cellulase such as improving the efficiency of cellulase producing microorganisms and cellulase efficiency (Garvey et al. 2013). Figure 4.4 explains the efficient destruction of biomass via cellulase to release sugar and produce biofuels. Additionally, non-efficient cellulase is incapable to release sugars from this biomass and hydrolysis becomes incomplete (Fig. 4.4). Although commercial cellulases are manufactured from native microorganism, economic viability seems far away. Currently, for efficient enzymatic hydrolysis of biomass, different cellulases (EG, CBH, and BGL) have been achieved from different microbial sources, which makes the process economically sustainable. Therefore, intense research is under way to find out novel

Fig. 4.4 Cartoon depiction of enzymatic cellulose deconstruction. Processive reducing (orange) and nonreducing end (green) cellulases move along the cellulose fibers liberating cellobiose units. The cellobiose released is then converted to individual glucose sugars by β-glucosidases (pink). Endo-acting cellulases (red) introduce chain breaks that the exo-acting enzymes can act upon. All of the aforementioned enzymes use hydrolytic mechanisms, while the new players in this process, LPMOs (blue), further potentiate their action using an oxidative mechanism to introduce further chain breaks on which the processive cellulases can initiate further degradation (Hemsworth et al. 2015)

cellulase producing systems in order to increase the flexibility of available microbial strain, yield, and processing. The current enzymatic hydrolysis research is aimed to improve the production of cellulase via screening of effective microorganisms that would provide new opportunities to face this challenges (Peterson and Nevalainen 2012; Chandel et al. 2012).

The fermentable sugars obtained from the effective hydrolysis via cellulase enzymes are used for the production of renewable energy (Fig. 4.5). In addition, fermentable sugars can be converted into bioethanol, biohydrogen, and methane through different processes with the help of specific microorganisms. Effective production and yield of biofuels are directly dependent on the amount of fermentable sugars present in reaction medium, and the fair sugar amount is directed by viable cellulase system. Apart from cellulase system, the targeted process can be improved by removing the structural weakness of biomass via effective pretreatment strategies for biofuel production.

4.5 Conclusion

Demand of cellulase enzyme is increasing tremendously in global enzyme market due to potential applications in degradation of lignocellulosic biomass. Additionally, fungal cellulases production is cost effective for degradation of

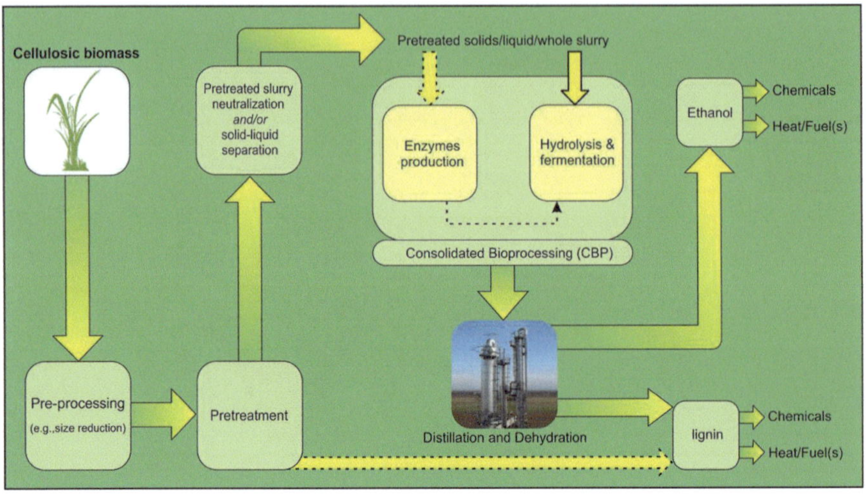

Fig. 4.5 Schematic of second generation ethanol production process. [Adopted from Kumar et al. 2016]

these kinds of wastes through SSF. A number of fungi have been reported for production of cellulase and considered as the potential groups over other known microorganisms. However, it is challenging to bring down the production cost of enzyme and bioconversion process of wastes. Therefore, efforts should be made to develop more suitable and optimized combination process to bring down the cost and improved quality of cellulase.

Acknowledgments N.S. thankfully acknowledges IIT (BHU) Varanasi, India for providing Institute Post-Doctoral Fellowship. P. K. Mishra and A. M. acknowledge Department of Chemical Engineering and Technology IIT (BHU) Varanasi. M. S. acknowledges the Department of Science and Technology, Govt. of India for DST INSPIRE Faculty [IFA13-MS-02] 2014 award.

References

Adrio JL, Demain AL (2014) Microbial enzymes: tools for biotechnological processes. Biomolecules 4:117–139. https://doi.org/10.3390/biom4010117.

Agrawal D, Matkar K (2016) Enhancement of cellulase production by substrate manipulation in three Aspergillus sp. Indian J Appl Res 6(3). ISSN – 2249-555X

Babu KR, Satyanarayana T (1996) Production of bacterial enzymes by solid state fermentation. J Sci Ind Res 55:464–467. ISSN 0022-4456

Bansal N, Tewari R, Soni R, Soni SK (2012) Production of cellulases from Aspergillus niger NS-2 in solid state fermentation on agricultural and kitchen waste residues. Waste Manag 32:1341–1346. https://doi.org/10.1016/j.wasman.2012.03.006.

Behera SS, Ray RC (2016) Solid state fermentation for production of microbial cellulases: recent advances and improvement strategies. Int J Biol Macromol 86:656–669. https://doi.org/10.1016/j.ijbiomac.2015.10.090.

Behera S, Arora R, Nandhagopal N, Kumar S (2014) Importance of chemical pretreatment for bio-conversion of lignocellulosic biomass. Renew Sust Energ Rev 36:91–106. https://doi.org/10.1016/j.rser.2014.04.047

Behera BC, Sethi BK, Mishra RR, Dutta SK, Thatoi HN (2016) Microbial cellulases diversity & biotechnology with reference to mangrove environment: a review. J Genet Eng Biotechnol. https://doi.org/10.1016/j.jgeb.2016.12.001

Camassola M, Dillon AJP (2014) Effect of different pretreatment of sugar cane bagasse on cellulase and xylanases production by the mutant penicillium echinulatum 9A02S1 grown in submerged culture. Biomed Res Int. https://doi.org/10.1155/2014/720740

Cavacopaulo A (1998) Mechanism of cellulase action in textile processes. Carbohydr Polym 37:273–277. https://doi.org/10.1016/S0144-8617(98)00070-8

Chahal OS (1983) Growlh characteristics of microorganisms in solid state fermentation for upgrading of protein value of lignocelluloses and cellulase production. In: Blanch HW, Poputsakis ET, Stephanopoulos G (eds) Foundation of biochemical engineering kinetics and thermodynamics in biological systems, ACS Symp. Vol 207, Chapter 20. American Chemical Society, Washington, DC, pp 421–442. https://doi.org/10.1021/bk-1983-0207.ch020

Chakraborty S, Gupta R, Jain KK, Kuhad RC (2016) Cost-effective production of cellulose hydrolysing enzymes from Trichoderma sp. RCK65 under SSF and its evaluation in saccharification of cellulosic substrates. Bioprocess BiosystEng 39:1659–1670. https://doi.org/10.1007/s00449-016-1641-6

Chandel AK, Chandrasekhar G, Silva MB, Silva SSD (2012) The realm of cellulases in biorefinery development. Crit Rev Biotechnol 32:187–202. https://doi.org/10.3109/07388551.2011.595385

Cherian E, Kumar MD, Baskar G (2016) Production and optimization of cellulase from agricultural waste and its application in bioethanol production by simultaneous saccharification and fermentation. Manag Environ Qual: Int J 27:22–35. https://doi.org/10.1108/MEQ-07-2015-0128

Conesa C, Seguí L, Laguarda-Miró N, Fito P (2016) Microwaves as a pretreatment for enhancing enzymatic hydrolysis of pineapple industrial waste for bioethanol production. Food Bioprod Process 100:203–213. https://doi.org/10.1016/j.fbp.2016.07.001

Couto SRG, Sanromán MA (2006) Application of solid-state fermentation to food industry—A review. J Food Eng 76:291–302. https://doi.org/10.1016/j.jfoodeng.2005.05.022

Das A, Paul T, Halder SK, Jana A, Maity C, Mohapatra PKD, Pati BR, Mondal KC (2013) Production of cellulolytic enzymes by Aspergillus fumigatus ABK9 in wheat bran-rice straw mixed substrate and use of cocktail enzymes for deinking of waste office paper pulp. Bio resource Technology 128:290–296. https://doi.org/10.1016/j.biortech.2012.10.080

Dashtban M, Schraft H, Qin W (2009) Fungal bioconversion of lignocellulosic residues; opportunities & perspectives. Int J BiolSci 5(6):578–595. https://doi.org/10.7150/ijbs.5.578

Delabona PDS, Rosangela DP, Pirota B, Codima CA, Tremacoldi CR, Rodrigues A, Farinas CS (2012) Using Amazon forest fungi and agricultural residues as a strategy to produce cellulolytic enzymes. Biomass Bioenergy 37:243–250. https://doi.org/10.1016/j.biombioe.2011.12.006

Dhillon GS, Kaur S, Brar SK (2013) Perspective of apple processing wastes as low-cost substrates for bioproduction of high value products: a review. Renew Sust Energ Rev 27:789–805. https://doi.org/10.1016/j.rser.2013.06.046

El-Bakry M, Abraham J, Cerda A, Barrena R, Ponsa S, Gea T, Sanchez A (2015) From wastes to high value added products: novel aspects of SSF in the production of enzymes. Crit Rev Environ Sci Technol 45:1999–2042. https://doi.org/10.1080/10643389.2015.1010423

El-Morsy E (2000) Fungi isolated from the endorhizosphere of halophytic plants from the red sea coast of Egypt. Fung Divers 5:43–54

Gao JH, Zhu WD, Yuan M, Guan F, Xi Y (2008) Production and characterization of cellulolytic enzymes from the thermos acidophilic fungal Aspergillus terreus M11 under solid-state cultivation of corn stover. Bioresour Technol 99:7623–7629. https://doi.org/10.1016/j.biortech.2008.02.005

Garvey M, Klose H, Fischer R, Lambertz C, Commandeur U (2013) Cellulases for biomass degradation: comparing recombinant cellulase expression platforms. Trends Biotechnol 31(10):581–593. https://doi.org/10.1016/j.tibtech.2013.06.006

Hemsworth GR, Johnston EM, Davies GJ, Walton PH (2015) Lytic polysaccharide monooxygenases in biomass conversion. Trends Biotechnol 33:747–761 http://dx.doi.org/10.1016/j.tibtech.2015.09.006.

Hendriks ATWM, Zeeman G (2009) Pretreatments to enhance the digestibility of lignocellulosic biomass. Bioresour Technol 100(1):10–18. https://doi.org/10.1016/j.biortech.2008.05.027

Ibrahim MF, Abd-Aziz S, Yusoff MEM, Phang LY, Hassan MA (2015) Simultaneous enzymatic saccharification and ABE fermentation using pretreated oil palm empty fruit bunch as substrate to produce butanol and hydrogen as biofuel. Renew Energy 77:447–455. https://doi.org/10.1016/j.renene.2014.12.047

Jose LA, Arnold LD (2014) Microbial enzymes: tools for biotechnological processes. Biomolecules 4:117–139. https://doi.org/10.3390/biom4010117

Juturu V, Wu JC (2014) Microbial cellulases: engineering, production and applications. Renew Sust Energ Rev 33:188–203. https://doi.org/10.1016/j.rser.2014.01.077

Kang SW, Park YS, Lee JS, Hong SI, Kim SW (2004) Production of cellulases and hemicellulases by Aspergillus niger KK2 from lignocellulosic biomass. Bioresour Technol 91:153–156. https://doi.org/10.1016/S0960-8524(03)00172-X

Kapoor M, Panwar D, Kaira GS (2016) Bioprocesses for enzyme production using agro-industrial wastes: technical challenges and commercialization potential (Chapter 3). In: Dhillon, Singh G, Kaur S (eds) Agroindustrial wastes as feedstock for enzyme production: apply and exploit the emerging and valuable use options of waste biomass. Academic Press/Elsevier, pp 61–93. isbn:9780128023921

Kasana RC, Salawan R, Dhar H, Dutt S, Gulati A (2008) A rapid and easy method for the detection of microbial celluloses on agar plates using gram's iodine. Curr Microbiol 57(5):53–507. https://doi.org/10.1007/s00284-008-9276-8

Kathiresan K, Manivannan S (2006) α-Amylase production by Penicillium fellutanum isolated from mangrove rhizosphere soil. Afr J Biotechnol 5:829–832

Kuhad R, Singh A (2007) Lignocellulose biotechnology: future prospects. IK International Publishing House Pvt. Ltd, New Delhi

Kuhad RC, Deswal D, Sharma S, Bhattacharya A, Jain KK, Kaur A, Pletschke BI, Singh A, Karp M (2016) Revisiting cellulase production and redefining current strategies based on major challenges. Renew Sust Energ Rev 55:249–272. https://doi.org/10.1016/j.rser.2015.10.132

Kumar S, Sharma HK, Sarkar BC (2011) Effect of substrate and fermentation conditions on pectinase and cellulase production by aspergillus niger NCIM 548 in submerged (SmF) and solid state fermentation (SSF). Food Sci Biotechnol 20(5):1289–1298. https://doi.org/10.1007/s10068-011-0178-3

Kumar D, Murthy GS (2013) Stochastic molecular model of enzymatichydrolysis of cellulose for ethanol production. Biotechnol Biofuels 6:63–82. https://doi.org/10.1186/1754-6834-6-63

Kumar R, Tabatabaei M, Karimi K, Horváth IS (2016) Recent updates on lignocellulosic biomass derived ethanol – A review. Biofuel Res J 9:347–356. https://doi.org/10.18331/BRJ2016.3.1.4

Kupski L, Pagnussatt FA, Buffon JG, Furlong EB (2014) Endoglucanase and total cellulase from newly isolated rhizopus oryzae and trichoderma reesei: production, characterization, and thermal stability. Appl Biochem Biotechnol 172:458–468. https://doi.org/10.1007/s12010-013-0518-2

Liang Y, Feng Z, Yesuf J, Blackburn JW (2010) Optimization of growth medium and enzyme assay conditions for crude cellulases produced by a novel thermophilic and cellulolytic bacterium, Anoxybacillus sp. 527. Appl Biochem Biotechnol 160:1841–1852. https://doi.org/10.1007/s12010-009-8677-x

Liu HQ, Feng Y, Zhao DQ, Jiang JX (2012) Evaluation of cellulases produced from four fungi cultured on furfural residues and microcrystalline cellulose. Biodegradation 23:465–472. https://doi.org/10.1007/s10532-011-9525-6

Liu J, Ju M, Wu W, Liu B, Zhan L, Wu M (2014) Lignocellulolytic enzyme production in solid-state fermentation of corn stalk with ammoniation pretreatment by Lentinusedodes L-8. Bioresources 9(1):1430–1444

Liu Z, Ho S-H, Sasaki K, Haan RD, Inokuma K, Ogino C, Van Zyl WH, Hasunuma T, Kondo A (2016) Engineering of a novel cellulose-adherent cellulolytic Saccharomyces cerevisiae for Cellulosic biofuel production. Sci Rep 6:24550. https://doi.org/10.1038/srep24550. http://www.novozymes.com

Luo W, Vrijmoed LLP, Jones EBG (2005) Screening of marine fungi for lignocellulose-degrading enzyme activities. Bot Mar 48(5-6):379–386. https://doi.org/10.1515/bot.2005.051

Maria GL, Sridhar KR, Raviraja NS (2005) Antimicrobial and enzyme activity of mangrove endophytic fungi of southwest coast of India. J Agric Technol 1:67–80

Megan G, Holger K, Rainer F, Camilla L, Ulrich C (2013) Cellulases for biomass degradation: comparing recombinant cellulase expression platforms. Trends Biotechnol 31(10):581–593. https://doi.org/10.1016/j.tibtech.2013.06.006

Menon V, Rao M (2012) Trends in bioconversion of lignocellulose: biofuels, platform chemicals & biorefinery concept. Prog Energy Combust Sci 38:522–550. https://doi.org/10.1016/j.pecs.2012.02.002

Mood SH, Golfeshan AM, Tabatabaei M, Jouzani GS, Najafi GH, Gholami M, Ardjmand M (2013) Lignocellulosic biomass to bioethanol, a comprehensive review with a focus on pretreatment. Renew Sust Energ Rev 27:77–93. https://doi.org/10.1016/j.rser.2013.06.033

Mosier N, Wyman C, Dale B, Elander R, Lee YY, Holtzapple M, Ladisch M (2015) Features of promising technologies for pretreatment of lignocellulosic biomass. Bio Res Technol 96(6):673–668. https://doi.org/10.1016/j.biortech.2004.06.025

Narra M, Dixit G, Divecha J, Madamwar D, Shah AR (2012) Production of cellulases by solid state fermentation with Aspergillus terreus and enzymatic hydrolysis of mild alkali-treated rice straw. Bioresour Technol 121:55–361. https://doi.org/10.1016/j.biortech.2012.05.140

Nathan VK, Rani ME, Rathinasamy G, Dhiraviam KN, Jayavel S (2014) Process optimization and production kinetics for cellulase production by Trichodermaviride VKF3. Springer Plus 3:92. https://doi.org/10.1186/2193-1801-3-92

Pandey A (2003) Solid-state fermentation. Biochem Eng J 13:81–84. https://doi.org/10.1016/S1369-703X(02)00121-3

Pandey A, Soccol CR, Mitchell D (2000) New developments in solid state fermentation: i-bioprocesses and products. Process Biochem 35:1153–1169. https://doi.org/10.1016/S0032-9592(00)00152-7

Parmar I, Rupasinghe HPV (2012) Optimization of dilute acid-based pretreatment and application of laccase on apple pomace. Bioresour Technol 124:433–439. https://doi.org/10.1016/j.biortech.2012.07.030

Pensupa N, Jin M, Kokolski M, Archer DB, Du C (2013) A solid state fungal fermentation-based strategy for the hydrolysis of wheat straw. Bioresour Technol 149:261–267. https://doi.org/10.1016/j.biortech.2013.09.061

Pereira JDC, Marques NP, Rodrigues A, Oliveira TBD, Boscolo M, Silva RD, Gomes E, Martins DAB (2015) Thermophilic fungi as new sources for production of cellulases and xylanases with potential use in sugarcane bagasse saccharification. J Appl Microbiol 118:928–939. https://doi.org/10.1111/jam.12757

Peterson R, Nevalainen H (2012) Trichoderma reesei RUT-C30 – thirty years of strain improvement. Microbiology 158:58–68. https://doi.org/10.1099/mic.0.054031-0

Phitsuwan P, Laohakunjit N, Kerdchoechuen O, Kyu KL, Ratanakhanokchai K (2013) Present and potential applications of cellulases in agriculture, biotechnology, and bioenergy. Folia Microbiol 58:163–176. https://doi.org/10.1007/s12223-012-0184-8

Raghuwanshi S, Deswal D, Karp M, Kuhad RC (2014) Bioprocessing of enhanced cellulase production from a mutant of Trichoderma asperellum RCK2011 and its application in hydrolysis of cellulose. Fuel 124:183–189. https://doi.org/10.1016/j.fuel.2014.01.107

Rasmussen ML, Shrestha P, Khanal SK, Pom-etto III, Van AL, Leeuwen J (2010) Sequential saccharification of corn fiber and ethanol production by the brown rot fungus Gloeophyllumtrabeum. Bioresour Technol 101(10):3526–3533. https://doi.org/10.1016/j.biortech.2009.12.115

Rasmussen H, Mogensen KH, Jeppesen MD, Sørensen HR, Meyer AS (2016) 4-Hydroxybenzoic acid from hydrothermal pretreatment of oil palm empty fruit bunches -its origin and influence on biomass conversion. Biomass Bioenergy 93:209–216. https://doi.org/10.1016/j.biombioe.2016.07.024. http://www.genencor.com

Rasmussen H, Tanner D, Sørensen HR, Meyer AS (2017) New degradation compounds from lignocellulosic biomass pretreatment: routes for formation of potent oligo phenolic enzyme inhibitors. Green Chem, Advance Article. doi:https://doi.org/10.1039/C6GC01809B

Ryu DDY, Mandels M (1980) Cellulases biosynthesis and applications. Enzym Microb Technol 2:91–102. https://doi.org/10.1016/0141-0229(80)90063-0

Sadhu S, Maiti TK (2013) Cellulase production by bacteria: a review. Br Microbiol Res J 3(3):235–258. http://imsear.hellis.org/handle/123456789/162913

Sajith S, Priji P, Sreedevi S, Benjamin S (2016) An overview on fungal cellulases with an industrial perspective. J Nutr Food Sci 6: 661. http://dx.doi.org/10.4172/2155-9600.1000461

Sánchez C (2009) Lignocellulosic residues: biodegradation and bioconversion by fungi. Biotechnol Adv 27:185–194. https://doi.org/10.1016/j.biotechadv.2008.11.001

Sarkar N, Aikat K (2012) Cellulase and xylanase production from rice straw by a locally isolated fungus aspergillus fumigatus NITDGPKA3 under solid state fermentation – statistical Optimization by response surface methodology. J Technol Innov Renew Energy 1:54–62. ISSN: 1929-6002

Schmidt O (2006) Wood and tree fungi: Biology, damage, protection, and use. Springer-Verlag, Berlin/Heidelberg

Schwarze F, Engels J, Mattheck C (2000) Fungal strategies of wood decay in trees. Springer-Verlag, Berlin/Heidelberg

Singh R, Manoj K, Anshumali M, Mehta PK (2016) Microbial enzymes: industrial progress in 21st century. Biotech 6:174. https://doi.org/10.1007/s13205-016-0485-8

Singhania RR, Patel AK, Soccol CR, Pandey A (2009) Recent advances in solid-state fermentation. Biochem Eng J 44:13–18. https://doi.org/10.1016/j.bej.2008.10.019

Singhania RR, Sukumaran RK, Patel AK, Larroche C, Pandey A (2010) Advancement and comparative profiles in the production technologies using solid-state and submerged fermentation for microbial cellulases. Enzym Microb Technol 46:541–549. https://doi.org/10.1016/j.enzmictec.2010.03.010

Srivastava N, Jaiswal P (2016) Production of cellulases using agriculture waste: application in biofuels production. Environmental biotechnology a new approach. Daya Publishing House® A Division of Astral International Pvt. Ltd. New Delhi–110 002.233-425 244

Srivastava N, Singh J, Srivastava M, Ramteke PW, Mishra PK (2015a) Improved production of reducing sugars from rice straw using crude cellulase activated with Fe_3O_4/Alginate nanocomposite. Bioresour Technol 183:262–266. https://doi.org/10.1016/j.biortech.2015.02.059

Srivastava N, Rawat R, Oberoi HS, Ramteke PW (2015b) A review on fuel ethanol production from lignocellulosic biomass. Int J Green En 12:949–960. https://doi.org/10.1080/15435075.2014.890104

Srivastava N, Srivastava M, Mishra PK, Singh P, Ramteke PW (2015c) Application of cellulases in biofuels industries: an overview. J Biofuels Bioenergy 1(1):55–63. https://doi.org/10.5958/2454-8618.2015.00007.3

Streffer F (2014) Lignocellulose to Biogas and other Products. JSM Biotechnol Bioeng 2(1): 1023–1030

Subramaniyam R, Vimala R (2012) Solid state and submerged fermentation for the production of bioactive substances: a comparative study. Int J Sci Nat 3(3):480–486

Sukumaran RK, Singhania RR, Mathew GM, Pandey A (2009) Cellulase production using biomass feed stock and its application in lignocellulose saccharification for bio-ethanol production. Renew Energy 34:421–424. https://doi.org/10.1016/j.renene.2008.05.008

Swain MR, Ray RC (2007) Alpha-amylase production by Bacillus subtilis CM3 in solid state fermentation using cassava fibrous residue. J Basic Microbiol 47:417–425. https://doi.org/10.1002/jobm.200710132

Taherzadeh MJ, Karimi K (2007) Enzyme-based hydrolysis processes for ethanol from lignocellulosic materials: a review. Bioresources 2:707–738. ISSN 1930-2126

Thota SP, Badiya PK, Yerram S, Vadlani PV, Pandey M, Golakoti NR, Belliraj SK, Dandamudi RB, Ramamurthy SS (2017) Macro-micro fungal cultures synergy for innovative cellulase enzymes production and biomass structural analyses. Renew Energy 103:766–773. https://doi.org/10.1016/j.renene.2016.11.010

Wang M, Liu K, Dai L, Zhang J, Fang X (2013) The structural and biochemical basis for cellulose biodegradation. J Chem Technol Biotechnol 88:491–500. https://doi.org/10.1002/jctb.3987

Wang Y, Fan C, Hu H, Li Y, Suna D, Wang Y, Peng L (2016) Genetic modification of plant cell walls to enhance biomass yield and biofuel production in bioenergy crops. Biotechnol Adv 34:997–1017. https://doi.org/10.1016/j.biotechadv.2016.06.001

Wilson DB (2009) Cellulases and biofuels. Curr OpinBiotechnol 20:295–299. https://doi.org/10.1016/j.copbio.2009.05.007

Xia L, Cen P (1994) Cellulase production by solid state fermentation on lignocellulosic waste from the xylose industry. Process Biochem 34:909–912. https://doi.org/10.1016/S0032-9592(99)00015-1

Yadav PS, Shruthi K, Prasad BVS, Chandra MS (2016) Enhanced production of β-glucosidase by new strain aspergillus protuberus on solid state fermentation in rice husk. Int J Curr Microbiol App Sci 5(12):551–564. http://dx.doi.org/10.20546/ijcmas.2016.512.061

Yoon LW, Ang TN, Ngoh GC, Chua ASM (2014) Fungal solid-state fermentation and various methods of enhancement in cellulase production. Biomass Bioenergy 67:319–338. https://doi.org/10.1016/j.biombioe.2014.05.013

Zabeda H, Sahu JN, Boyce AN, Faruq G (2016) Fuel ethanol production from lignocellulosic biomass: an overview on feed-stocks and technological approaches. Renew Sust Energ Rev 66:751–774. https://doi.org/10.1016/j.rser.2016.08.038

Zhang YHP, Lynd LR (2004) Toward an aggregated understanding of enzymatic hydrolysis of cellulose: non-complexed cellulase systems. Biotechnol Bioeng 88:797–824. https://doi.org/10.1002/bit.20282

Zhang YHP, Himmel ME, Mielenz JR (2006) Outlook for cellulase improvement: screening and selection strategies. Bio technol Adv 24(5):452–481. https://doi.org/10.1016/j.biotechadv.2006.03.003

Zhuang J, Marchant MA, Nokes SE, Strobel HJ (2007) Economic analysis of cellulase production methods for bio-ethanol. Appl Eng Agric 23(5):679–687. 10.13031/2013.23659

Prospects of Arbuscular Mycorrhizal Fungi for Heavy Metal-Polluted Soil Management

5

Gomathy Muthukrishnan, Sabarinathan Kuttalingam Gopalasubramaniam, and Pandiyarajan Perumal

Abstract

The entire ecosystem has been contaminated by heavy metals. They are not only toxic to humans but also the microorganisms in the soil which make the land uncultivable for crop growth. The efficiency of the microorganisms lies in their ability to make them available to the crop plants. Among the microorganisms, mycorrhizal fungi are the only ones which provide a direct link between soil and root of the crops. The mechanism behind the metal tolerance nature of the AM fungi is studied and reported. As the health of plants and mycorrhizal symbiosis are linked very closely, it is much needed to understand the response of plants to heavy metals.

Keywords

AM fungi · Heavy metals · Mechanisms · Pollution · Soil reclamation

5.1 Heavy Metals

During the last few decades, biosphere pollution due to heavy metals such as Cu, Zn, Pb, Cr, Cd, Ag, Se, Hg, etc. was accelerated severely due to mining, smelting, tanning, treatment of agricultural soils with agrochemicals and soil sludge, etc. Many problems are related to the contamination of soil and water and disruptions of

G. Muthukrishnan (✉) · S. K. Gopalasubramaniam
Agricultural College and Research Institute, Killikulam, India
e-mail: mg2003@rediffmail.com; sabarimicro@hotmail.com

P. Perumal
Anbil Dharmalingam Agricultural College and Research Institute, Trichy, India
e-mail: pprtnau@gmail.com

D. G. Panpatte et al. (eds.), *Microorganisms for Green Revolution*,
Microorganisms for Sustainability 7,
https://doi.org/10.1007/978-981-10-7146-1_5

natural ecosystems (He et al. 2005). The soil ecosystem is contaminated with heavy metals in various ways like agricultural, industrial, urban sectors, etc., and they remained in the soil for longer period of time and contaminate soil, water and air. Humans were much affected due to the heavy metal exposure through ingestion of metal-polluted food or uptake of drinking water which leads to complicated health problems. In soils, due to the higher mobility and solubility of heavy metals, viz. Cr, Pb, Co, Zn and Cd, cause significant damage to ecosystem and human health. These metals occur in several forms in soil and water. In soil the speciation and mobility of heavy metals are determined by sequential extraction, which solubilizes different phases of metals (Kabata-Pendias and Pendias 1992).

Many soil microbes in the soil and in the soil environment have substantial impact on mankind (Doyle and Lee 1986). The information about the diversity of soil microbes, their behaviour in the rhizosphere and soil environment and influence of microbes in the plant is needed elaborately. Microorganisms inhabiting in the rhizosphere region interact with plant roots and mediate nutrient availability. Those forming useful symbiotic associations with the roots and microorganisms contributing for plant nutrition and enhancing the soil properties need to be completely exploited, and inoculation of these beneficial microbes to the nutrient poor agricultural soils has to be encouraged greatly (Khan 2002).

Plant and soil health are dependent upon the interactions of biological, physical and chemical components of both. The rhizosphere is the location in and around the root where the greatest flow of energy and minerals is abundantly found (Wright and Miller 1994). In this highly productive region, a vital symbiosis exists between roots of 80% of all vascular plant species and soilborne arbuscular mycorrhizal (AM) fungi (Smith and Read 1997). Almost all the plant families except a few (Smith and Read 1997) were successfully colonized by AM fungi, performing the nature's role of mineral and nutrient transfer to plants (Jeffries and Barea 1994) that favours the biogeochemical cycling of elements.

This mutualistic association has existed for more than 400 million years or since plants first moved from an aquatic to a terrestrial environment (Simon et al. 1993; Taylor et al. 1995). In this symbiosis, plants benefit by increased uptake of immobile nutrients in the soil, while the fungus receives photosynthetic carbon and other essential nutrients from the host (Smith and Read 1997). These fungal endosymbionts are universal in their association with flowering plants including agriculturally important crop species (Jeffries et al. 2003). The process of using mycorrhizal fungi for remediation or to ameliorate the soil from pollution is referred as mycoremediation.

5.2 Early Origin of AM Fungi

The name vesicular-arbuscular mycorrhizal (VAM) fungus was given by Dangeard (1900). Plants growing in coal mine wastes were found to be colonized with VAM fungi (Daft and Nicolson 1974). In halophytes and xerophytes, the symbiotic association occurs (Khan 1974) and also in artic, temperate and tropical herbs

(Mosse et al. 1981). The subterranean land plants fossil records showed the presence of VAM fungi (Pirozynski and Malloch 1975). Even in the rootless plants of gametophytes, some mosses, lycopods and psilotales were associated with VAM fungi (Mosse et al. 1981) and also reported in aquatic plants and wide range of habitats usually in the roots of angiosperms, gymnosperms and pteridophytes (Quilambo 2003). In dicotyledons and monocotyledons, 83% and 79% of plants respectively, form mutualistic symbiosis with AM fungi (Bonfante-Fasolo 1987). VAM association rarely occurs in families like Betulaceae, Chenopodiaceae, Cyperaceae, Amaranthaceae and Polygonaceae (Jagpal and Mukerji 1988). In some families viz., Ericaceae, Fumariaceae, Orchidaceae, Pinaceae and Urticaceae, the AM association will not occur. During the early Cretaceous era, AM associations are ubiquitous in both living angiosperms and Triassic fossil cycads (Taylor and Taylor 1993). In the early Devonian period of more than 400 million years ago, evidence of mycorrhizal symbiosis was established by the discovery of arbuscules in *Aglaophyton major* (Remy et al. 1994). In ancient plants, inheritable and highly influential colonization of VAM fungi exists worldwide (angiosperms, gymnosperms, ferns) (Simon et al. 1993). Before the existence of land plants, Glomales are the very oldest group, and Phipps and Taylor (1996) reported that AM infections were found even before the occurrence of roots.

The unique structures of VAM fungi, viz. hyphae, vesicles and arbuscules, are found in mosses, liverworts and hornworts (Schüßler 2000). AM fungal associations are reported in many plant families including the vascular plant families such as *Sphenophytes*, *Lycopodophytes*, *Pteridophytes*, etc. (Brundrett 2002). Universally, these ubiquitous symbiotic AM fungi form symbiotic relationships with halophytes, hydrophytes and xerophytes (Khan 2003). So the mycorrhizal diversity and plant community structure influence the below-ground networks. The extent of benefit not only depends on the particular plant and AM fungi involved but also on the other microbiota and soil abiotic factors (Chaudhry et al. 2005).

5.3 Mycorrhizal Symbiosis

'Mycorrhiza' is the term coined by Frank (1885) to define the association between fungi and higher plants. This was the very old symbiosis even before or more than 460 million years. VAM fungi will not grow and complete their life cycle in the absence of the plants that provide the photosynthetic carbon and in turn receive nutrients. Later this photosynthetic carbon is translocated in the form of lipids and sugar into the external mycelium (Douds and Millner 1999).

During symbiotic interactions, the spore of AM fungi germinates after recognizing the host through signals and develops a presymbiotic mycelium with limited growth in the proximity of the host root (Giovannetti et al. 1994) followed by appressorium formation on the root epidermis. After the formation of appressorium, roots can be colonized either by Arum-type or by Paris-type of colonization. Then the fungal hyphae penetrate the root cortical cell walls and form haustoria-like structures called arbuscules (Smith and Read 1997). This tree-like structured arbuscules are the site for exchange of nutrients between the plant and the fungus. These arbuscules are known for their increased surface area, and the hexose sugars from the plants are exchanged and transported by the fungus. Vesicles are the structures produce by some AM fungi and not by all, and that's why the name has been changed from VAM to AM fungi, believed to function as storage organs (Smith and Read 1997). In this symbiosis AM fungi allot around 20% of photosynthetically fixed carbon in the fungal partners (Johnson et al. 2002) in many natural environments. The carbon present in the fungal partner is utilized by the other microorganisms in the rhizospheric region. More than 160 fungal taxa were described in the order Glomales alone based on spore morphology (Schussler et al. 2001) and recent literatures confirmed through molecular studies and stated that the number may be very high than the already stated information, although recent molecular analyses indicated that the actual number of AM taxa may be much higher (Daniell et al. 2001; Vandenkoornhuyse et al. 2002).

Arbuscular mycorrhizal fungi also interface directly with the soil by producing extraradical hyphae of AM fungi that extend several centimetres into the soil and explore larger volume of soil in a greater magnitude than the root zone that increases the efficiency of uptake of nutrients (Rhodes and Gerdemann 1975; Auge 2001). Extraradical hyphae of AM fungi played an important role in soil stabilization through formation of soil aggregates and by the production of glomalin (Tisdall and Oades 1979; Wright et al. 1998). Since the extraradical hyphae are well known for acquisition of nutrients, phosphorus (P) is solubilized and mobilized to the plants (Read and Moreno 2003). Not only P but also other minerals and organically bound nitrogen from plant litter can also be mobilized well (Hodge et al. 2001). For this purpose 12–27% of photo-assimilated carbon is utilized by AM fungi Gildon and Tinker (1983). Fluctuations in the carbon level in the plant are balanced with the help of extraradical hyphae of AM fungi by changing host plant physiology. AM fungi interact with other microorganisms in the soil environment and remove the pathogenic microbes which affect the plants through negative effects, and they also alleviate the metal toxicity (Khan et al. 2000). AM symbiosis also influences the exudates of roots by modifying the root physiology and mycorrhizosphere.

The uptake of P needs higher quantity of carbon, and this was met out by utilizing the larger surface area of a leaf which ultimately increased photosynthetic efficiency by utilizing the larger surface area of a leaf which in turn increased the photosynthesis (Bolan 1991; George et al. 1992). The high carbon cost of P uptake is compensated by an increase in photosynthetic capability of the host through increased leaf surface area, and in addition, the fungal cell membrane is capable of concentrating solutes against a gradient (Bolan 1991; George et al. 1992). Even under stress condition, mycorrhiza efficiently acquires P and other nutrients. Almost all plants are colonized by AM fungi except a few families, and they are not much host specific; the nature of symbiosis is influenced by several factors like plant, fungus and soil biotic and abiotic factors (Brundrett 1991; Varma 1995).

AM fungi render many benefits to the host plant. AM fungi can efficiently take up nutrients (P, N, Fe, Cu and Zn) with the help of ephemeral hyphae in microsites (Douds and Millner 1999; Pawlowska et al. 2000). This symbiosis could reduce the pathogenic fungal infections by triggering a defence mechanism which alters the production of root exudates and stimulates the beneficial microbes in the rhizosphere, creating a competition for niche and nutrients (Hooker and Black 1995; Borowicz 2001). It also depends on the factors like host genotype, species of AM fungi and other abiotic factors. It also alleviates the metal toxicity through several mechanisms and keeps the soil in a low concentration of metals by hyperaccumulating the metals in the root and shoot portion of the plant and keeping the reproductive part free of metal contamination (Diaz et al. 1996; Gonzalez-Chavez et al. 2002; Subramanian et al. 2008; Gomathy et al. 2011). Metal uptake depends on the AM species, metal concentration, soil fertility and pH, and this metal uptake alters the uptake and translocation of P in the host plant (Diaz et al. 1996; Gonzalez-Chavez et al. 2002). The soil organic matter and biomass carbon content also have a high dependency on AM fungal population (Tisdall et al. 1997; Rillig and Steinberg 2002).

5.4 Heavy Metals and Their Effect on Plants

The ecosystem is polluted due to various factors. Heavy metals are one among them, and their entry into the environment depends on wide reasons, such as waste disposal, mining, smelting, tanning, paintings, agricultural pesticides, sewage sludge, etc. Metals are part of the soil; some are needed by the plants as micronutrients; and some don't have any biological activity and are merely a pollutant. These metals cause so many diseases to human when they enter into the food chain. In the soil they persist over thousands of years which impose serious threats to all the living things and affect or exclude even the population of microorganisms.

Root and shoot growth is greatly affected by the heavy metals which reduce the growth of the plants. When Ni concentration gets increased beyond 50 mg/kg dry weight, it suffered from Ni toxicity even in lower concentrations needed by plants (Aydinalp and Marinova 2009).

Nichols et al. (2000) observed that chlorophyll a and b and carotenoid content were decreased in the plants when they are exposed to 2 ppm of Cr. The decrease was due to the degradation of protein present in the peripheral part of the leaves (Shanker 2003). Seed germination is also affected and reduced due to heavy metals. Sugarcane bud germination was lowered because of the exposure to 80 ppm of Cr (Jain et al. 2000). Seed germination of vegetables got decreased in the Pb- and Ni-treated plants (Srinivas et al. 2013). Study on the leafy vegetables revealed that varying concentrations of nickel, chromium, lead and mercury significantly reduced the biomass, photosynthesis and chlorophyll content compared to control (Sheetal et al. 2016).

Root length was much affected by Cr (Prasad et al. 2001), and reduced growth was observed when the wheat was grown in 20 ppm of Cr (Chen et al. 2001). Chibuike and Obiora (2014) recorded reduction in number of tillers and panicles in wheat. Leaf growth and leaf size were greatly reduced in the concentration of 60 ppm (Singh 2001). Cadmium was found to reduce the plant growth, chlorophyll content and number of nodules in chickpea. The combination of cadmium with other metals such as nickel and lead reduced the protein content in grains also (Chibuike and Obiora 2014).

5.5 Heavy Metal Pollution and Importance of AM Fungi

Heavy metal is the general collective term which strictly refers to metals and metalloids with a specific mass higher than 5 g cm^{-3} (Baker 1987). Soil is contaminated with heavy metals that are released from various industries. Some of the metals were micronutrients (Zn, Cu, Mn, Ni and Co), and some don't have any biological role (Cd, Cr, Pb and Hg) (Marschner and Romheld 1995). The contamination and accumulation of these heavy metals might be through various ways, viz. mining, tanning, disposal of high metal wastes, leaded gasoline and paints, coal combustion, petrochemicals, etc. Soils are the main deposit of all the heavy metals released from all the sources. These heavy metals were migrated from polluted areas to unpolluted areas through air, water, sewage sludge and leachates. They persist in the environment for a longer period of time, and they could not be completely removed physically or degraded chemically (Kroopnick 1994). Remediation measures are needed to reclamate the soil for crop cultivation. Some of the techniques like excavation of soil, disposing to a landfill site (Parker 1994), on-site management and soil washing (Elliott et al. 1989) are practiced to reclamate the soil. But these techniques are not much reliable, and they are costly which need further treatment of soil. Doing the aforementioned techniques removes the microbial flora and other biological activities in the soil and makes the land worthless for plant growth (Zak and Parkinson 1982). To make it fit for the cultivation of crops, phytoremediation techniques are working well. Plants used for this technique should have high levels of root and shoot biomass so that the heavy metals could effectively be removed from the soil and stored in the above- and below-ground plant parts (Leyval et al. 1997). Plants growing in the devastated lands were devoid of mycorrhizal association and have very poor root and shoot growth (Pawlowska et al. 1996). But rather going for phytoremediation, bioremediation techniques are promising where the previous technique again contaminates the soil and the latter captures or sequesters the metals in the microbial system. So as to improve the vegetation, AM fungi should be inoculated in the affected areas.

Mycorrhizae-inoculated plants tend to accumulate more amounts of Zn and Cu (Heggo et al. 1990). Leyval et al. (1997) suggested that different AM fungal species have varied ability to translocate heavy metals. Mycorrhizae-treated plants accumulated huge quantities of metals than control plants (Joner and Leyval 1997). AM fungi play a pivotal role in metal tolerance (Del Val et al. 1999) and metal accumulation (Zhu et al. 2001; Jamal et al. 2002) as the external mycelium explored a larger volume of soil since it unravelled beyond the root zone (Khan et al. 2000; Joner et al. 2000; Malcova et al. 2003). Chen et al. (2001) found the storage of metals in the structures like roots and spores, and they found 1200 ppm of Zn in the fungal tissue of *Glomus mosseae* and 600 ppm in *G. versiforme*.

In the devastated lands, the suitable bioremediation technique is the introduction of AM fungi that establishes better plant growth (Enkhtuya et al. 2002), due to aggregation of soils by production of glomalin (Kabir and Koide 2000), water availability (Auge 2001), better nutrition (Feng et al. 2003), huge mycelium and glomalin production, etc. (Leal et al. 2016). The arbuscular mycorrhizal fungus *Rhizophagus irregularis* could tolerate Al and Pb at higher concentrations (Gavito et al. 2014).

5.6 Incidence of AM Fungi at Metal-Contaminated Sites

Mycorrhizal species which tolerated 100 mg kg^{-1} of Zn and 863 μg g^{-1} Cd in the soil was reported by Gildon and Tinker (1981) in the mine-spoil area with the mycorrhizal colonization percentage of 35 in the roots of clover. Bohn and Liberia (1982) found only 11 spores of *Glomus* g^{-1} of sludge soil. From the metal-polluted sites, Dueck et al. (1986) isolated *Glomus fasciculatum*, and *Glomus mosseae* was isolated by Weissenhorn and Leyval (1995). Metals such as Zn, Cu, Pb, Ni and Cr that contaminated soils were abundantly found with the mycorrhizal species such as *Glomus*, *Gigaspora*, *Acaulospora* and *Sclerocystis* (Sambandan et al. 1992). Extensive mycorrhizal colonization that occurred in metal-devastated areas was illustrated by Diaz and Honrubia (1994). In coal mine areas, metal-tolerant AM fungus, *Scutellospora dipurpurescens*, was isolated (Griffioen 1994). Weissenhorn et al. (1995) recorded profuse colonization in smelter- and sludge-contaminated soils and tolerate 1220 ppm of Cd and 895 ppm of Pb. Leyval et al. (1995) found 46 spores of *Glomus* in metal-polluted area. Raman and Sambandan (1998) found spores of *Glomus* and *Gigaspora* in various polluted environments. Leyval et al. (1995) recorded the germination of AM spores even in 6060 ppm Pb, 24,410 ppm Zn and 1630 ppm Cu. *Glomus* isolate obtained from *Viola calaminaria* well colonized in the maize-growing areas and helped the plant to tolerate higher metal concentration. Two identified species of *Glomus*, *viz. Glomus claroideum* and *Glomus mosseae*, and two unidentified species were isolated from sewage soils (Del Val et al. 1999). Pawlowska et al. (2000) recorded the presence of *Glomus constrictum*, *Glomus mosseae* and an unidentified *Glomus* species from polluted landfill. Many *Glomus* species (*G. caledonium*, *G. claroideum*, *G. constrictum*, *G. fasciculatum*, *G. intraradices*), *Acaulospora* (*A. delicata* and *A. undulata*) and *Entrophospora* (*E. infrequens*) were noted in arsenic mine spoils (Gonzalez-Chavez et al. 2002). In Zn-contaminated soil, *Fragaria vesca* plants inoculated with mycorrhizae were found to have 70% colonization by *G. mosseae* (Turnau et al. 2001). In the mycorrhizae inoculated soil, spores were collected, and characterization was done by polymerase chain reaction.

Koomen et al. (1990) reported that higher concentrations of heavy metals reduce or completely eliminate the AM fungal population. But controversy to the above point, many reports suggested the presence of mycorrhizal population even at elevated concentrations of heavy metal providing a way to rehabilitation of contaminated areas for potential plant growth (Subramanian et al. 2008). The presence of different species of AM fungi varied based on different locations, different soils, prevailing environment, presence of crop, etc. (Leal et al. 2016).

5.7 Mycorrhizal Mechanisms Involved in Tolerance of Heavy Metals

The polluted and highly contaminated areas of land having diversified mycorrhizal colonization indicates that the selective mechanisms are present in the fungal systems for specific ecosystem to withstand the pollutants (Meharg 2003; Subramanian et al. 2008). AM fungi helped the plants in the contaminated area to grow and establish well and detoxified the metal in the soil and accumulate the metals in their structures (Maier et al. 2009). The secretion of organic acids in the rhizosphere may have a key role in the tolerance of heavy metals.

5.7.1 Biosorption of Heavy Metals

5.7.1.1 Adsorption

AM fungal mycelium is having high metal-adsorbing capacity when compared to other ions (Berthelin et al. 1993). Colpaert and Van Assche (1992) reported that extraradical mycelium adsorbed more quantity of Zn and avoided the uptake and translocation of Zn in the shoot portion of *Pinus sylvestris*. In the metal solution, the metal concentration was lower in the area where the hyphal mats were present and the remaining solution having higher concentration of metals (Griffiths et al. 1994). Not only the AM hyphae but also the cell walls have a high adsorption affinity for the metal Pb (Colpaert and Van Assche 1992). The mycorrhizal cell wall is loaded with more negatively charged ligands such as phosphoryl, carboxyl, hydroxyl and phenolic groups and can absorb Cu^{2+} ions (Zhou 1999). Compared to other fungi, these AM fungal hyphae have more surface area for accumulation of toxic metals, and they serve as an excellent adsorptive site for accumulation of cations that exclude the entry in plants (Joner et al. 2000). Christie et al. (2004) reported that *G. mosseae* hyphae adsorb more Zn ions in the Zn-contaminated areas.

For trace element adsorption, the main site was the extraradical mycelium of AM fungi (Kaldorf et al. 1999), and the cell wall is the main adsorptive site for all the toxic elements (Zhou 1999). All these together combinedly make the environment metal-free and assure for proper plant growth. Cu and As in the soil were reported to be accumulated and localized in the extraradical mycelium of AM fungi (Gonzalez-Chavez et al. 2002).

5.7.1.2 Ion Exchange

The metal ions in the contaminated soil exchanged with the bivalent metal ions and polysaccharides present in the cell walls of AM fungi. Metabolic binding was observed inbetween the metal ion and the fungal hyphae site was metabolic binding comparing with non-metabolic binding (Gadd 1990). The cell wall of fungi has negatively charged groups, and this determines the cation exchange capacity (CEC), and this CEC differed between AM fungal species, and fluctuates based on the pH (Grauer 1992). If the pH increases, it might also increase the cation sorption in the cell wall of AM fungi (Cronan 1991).

5.7.1.3 Complexation

Many organic acids are released from the symbiotic roots in terms of root exudates, and these exudates contain citric acid, malic acid, lactic acid, etc., and these organic acids complexed the metals in the soil and reduce the metal concentration in the soil system (Rauser 1999; Gomathy et al. 2011).

Dehn and Schuepp (1989) reported the cystine ligands that are present in AM fungi hyphal protein complexed the metals in the fungal hyphal region. These metallo-organic complexes are important in reducing the availability of metals in the root zone (Shanker et al. 2005).

5.7.1.4 Precipitation

Precipitation is also one of the mechanisms where AM fungi adapt to reclamate the soil from heavy metals. Kaldorf et al. (1999) reported that AM fungi intracellularly precipitate the metallic cations in their hyphae. AM fungi are known to produce abundant quantity of protein called glomalin from its extraradical hyphae where the protein precipitates the heavy metals (Wright and Upadhyaya 1998). Turnau et al. (2001) have explained precipitation of metals (Cd, Ti and Ba) in fungal structures like vesicles and in intracellular space of fungi than in plant cells. Metal precipitation is also noted in polyphosphate granules (Kaldorf et al. 1999).

5.7.1.5 Solubilization of Minerals Containing Heavy Metals

The acidification of the rhizosphere due to the organic acids secretion by AM fungi leads to weathering of rocks and minerals. Leyval et al. (1990) suggested that due to the AM symbiosis, solubilization and mobilization of biotite happened in the root zone. Water-soluble inorganic phosphates have been solubilized by AM fungi which in turn weathered the minerals (Paris et al. 1994). Becerril et al. (2005) also observed the same that mycorrhizal fungi solubilized and accumulated greater quantity of Cd and it remained in the solution.

5.7.1.6 Hyphal Transport of Heavy Metals by AM Fungi

Transport of metal elements by AM fungi to the plant happened in three steps, uptake, translocation and transfer. Uptake refers the movement of ion from soil solution to fungal cytoplasm that is an energy-dependent process. The next step is the translocation where the heavy metals are translocated through cytoplasmic streaming (Cooper and Tinker 1978). The third step is the transfer of the heavy metals into the plant system from root to shoot. Ectomycorrhizal fungus that moved the Pb into the cortex of *Laccaria bicolor* roots was observed by Marschner et al. (1998). In barley root to shoot transfer of Cd was low in mycorrhizae-inoculated plants compared to uninoculated plants (Tullio et al. 2003).

5.7.2 Detoxification of Heavy Metals by AM Fungi

5.7.2.1 Avoidance Mechanism

Microorganisms and plants are adapting a specific mechanism called avoidance mechanism where the pollutants or toxicants are being avoided from entering into the cell (Meharg 1993). Marschner et al. (1998) observed that the metal uptake has declined in spruce seedlings inoculated with AM fungi due to the efficient filtration mechanism of fungi. For successful avoidance of metals to the cells, biochemical adaptations were being tuned such as suppressing the toxicant in flux transporters, enhancing efflux of a toxicant, releasing toxicant-complexing agents into the soil, secretion of enzymes, etc. (Meharg 2003).

5.7.2.2 Compartmentalization Strategies

In compartmentalization strategy, the pollutant or the toxic metal was compartmentalized in specific organs either in plants or in fungus. In those compartments, the metabolic functions will be less, and the toxic metal is changed to nontoxic. The compartmental organs may be vacuole or subcellular compartments that are stored away from the cytoplasm (Meharg 1993, 1994). Overcoming the compartments, the toxic metal which entered into the cell may be changed to other forms to ease the transport within the plant or to store the metal. To carry out the transport of specific metals, chelators, namely, dicarboxylic organic acids, amino acids, phytochelatins (PCs) and metallothioneins (MTs), are involved in this. These metal toxicants alter the metabolic pathways such as transformation, complexation, or reduced transport into the cell or localization within the cell. Metal speciation may be changed due to alterations in pH, changing the reduction-oxidation reactions, complexing the biomolecules, etc. (Amor et al. 1998). Kaldorf et al. (1999) explained that metal compartmentalization also occurred in plastids or in other organelles. Protein metal complex and phytochelatins in the tonoplast transfer the toxic metals into the vacuole (Ortiz et al. 1995; Cobbett and Goldsburough 2002).

5.7.2.3 Dilution

Mycorrhizae-inoculated plants have luxurious growth of roots and shoots compared to nonmycorrhizal plants (Gomathy et al. 2011). Moreover the external mycelium explores larger volume of soil than root which resulted in higher uptake of water and nutrients especially phosphorus required for root growth (Smith and Read 1997; Chen et al. 2003) with root branching and higher number of lateral root hairs (Turnau et al. 1993; Jentshke et al. 1998). Owing to profuse growth of infected AM fungal roots and shoots, the heavy metals taken up by the roots translocated to shoots get diluted than nonmycorrhizal plants.

5.7.2.4 Effect of AM Fungi on Heavy Metal Tolerance in Plants

Tolerance is the change in sensitivity to a specific toxicant or pollutant. Tolerance nature of plants could be adaptive or induced or constitutive (Meharg 2003). In the areas like polluted mine and smelter areas, many plants are growing by tolerating the pollutant by the mechanism of evolution of tolerant genotypes or induced tolerance (Antonovics et al. 1971; Schat and Ten Bookum 1992). Gildon and Tinker (1981) noted the resistance of AM fungi to cadmium and zinc in several plants. Killham and Firestone (1983) stated that mycorrhizal plants take more amounts of heavy metals from the soil compared to nonmycorrhizal plants that are more toxic for them.

Dueck et al. (1986) studied the influence of *Glomus fasciculatum*-inoculated plants such as *Festuca rubra* and *Calamagrostis epigejos* spiked with Zn at various levels and noted that the inoculated *Glomus* species reclamate and remove the metal contaminant from the soil. Different *Glomus* species were subjected to varying level of Mn-affected soil, and sorghum plants were grown in the soil, and the results indicated that among the species, *Glomus macrocarpum* accumulated more amount of Mn than uninoculated control plants (Raju et al. 1990).

They also found that the plants inoculated with *Glomus intraradices* showed decreased Mn uptake compared to uninoculated plants. Reduced concentrations of Zn and Cu in mycorrhizal plants were also observed (Xiong 1993; Audet and Charest 2006). Diaz et al. (1996) studied the influence of Zn and Pb uptake by *Lygeum spartum* and *Anthyllis cytisoides* plants inoculated with *G. mosseae* and *G. macrocarpum* in soils with different levels of Zn and Pb. They reported that, at lower doses, mycorrhizal plants had equal or higher Zn or Pb concentrations than nonmycorrhizal controls. Hildebrandt et al. (1999) noted that in the roots of *Viola calaminaria*, the identified *Glomus* sp. improved maize growth and reduced the translocation of heavy metals to shoot. Likewise *G. mossae* also proved to enhance the growth of maize, alfalfa, barley, etc. in cadmium-rich soils (Kaldorf et al. 1999). Meharg and Cairney (2000) explained that genes encoding increased metal resistance are rapidly transferred through AM populations in the presence of selection pressures. When such selection pressure was removed, the genes were rapidly lost and return to low frequencies. Gonzalez-Chavez et al. (2002) reported that AM fungi have conferred arsenate resistance to *Holcus lanatus*. Liao et al. (2003) studied the interactions between arbuscular mycorrhizae (*Acaulospora laevis*, *Glomus caledonium* and *Glomus manihotis*) and heavy metals (Cu and Cd) and reported that the mycorrhizal colonization rates of *Glomus caledonium* were the highest among the treatments. Bi et al. (2003) compared the effects of two arbuscular mycorrhizal fungi, viz. *Glomus mosseae* and *Glomus versiforme*, on the growth and nutrient uptake of maize and reported that colonization by both AM fungi increased plant growth as compared to nonmycorrhizal controls, with *Glomus mosseae* giving higher yield of maize than *Glomus versiforme*. They further reported that the translocation of Na to the shoots was less, perhaps protecting the plants from excessive Na accumulation, indicating the contribution of AM fungi in rhizofiltration mechanism. In the cadmium-contaminated soils, varied response was observed by Janouskova et al. (2005) when the plants were inoculated with *Glomus intraradices* and *Gigaspora margarita*.

5.7.3 Influence of VAM on Glomalin Production

5.7.3.1 Origin of Glomalin

Glomalin a glycoprotein was identified at USDA during the 1990s when they were looking for monoclonal antibodies reactive with AM fungi. During the analysis, one of the antibodies reacted with a substance on AM fungal hyphae, and it was present in almost all AM fungal species and named as glomalin to remember the order Glomales in which AM fungi present (Wright et al. 1996). Many other fungi (*Rhizoctonia*, *Gaeumannomyces*, *Endogone*, *Mucor* and *Phytophthora*) present in the soil were subjected to cross-reaction against glomalin; none were found to be immunoreactive (Wright et al. 1996). By following a series of extraction procedure, glomalin was extracted and characterized through total and immunoreactive protein assays. Presence of glomalin is very wide and abundant in all types of soil like acidic, calcareous, grassland and cropland soil regardless of various environments (Wright and Upadhyaya 1998) similar to the presence of AM fungi (Carlile and Watkinson 1996; Olsson et al. 1999; Wright and Upadhyaya 1998). Wright et al. (1998) observed the presence of glomalin in the range from 2 to >60 mg g^{-1} of soil. Glomalin was plentifully produced and present on the extraradical hyphae of all AM fungi regardless of the species (Wright et al. 1996). Wright and Upadhyaya (1998) explained the possible mechanisms of the presence of glomalin in the soil by the secretion of glomalin into the environment or the medium in which it presents other by means of hyphal wall incorporation and further release to the surrounding area. Rillig et al. (2001) noted the presence of tightly bound iron molecule in the range of 0.04–8.8% and devoid of phenolic compounds such as tannins. Glomalin strongly adhere to the fungal hyphae and after degradation of hyphae, it started moving away from the hyphae to the adhering soil particles or organic matter. It also protects the roots and other organic particles from degradation by other microorganisms which may overprotect the structures of AM fungi inside the root (Wright et al. 1996).

5.7.3.2 Glomalin and Heavy Metal Sequestration

The bioavailability of heavy metals in the soil is mainly based on the speciation and the soil components (Boruvka and Drabek 2004). The free amino, hydroxyl and carboxyl groups present in the cell wall of AM fungi can bind with the toxic metals like Cu, Pb, Cd and Cr (Kapoor and Viraraghavan 1995). This adsorbing ability of the metals in the AM fungi could commercially be used for biosorbents (Morley and Gadd 1995). Glomalin protein is having the ability to sequester the metals in the cell wall of AM fungi. Glomalin plays a great role in phytostabilization and maintaining a balance in the soil ecosystem by sequestering the metal and rendering the plant to grow in a metal-free environment. In the interface of soil and hyphae, glomalin performs the role of immobilization of heavy metals (Wright and Upadhyaya 1996) and reduces the bioavailability (Gonzalez-Chavez et al. 2004). Many species of AM fungi were tested for their ability to produce glomalin and found that the higher the quantity of hyphae, the higher the production of glomalin (Joner et al. 2000). Lovelock et al. (2004) found that even in the sand cores of forest soils, glomalin production was observed from the AM fungal hyphae. Khan (2005) explained that sequestration is an important mechanism for biostabilization of heavy metals in the polluted soils. *G. intraradices*-infected Ri T-DNA carrot roots were proved for glomalin in hyphae and in walls of spores (Driver et al. 2005).

Stable aggregates were formed in the soil which prevents soil erosion (Rillig et al. 2003). This mycorrhizoremediation (use of mycorrhizal plants in the phytoremediation of metal-contaminated soils) technology is found to be the best one for remediation of polluted sites (Gomathy et al. 2011).

5.7.4 Influence of Mycorrhizal Root Exudates in Heavy Metal Sequestration

During environmental stress, roots release greater quantities of organic material either passively or actively into the rhizosphere (Curl and Trueglove 1986; Whipps 1990) where some of the compounds were involved in the relieving of external stress (Mori et al. 1991).

Root exudates are the first plant signals that AM fungi perceive and thought to play an important role in the establishment of AM symbiosis. The root exudates of AM contained many organic compounds such as amino compounds, organic acids, fatty acids, sterols, flavones, etc. which helped in the heavy metal sequestration. Heavy metals can be sequestered through the production of metal-binding compounds like amino acids (Reilly 1972), citric acid (Thurman and Rankin 1982), malic acid (Brookes et al. 1981), phytochelatins (PCs) (Cobbett 2002; Hall 2002), etc. Organic acids solubilized the metals in acid soils (Jones and Darrah 1994). Organic acids can bind elements such as metals, and their role as detoxification agents has been widely discussed by Jones (1998). The extrametrical mycelium plays an important role in increasing the absorptive surface area for nutrient uptake (Smith and Read 1997) and making available microsites that are inaccessible to fine roots and increase the root exudate production (Jongmans et al. 1997). Organic acid excretion varies among the mycorrhizal species, and differences were found between the isolates of the same species (Jonnarth et al. 2000). AM fungi excreted organic acids such as citric, malic and oxalic acids into the rhizosphere (Landeweert et al. 2001).

Organic acids are having lower molecular weight, and microorganisms were also producing them when they are degrading organic materials (Fox and Comerford 1990; Young et al. 1998). These are used as nutrients for microorganisms (Lundstrom 1994; Jones and Darrah 1994). It also solubilizes scarcely soluble metal to growing plants (Rengel and Romheld 2000). Higher quantities of metals are solubilized by aromatic and aliphatic organic acids (Williams and Nascimento 2006). Highly insoluble minerals also tend to solubilize with the help of organic acids such as malate, citrate, etc. (Jones and Darrah 1994). The complex reaction of metals with organic acids not only based on complexation ability of organic compounds but on hydrolysation of organic acids, adsorption, desorption reactions, etc. (Cline et al. 1987). These acids also detoxify the excess heavy metals (Harmens et al. 1993; Ma et al. 1997). Mycorrhizal application could alter the root exudate production pattern (Vierheilig et al. 2003).

5.7.5 Organic Acids and Metal Sequestration

Among the organic acids produced, ectomycorrhizal fungi that released oxalic acid in higher quantities were noted by many scientists, and they observed the mantles of ectomycorrhiza and found calcium oxalate crystals in *Pinus radiata* and *Eucalyptus marginata* (Griffiths et al. 1994) and also in mycorrhizas of *Monotropa uniflora* (Snetselaar and Whitney 1990). Even the water droplets found in the mycelium of *Suillus bovinus* were observed for oxalic acid (Sun et al. 1999). Seedlings of birch, *Picea abies* and *Paxillus involutus*, were found with citrate and malate organic acids (Blaudez et al. 1998; Nowotny et al. 1998). Dutton and Evans (1996) reported that mycelium of mycorrhizae-inoculated plants retained calcium oxalate in their hyphae and act as a reservoir for calcium source for the plants and indirectly enhanced the sulphate availability in the root zone. Moreover the production of this oxalic acid was reported to suppress the diseases of plants (Duchesne et al. 1989). Mycorrhizal fungi in *Paxillus involutus* produced huge quantities of organic acids and tiny amounts of oxalic acid (Rasanayagam and Jeffries 1992).

These organic acid productions increase the cation uptake from the soil and benefit the plant growth and potentially inactivate the toxic inorganic aluminium (Hale and Griffin 1974). Toxic cations could be potentially detoxified by the organic acids; there are three categories in which strong detoxifiers are dicarboxylic acids, oxalic, citric and tartaric acids; intermediate detoxifiers are malic, malonic and salicylic acids; and weak detoxifiers include monocarboxylic acids, succinic, lactic, formic, acetic acids, etc. (Hue et al. 1986). Meharg (2003) reported that mycorrhizal plants of *Pinus sylvestris* efficiently detoxify Ni and Cd compared to nonmycorrhizal plants.

In the rhizosphere region, root exudates played a major role in metal solubilization as well as altered the level of ions and molecules (Caltado et al. 1988). Lundstrom (1993) explained the complexation process of organic acids where Cr (III) gets solubilized well in various plant species. Organic acids are negatively charged ions, and they interact with positively charged metal ions under various soil conditions (Jones and Darrah 1994). Tolra et al. (1996) reported higher concentrations of Ni and Zn resulted in elevated quantities of organic acids such as citric, malic, malonic and oxalic acids. In Cd-contaminated soil, mycorrhizae-inoculated *Medicago truncatula* roots excreted isoflavonoids in the root zone. Root exudates influence various characters in the soil such as metal solubility and availability, nutrient availability, microbial activity, soil properties, etc. (Zhang et al. 1989). Metal solubilization and mobilization mainly depend on the ligand to which it gets attached, the solid phase, metal involved (Treeby et al. 1989), plant species (Morel et al. 1986), mycorrhizal species involved, etc. (Romheld and Marschner 1990). Root exudates of some plants particularly bind with the specific metal. McGrath (1989) pointed that root exudates of maize and tobacco particularly have high affinity towards Cd. Root exudates of pea plants preferably solubilize Mn and Cu and form metallo-organic complexes which further enhanced the metal absorption (Mench et al. 1989). Yang et al. (2003) demonstrated Al (III) and Ni (II) have been detoxified by citric acid.

5.7.5.1 Organic Acids and Heavy Metals

Root exudates are influencing the availability of heavy metals, and they alter the redox behaviour of the metal studied by Caltado et al. (1988). In tomato, the uptake of chromium depends upon the presence of carboxylic acid and amino acids (Srivastava et al. 1999). Cr was found in the soil solution due to the action of organic acids, viz. citric, aspartic and oxalic acids, that convert the inorganic Cr to organically bound Cr (James and Bartlett 1983). Mench and Martin (1991) stated that the metal-binding ability depends on the metal-solubilizing ability that is compared with the dissociation constants, uptake pattern and presence of various organic acids (Selvaraj 1998).

Increasing concentrations of organic acids increased the uptake of Cr without affecting the distribution in plant parts (Srivastava et al. 1999). In wheat when the metal concentration increased, the production of organic acids also increased, and in turn translocation of heavy metal to shoot also gets enhanced (Chaney et al. 1997). Shahandeh and Hossner (2000) have showed that organic acids increased the translocation of Cr. Davis et al. (2001) have explained that phytoremediation of soils has a check on the presence of mycorrhizae and organic acids.

References

Amor YE, Inze BD, Levine A (1998) The involvement of poly (ADP-ribose) polymerase in the oxidative stress responses in plants. FEMS Lett 440:1–7

Antonovics JA, Bradshaw D, Turner RG (1971) Heavy metal tolerance in plants. Adv Ecol Res 7:1–85

Audet C, Charest C (2006) Effects of AM colonization on wild tobacco plants grown in zinc contaminated soil. Mycorrhiza 16:277–283

Auge RM (2001) Water relations, drought and vesicular- arbuscular mycorrhizal symbiosis. Mycorrhiza 11:3–42

Aydinalp C, Marinova S (2009) The effects of heavy metals on seed germination and plant growth on alfalfa plant (*Medicago sativa*). Bulg J Agric Sci 15:347–350

Baker AJM (1987) Metal tolerance. New Phytol 106:93–111

Becerril F, Tuinen D, Laurent FM, Metwally A, Dietz KJ, Gianinazzi V (2005) Molecular changes in *Pisum sativum* roots during Arbuscular mycorrhiza buffering of cadmium stress. Mycorrhiza 16:51–60

Berthelin J, Belgy G, Boymond D, Leyval C (1993) Microbial biosorption and ion accumulation of metals in filed conditions in heavy metal polluted soils and around uranium mining wastes. In: FEMS symposium metals, Microorganisms relationships and applications France, pp 61–70

Bi YL, Li XL, Christie P, Hu ZQ, Wong MH (2003) Growth and nutrient uptake of arbuscular mycorrhizal maize in different depths of soil overlying coal fly ash. Chemosphere 50:863–869

Blaudez D, Chalot M, Dizengremel P, Botton B (1998) Structure and function of the ectomycorrhizal associations between *Paxillus involutus* and *Betula pendula*. New Phytol 138:543–552

Bohn KS, Liberia AE (1982) In: Symposium on surface mining hydrology, sedimentology and reclamation. Graves DH (ed). University of Kentucky, Lexington

Bolan NS (1991) A critical review on the role of mycorrhizal fungi in the uptake of phosphorus by plants. Plant Soil 134:189–207

Bonfante-Fasolo P (1987) Vesicular-arbuscular mycorrhizal fungus plant interactions at the cellular level. Symbiosis 3:249–254

Borowicz VA (2001) Do arbuscular mycorrhizal fungi alter plant-pathogen relations? Ecology 82:3057–3068

Boruvka L, Drabek O (2004) Heavy metal distribution between fractions of humic substances in heavy polluted soils. Plant Soil Environ 50:339–345

Brookes AJ, Collins JC, Thurman DA (1981) The mechanism of zinc tolerance in grasses. J Plant Nutr 3:695–700

Brundrett M (1991) Mycorrhizas in natural ecosystems. Adv Ecol Res 2:171–220

Brundrett MC (2002) Coevolution of roots and mycorrhizas of land plants. New Phytol 154(2):275–304

Caltado DA, Mcfadden KM, Thomas RE (1988) Wilddung. Organic constituent and complexation of nickel (II), Cadmium (II) and Plutonium (VI) in soybean xylem exudates. Plant Physiol 86:734–739

Carlile MJ, Watkinson SC (1996) Parasites and mutualistic symbionts. In: Carlile MJ, Watkinson SC (eds) The Fungi. Academic Press, New York, pp 307–371

Chaney RL, Malik M, Li YM, Brown SL, Angle JS, Baker AJM (1997) Phytoremediation of soil metals. Curr Opin Biotechnol 8:279–284

Chaudhry MS, Batool Z, Khan AG (2005) Preliminary assessment of plant community structure and arbuscular mycorrhizas in rangeland habitats of Cholistan Desert, Pakistan. Mycorrhiza 15(8):606–611

Chen B, Christie P, Li L (2001) A modified glass bead compartment cultivation system for studies on nutrient and trace metal uptake by arbuscular mycorrhiza. Chemosphere 42:185–192

Chen BD, Li XL, Tao HQ, Christie P, Wong MH (2003) The role of arbuscular mycorrhiza in zinc uptake by red clover growing in a calcareous soil spiked with various quantities of zinc. Chemosphere 50:839–846

Chibuike GU, Obiora SC (2014) Heavy metal polluted soils: effect on plants and bioremediation methods. App Environ Soil Sci 12

Christie P, Li XL, Chen BD (2004) Arbuscular mycorrhiza can depress translocation of zinc to shoots of hostplants in soils moderately polluted with zinc. Plant Soil 261:209–217

Cline GR, Powell PE, Szaniszlo PJ, Reid C (1987) Comparison of the abilities of hydroxamic, synthetic and other natural organic acids to chelate Fe and other ions in nutrient solution. Soil Sci Am J 46:1158–1164

Cobbett C, Goldsburough P (2002) Phytochelatins and metallothioneins: roles in heavy metal detoxification and homeostasis. Annu Rev Plant Biol 53:159–182

Colpaert JV, Van Assche JA (1992) Zinc toxicity in ectomycorrhizal Pinus sylvestris. Plant Soil 143:201–211

Cooper KM, Tinker PB (1978) Translocation and transfer of nutrients in vesicular Arbuscular mycorrhizas Uptake and translocation of phosphorus, zinc and sulphur. New Phytol 81:43–52

Cronan C (1991) Differential adsorption of Al, Ca and Mg by roots of red spruce. Tree Physiol 8:227–237

Curl EA, Trueglove B (1986) The rhizosphere. Springer, Berlin

Daft MJ, Nicolson TH (1974) Arbuscular mycorrhizae in plants colonizing coal wastes in Scotland. New Phytol 72:1129–1138

Dangeard PA (1900) In: Progress in microbial ecology. Mukherji KG, Agnihorti VP, Singh RP (eds). Print House, Lucknow, pp 489–524

Daniell TJ, Husband R, Fitter AH, Young JPW (2001) Molecular diversity of arbuscular mycorrhizal fungi colonizing arable crops. FEMS Microbiol Ecol 36:203–209

Davis FT, Puryear JD, Newton RJ, Egilla JN, Grossi JAS (2001) Mycorrhizal fungi enhances accumulation and tolerance of chromium in sunflower (*Helianthus annuus*). J Plant Physiol 158:777–786

Dehn B, Schuepp H (1989) Influence of VA mycorrhizae on the uptake and distribution of heavy metal in plants. Agric Ecosyst Environ 29:79–83

Del Val C, Barea JM, Azcòn-Aguilar C (1999) Assessing the tolerance to heavy metals of arbuscular mycorrhizal fungi isolated from sewage sludge-contaminated soils. Appl Soil Ecol 11:261–269

Diaz G, Honrubia M (1994) A mycorrhizal survey of plants growing on mine wastes in Southeast Spain. Arid Soil Res Rehabil 8:59–68

Diaz G, Azcon-Aguilar C, Honrubia M (1996) Influence on mycorrhizae on heavy metal (Zn and Pb) uptake and growth of *Anthyllis cytisoides*. Plant Soil 180:241–249

Douds DD, Millner PD (1999) Biodiversity of arbuscular mycorrhizal fungi in agroecosystems. Agric Ecosys Environ 74:77–93

Doyle RJ, Lee NC (1986) Microbes, warfare, religion and human institutions. Can J Microbiol 32:193–200

Driver JD, Holben WE, Rilling MC (2005) Characterization of glomalin as a hyphal wall component of Arbuscular mycorrhizal fungi. Soil Biol Biochem 37(1):101–106

Duchesne LC, Ellis BE, Peterson RL (1989) Disease suppression by ectomycorrhizal fungus *Paxillus involutus*: contribution of oxalic acid C. J Bot 67:2726–2730

Dueck TA, Visser P, Ernst WHO, Schat H (1986) Vesicular arbuscular mycorrhizae decrease zinc toxicity to grass growing in zinc polluted soil. Soil Biol Biochem 78:331–333

Dutton MV, Evans CS (1996) Oxalate production by fungi: its role in pathogenicity and ecology in the soil environment. Can J Microbiol 42:881–895

Elliott HA, Linn JH, Shields GA (1989) Role of Fe in extractive decontamination of Pb polluted soils. Hazard Waste Hazard Mater 6:223–228

Enkhtuya B, Rydlová J, Vosátka M (2002) Effectiveness of indigenous and non-indigenous isolates of arbuscular mycorrhizal fungi in soils from degraded ecosystems and man-made habitats. Appl Soil Ecol 14:201–211

Feng G, Song YC, Li XL, Christie P (2003) Contribution of arbuscular mycorrhizal fungi to utilization of organic sources of phosphorus by red clover in a calcareous soil. Appl Soil Ecol 22:139–148

Fox TR, Comerford NB (1990) Low-molecular-weight organic acids in selected forest soils of the southern eastern USA. Soil Sci Am J 54:1139–1144

Frank AB (1885) Uber die an Wursel symbionts beruhende Ernahung geiwieser Baume durch uterirdische pilze. Ber Drsch bot Ges 3:128–145

Gadd GM (1990) Heavy metal accumulation by bacteria and other microorganisms. Experimentia 46:834–840

Gavito ME, Abud YC, Santiz YM, Trujillo MM, Rivera E (2014) Effects of aluminium and lead on the development of Rhizophagus irregularis and roots in root cultures. Environ Eng Manag J 13(9):2357–2361

George E, Haussler K, Kothari SK, Ki XL, Marschner H (1992) Contribution of mycorrhizal hyphae to nutrient and water uptake by plants. In: Read DJ et al (eds) Mycorrhizas in ecosystems. CAB International, Cambridge, pp 42–47

Gildon A, Tinker P (1981) A heavy metal-tolerant strain of a mycorrhizal fungus. New Phytol 95:263–268

Gildon A, Tinker P (1983) Interactions of vesicular-arbuscular mycorrhizal infection and heavy metals in plants. II. The effect of infection on uptake of copper. New Phytol 95:263–268

Giovannetti M, Sbrana C, Logi C (1994) Early processes involved in host recognition by Arbuscular mycorrhizal fungi. New Phytol 127:703–709

Gomathy M, Sabarinathan KG, Thangaraju M, Subramanian KS, Sivashankari Devi T, Ananthi K (2011) The effect of mycorrhizae inoculated maize root exudates in alleviation of chromium toxicity in chromium polluted environments. Insight Micro 1(2):20–30

Gonzalez-Chavez C, Haen JD, Angronsveld JV, Dodd JC (2002) Copper sorption and accumulation by the extra radical mycelium of different *Glomus* sp. Isolated from the same polluted soil. Plant Soil 240:287–297

Gonzalez-Chavez MC, Carrillo Gonzalez R, Wright SF, Nichols K (2004) The role of glomalin, a protein produced by arbuscular mycorrhizal fungi, in sequestering potentially toxic elements. Environ Pollut 130(3):317–323

Grauer UE (1992) Faktoren der aluminium tolerance bei verschiedenen pflanzenarten. Verlag U E Gruer, Wendlingen

Griffioen WAJ (1994) Characterization of a heavy metal-tolerant endomycorrhizal fungus from the surroundings of a zinc refinery. Mycorrhiza 4:197–200

Griffiths RP, Baham JE, Caldwell BA (1994) Soil solution chemistry of ectomycorrhizal mats in forest soil. Soil Biol Biochem 26:331–337

Hale MG, Griffin GJ (1974) Effect of injury on exudation from immature and mature plant fruits (Abstr.) Plant Physiol 13:56–57

Hall JL (2002) Cellular mechanisms for heavy metal detoxification and tolerance. J Exp Bot 53:1–11

Harmens H, Koevnets PLM, Verkleij JAC, Ernst WHO (1993) The role of low molecular weight organic acids in the mechanism of increased zinc tolerance in *Silene vulgaris*. New Phytol 126:615–621

He ZL, Yang XED, Stoffella PJ (2005) Trace elements in agroecosystems and impacts on the environment. J Trace Elem Med Biol 19(2-3):125–140

Heggo A, Angle JS, Chaney RL (1990) Effects of vesicular Arbuscular mycorrhizal fungi on heavy-metal uptake by soybeans. Soil Biol Biochem 22:865–869

Hildebrandt U, Kaldorf M, Bothe H (1999) The zinc violet and its colonization by arbuscular mycorrhizal fungi. J Plant Physiol 154:709–717

Hodge A, Campbell CD, Fitter AH (2001) An Arbuscular mycorrhizal fungus accelerates decomposition and acquires nitrogen directly from organic matter. Nature 413:297–299

Hooker JE, Black KE (1995) Arbuscular mycorrhizal fungi as components of sustainable soil-plant systems. Crit Rev Biotechnol 15:201–212

Hue NV, Craddock GR, Adams F (1986) Effect of organic acids on aluminium toxicity in subsoils. Soil Sci Soc Am J 50:28–34

Jagpal R, Mukerji KG (1988) Distribution of VA-mycorrhizal association in old Delhi ridge. In: Verma AK, Mukerji KG, Tilak KG, Tilak KVBR, Rai J (eds) Mycorrhiza round table. International Development Centre, Canada, pp 257–267

Jain R, Srivastava S, Madan VK, Jain R (2000) Influence of chromium on growth and cell division of sugarcane. Indian J Plant Physiol 5:228–231

Jamal A, Ayub N, Usman M, Khan AG (2002) Arbuscular mycorrhizal fungi enhance zinc and nickel uptake from contaminated soil by soyabean and lentil. Int J Phytoremed 4:205–221

James BR, Barlett RJ (1983) Behavior of chromium in soils VI. Interactions between oxidation – reduction and organic complexation. J Environ Qual 12:173–176

Janouskova MD, Pavlikova T, Macek Vosatka M (2005) Arbuscular mycorrhiza decreases cadmium phytoextraction by transgenic tobacco with inserted metallothionein. Plant Soil 272:29–40

Jeffries P, Barea JM (1994) Biogeochemical cycling and Arbuscular mycorrhizas in the sustainability of plant soil systems. In: Gianinazzi S, Schuepp H (eds) Impact of arbuscular mycorrhizas on sustainable agriculture and natural ecosystems. Birkhauser, Basel, pp 101–115

Jeffries P, Gianinazzi S, Perotto S, Turnau K, Barea JM (2003) The contribution of arbuscular mycorrhizal fungi in sustainable maintenance of plant health and soil fertility. Biol Fertil Soils 37:1–16

Jentshke G, Winter S, Goldbold DL (1998) Ectomycorrhizas and Cd toxicity in Norway spruce seedlings. Tree Physiol 18:256–258

Johnson D, Leake JR, Ostle N, Ineson P, Read DJ (2002) In situ (CO2)-C-13 pulse-labelling of upland grassland demonstrates a rapid pathway of carbon flux from Arbuscular mycorrhizal mycelia to the soil. New Phytol 153:327–334

Joner EF, Leyval C (1997) Uptake of ^{109}Cd by roots and hyphae of a *G. mosseae*, *Trifolium subterraneum* mycorrhiza from soil amended with high and low concentrations of cadmium. New Phytol 135:353–360

Joner EJ, Leyval C, Briones R (2000) Metal binding capacity of arbuscular mycorrhizal mycelium. Plant Soil 226:227–234

Jones DL (1998) Organic acids in the rhizosphere a critical review. Plant Soil 205:25–40

Jones DL, Darrah PR (1994) Role of root derived organic acids in the mobilization of nutrients from the rhizosphere. Plant Soil 166:247–257

Jongmans AG, Van Breemen N, Lundstrom U, Finally RD, Van Hees PAW, Giesler R, Melkerud PA, Olsson M, Srinivasan M, Unestam T (1997) Rock-eating fungi. Nature 389:683–684

Jonnarth UA, Vanhees PAW, Lundstrom US, Finlay RD (2000) Organic acids produced by mycorrhizal *Pinus sylvestris* exposed to elevated aluminium and heavy metal concentrations. New Phytol 146:557–567

Kabata-Pendias A, Pendias H (1992) Trace elements in soils and plants, 2nd edn. CRC Press, Boca Raton

Kabir Z, Koide RT (2000) The effect of dandelion or a cover crop on mycorrhiza inoculum potential, soil aggregation and yield of maize. Agric Ecosyst Environ 78:167–174

Kaldorf M, Kuhn AJ, Schroder WH, Hildebrandt U, Bothe H (1999) Selective element deposits in maize colonized by a heavy metal tolerance conferring arbuscular mycorrhizal fungus. J Plant Physiol 154:718–728

Kapoor A, Viraraghavan T (1995) Fungal biosorption an alternative treatment option for heavy metal bearing wastewater: a review. Bioresour Technol 53(3):195–206

Khan AG (1974) The occurrence of mycorrhizae in halophytes, hydrophytes and xerophytes and of endogone spores in the adjacent soils. J Gen Microbiol 81:7–14

Khan AG (2002) The significance of microbes. In: Wong MH, Bradshaw AD (eds) The restoration and management of derelict land: modern approaches. World Scientific Publishing, Singapore, pp 80–92

Khan AG (2003) Mycotrophy and its significance in wetland ecology and wetland management. In: Wong MH (ed) Proceedings Croucher Foundation Study Institute: wetland ecosystems in Asia-function and management, Hong Kong, pp 11–15

Khan AG (2005) Role of soil microbes in the rhizospheres of plants growing on trace metal contaminated soils in phytoremediation. J Trace Elem Med Biol 18(4):355–364

Khan AG, Kuek C, Chaudhry TM, Khoo CS, Hayes WJ (2000) Role of plants, mycorrhizae and phytochelators in heavy metal contaminated land remediation. Chemosphere 41:197–207

Killham K, Firestone MK (1983) Vesicular-arbuscular mycorrhizal mediation of grass response to acidic and heavy metal depositions. Plant Soil 72:39–48

Koomen I, McGrath SP, Giller KE (1990) Mycorrhizal infection of clover is delayed in soils contaminated with heavy metals from past sewage sludge applications. Soil Biol Biochem 22:871–873

Kroopnick PM (1994) Vapor abatement cost analysis methodology for calculating life cycle costs for hydrocarbon vapor extracted during soil venting. In: Wise DL, Trantolo DJ (eds) Remediation of hazardous waste. Marcel Dekker, New York, pp 779–790

Landeweert R, Hoffland E, Finlay RD, Kuyper TW, Van Breemen N (2001) Linking plants to rocks: ectomycorrhizal fungi mobilize nutrients form minerals trends in. Ecol Evol 16:248–254

Leal PL, Siqueira JO, Stürmer SL (2016) Switch of tropical Amazon forest to pasture affects taxonomic composition but not species abundance and diversity of arbuscular mycorrhizal fungal community. Appl Soil Ecol 71:72–80

Leyval C, Laheurte F, Belgy G, Berthelin J (1990) Weathering of micas in the rhizospheres of maize, pine and beech seedlings influenced by mycorrhizal and bacterial inoculation. Symbiosis 9:105–109

Leyval C, Singh BR, Joner EJ (1995) Occurrence and infectivity of arbuscular mycorrhizal fungi in some Norwegian soils influenced by heavy metals and soil properties. Water Air Soil Pollut 84:203–216

Leyval C, Turnau K, Haselwandter K (1997) Effect of heavy metal pollution on mycorrhizal colonization and function: physiological, ecological and applied aspects. Mycorrhiza 7:139–153

Liao JP, Lin XG, Cao ZH, Shi YQ, Wong MH (2003) Interactions between Arbuscular mycorrhizae and heavy metals under sand culture experiment. Chemosphere 50:847–853

Lovelock CE, Wright SF, Clark DA, Ruess RW (2004) Soil stocks of glomalin produced by arbuscular mycorrhizal fungi across a tropical rain forest landscape. J Ecol 92:278–287

Lundstrom US (1993) The role of organic acids in the soil solution chemistry of a podsolized soil. J Soil Sci 44:121–133

Lundstrom US (1994) Significance of organic acids for weathering and the podsolization process. Environ Int 20:21–30

Ma JF, Zheng SJ, Matsumoto H (1997) Specific secretion of citric acid induced by Al stress in *Cassia tora* L. Plant Cell Physiol 38:1019–1025

Maier RM, Pepper IL, Gerba CP (2009) Environmental microbiology. Academic Press, San Diego, CA

Malcova R, Rydlova J, Vosatka M (2003) Metal free cultivation of *Glomus* sp. BEG 140 isolated from Mn contaminated soil reduces tolerance to Mn. Mycorrhiza 13:151–157

Marschner H, Romheld V (1995) Strategies of plants for acquisition of iron. Plant Soil 165:262–274

Marschner P, Jentschke G, Godbold DL (1998) Cation exchange capacity and lead sorption in ectomycorrhizal fungi. Plant Soil 205:93–98

McGrath SP (1989) Crop yield and composition In: de Haan FAM (ed) Soil quality assessment. Final report to the C.E.C., DG XII, Contrat EV4A/0008/NL, Agricultural University Wageningen, The Netherlands, pp 1–35

Meharg AA (1993) The role of the plasmalemma in metal tolerance in angiosperms. Physiol Plant 88:191–198

Meharg AA (1994) Integrated tolerance mechanisms, constitutive and adaptive plant responses to elevated metal concentrations in the environment. Plant Cell Environ 17:989–993

Meharg A (2003) The mechanistic basis of interactions between mycorrhizal associations and toxic metal cations. Mycol Res 107:1253–1265

Meharg AA, Cairney JWG (2000) Co-evolution of mycorrhizal symbionts and their hosts to metal-contaminated environments. Adv Ecol Res 30:69–112

Mench M, Martin E (1991) Mobilization of cadmium and other metals from two soils by root exudates of Zea mays. *Nicotiana tabacum* and *Nicotiana rustica*. Plant Soil 132:187–196

Mench MJ, Tancogne J, Gomez A, Juste C (1989) Cadmium bioavailability to *Nicotiana tabacum* L., *Nicotiana rustica* L. and *Zea mays* L., grown in soil amended or not amended with cadmium nitrate. Biol Fertil Soils 8:48–53

Morel JL, Mench M, Guckert A (1986) Measurement of Pb, Cu, Cd binding with mucilage exudates from maize (*Zea mays* L.) roots. Boil Fertil Soils 2:29–34

Morley GF, Gadd GM (1995) Sorption of toxic metals by fungi and clay minerals. Mycol Res 99:1429–1438

Mori SN, Nishizawa H, Hayashi M, Yoshimura E, Ishihara J (1991) Why are young rice plants highly susceptible to iron deficiency. Plant Soil 130:143–356

Mosse B, Stribley DP, Le Tacon F (1981) Ecology of mycorrhizae and mycorrhizal fungi. In: Alexander M (ed) Advance in microbial ecology. Plenum Press, New York, p 250

Nichols PB, Couch JD, Hamdani SH (2000) Selected physiological responses of *Salvinia minima* to different chromium concentrations. Aquat Bot 68:313–319

Nowotny J, Schwanz J, Rothe GM (1998) Influence of soil acidification and liming on selected enzymes of the carbohydrate metabolism and the contents of two major organic acids of mycorrhizal roots of Norway spruce (*Picea abies*). Plant Soil 199:41–51

Ortiz DF, Ruscuitti T, McCue KF, Ow DW (1995) Transport of metal-binding peptides by HMT1, a fission yeast ABC-type vacuolar membrane protein. J Biol Chem 270:4721–4728

Olsson PA, Thingstrup I, Jakobsen I, Baath E (1999) Estimation of the biomass of arbuscular mycorrhizal fungi in linseed field. Soil Biol Biochem 31:1879–1887

Paris F, Bonnaud P, Ranger J, Lapeyrie F (1994) Phyllosilicate alteration by ecto-mycorrhizal fungi in vitro. Acta Bot Gallica 141:529–532

Parker R (1994) Environmental restoration technologies. EMIAA year book, pp 169–171

Pawlowska TB, Blaszkowski J, Rühling A (1996) The mycorrhizal status of plants colonizing a calamine spoil mound in southern Poland. Mycorrhiza 6:499–505

Pawlowska TE, Chaney RL, Chin M, Charvat I (2000) Effects of metal phytoextraction practices on the indigenous community of arbuscular mycorrhizal fungi at a metal contaminated landfill. Appl Enivorn Microbiol 66(6):2526–2530

Phipps CJ, Taylor TN (1996) Mixed arbuscular-mycorrhizae from the Triassic of Antarctica. Mycologia 88:707–714

Pirozynski KA, Malloch DW (1975) The origin of land plants; a matter of mycotrophism. Biosystems 6:153–164
Prasad MNV, Greger M, Landberg T (2001) *Acacia nilotica* L bark removes toxic elements from solution: corroboration from toxicity bioassay using *Salix viminalis* L in hydroponic system. Int J Phytoremed 3:289–300
Quilambo OA (2003) The vesicular-arbuscular mycorrhizal symbiosis. Afr J Biotechnol 2:539–546
Raju PS, Clark RB, Ellis JR, Maranville JW (1990) Effects of species of VA mycorrhizal fungi on growth and mineral uptake of sorghum at different temperatures. Plant Soil 121:165–170
Raman N, Sambandan S (1998) Distribution of VAM fungi in tannery effluent polluted soils of Tamil Nadu, Indian. Bull Environ Contam Toxicol 60:142–150
Rasanayagam S, Jeffries P (1992) Production of acid is responsible for antibiosis by some ectomycorrhizal fungi. Mycol Res 96:971–976
Rauser WE (1999) The structure and function of metal chelators produced by plants. Cell Biol Biophys 31:19–48
Read DJ, Moreno JP (2003) Mycorrhizas and nutrient cycling in ecosystems – a journey towards relevance. New Phytol 157:475–492
Reilly C (1972) Amino acids and amino acid copper complexes in water soluble extracts of copper tolerant and non tolerance *Becium homblei*. Z. Pflanzenphysiol 66:294–296
Remy W, Taylor TN, Hass H, Kerp H (1994) Four hundred-million-year-old vesicular Arbuscular mycorrhizae. Proc Natl Acad Sci 91:11841–11843
Rengel Z, Romheld V (2000) Root exudation and Fe uptake and transport in wheat genotypes differing in tolerance to Zn deficiency. Plant Soil 222:25–34
Rhodes LH, Gerdemann JW (1975) Phosphate uptake zones of mycorrhizal and non-mycorrhizal onions. New Phytol 75:555–561
Rillig MC, Wright SF, Nichols KA, Schmidt WF, Torn MS (2001) Large contribution of arbuscular mycorrhizal fungi to soil carbon pools in tropical forest soils. Plant Soil 233:167–177
Rillig MC, Steinberg PD (2002) Glomalin production by an arbuscular mycorrhizal fungus: a mechanism of habitat modification? Soil Biol Biochem 34:1371–1374
Rillig MC, Ramsey PW, Morris S, Paul EA (2003) Glomalin, an arbuscular-mycorrhizal fungal soil protein, respond to land-use change. Plant Soil 253(2):293–299
Romheld V, Marschner H (1990) Genotypical differences among graminaceous species in release of phytosiderophores and uptake of iron phytosiderophores. Plant Soil 123:147–153
Sambandan K, Kannan K, Raman N (1992) Distribution of vesicular-arbuscular mycorrhizal fungi in heavy metal polluted soils of Tamil Nadu, India. J Environ Biol 73:159–167
Schat H, Ten Bookum WM (1992) Genetic control of copper tolerance in *Silene vulgaris*. Heredity 68:219–229
Schüßler A (2000) *Glomus claroideum* forms an arbuscular-mycorrhiza-like symbiosis with the hornwort *Anthoceros punctatus*. Mycorrhiza 10(1):15–21
Schussler A, Schwarzott D, Walker C (2001) A new fungal phylum, the Glomeromycota: phylogeny and evolution. Mycol Res 105:1413–1421
Selvaraj T (1998) Studies on mycorrhizal and rhizobial symbioses on tolerance on tannery effluent treated *Prosopis juliflora*. PhD thesis. University of Madras, Chennai, India, p 215
Shahandeh H, Hossner LR (2000) Enhancement of Cr (III) phytoaccumulation. Int J Phytoremed 2:269–286
Shanker AK (2003) Physiological, biochemical and molecular aspects of chromium toxicity and tolerance in selected crops and tree species. PhD Thesis. Tamil Nadu Agricultural University, Coimbatore, India
Shanker AK, Cerventes C, Herminia L, Avudainayagam S (2005) Chromium toxicity in plants. Environ Int 31:739–753
Sheetal KR, Singh SD, Anand A, Prasad S (2016) Heavy metal accumulation and effects on growth, biomass and physiological processes in mustard. Indian J Plant Physiol 21(2):219–223
Simon L, Bousquat J, Levesque RC, Lalonde M (1993) Origin and diversification of endomycorrhizal fungi and coincidence with vascular land plants. Nature 363:67–69

Singh AK (2001) Effect of trivalent and hexavalent chromium on spinach (*Spinacea oleracea* L). Environ Ecol 19:807–810
Smith SE, Read DJ (1997) Mycorrhizal symbiosis, 2nd edn. Academic, London, p 589
Snetselaar KM, Whitney KD (1990) Fungal calcium oxalate in mycorrhizae of *Monotropa uniflora*. Can J Bot 68:533–543
Srinivas J, Purushotham AV, Murali Krishna KV (2013) The effects of heavy metals on seed germination and plant growth on Coccinia, Mentha and Trigonella plant seeds in Timmapuram. Int Res J Environ Sci 2(6):20–24
Srivastava S, Nigam R, Prakash S, Srivastava MM (1999) Mobilization of trivalent chromium in presence of organic acids: a hydroponic study of wheat plant (*Triticum vulgare*). Bull Environ Contam Toxicol 63:524–530
Subramanian KS, Thangaraju M, Jegan RA (2008) Mechanisms involved in tolerance of mycorrhizal plants to heavy metals. In: Saikia R, Bezbaruah R, Bora T (eds) Microbial biotechnology. New India Publishing Agency, New Delhi, pp 299–317
Sun YP, Unestam T, Lucas SD, Johanson KJ, Kenne KJ, Finlay R (1999) Exudation reabsorption in a mycorrhizal fungus, the dynamic interface for interaction with soil and other microorganisms. Mycorrhiza 9:137–144
Taylor TN, Taylor EL (1993) Biology and evolution of fossil plants. Englewood Cliff, Prentice Hall, NJ, 59 p
Taylor TN, Remy W, Hasss H, Keep H (1995) Fossil arbuscular mycorrhizae from the early Devonion. Mycologia 87:560–573
Thurman DA, Rankin JL (1982) The role of organic acids in zinc tolerance in *Deschampsia cespitosa*. New Phytol 91:629–635
Tisdall JM, Oades JM (1979) Stabilization of soil aggregates by the root systems of rye grass. Aust J Soil Res 17:429–441
Tisdall JM, Smith SE, Rengasamy P (1997) Aggregation of soil by fungal hyphae. Aust J Soil Res 35:55–60
Tolra RP, Poschenrieder C, Barcelo J (1996) Zinc hyperaccumulation in *Thlaspi caerulescens*. II. Influence on organic acids. J Plant Nutr 19:1541–1550
Treeby M, Marschner H, Romheld V (1989) Mobilization of iron and other micronutrient cations from a calcareous soil by plant-borne, microbial and synthetic metal chelators. Plant Soil 112:153–158
Tullio M, Pierandrei F, Salerno A, Rea E (2003) Tolerance to cadmium of vesicular arbuscular mycorrhizae spores isolated from a cadmium-polluted and unpolluted soil. Biol Fertil Soils 37:211–214
Turnau K, Ryszka P, Gianinazzi-Pearson V, van Tuinen D (2001) Identification of arbuscular mycorrhizal fungi in soils and roots of plants colonizing zinc wastes in southern Poland. Mycorrhiza 10:169–174
Turnau K, Kottke I, Oberwinkler F (1993) Element localization in mycorrhizal roots of *Pteridium aquilinum* L Kuhh collected from experimental plots treated with cadmium dust. New Phytol 123:313–324
Vandenkoornhuyse P, Husband R, Daniell TJ, Watson IJ, Duck JM, Fitter AH, Young JPW (2002) Arbuscular mycorrhizal community composition associated with two plant species in a grassland ecosystem. Mol Ecol 11:1555–1559
Varma A (1995) Arbuscular mycorrhizal fungi: the state of the art. Crit Rev Biotech 15:179–199
Vierheilig H, Lerat S, Piche Y (2003) Systemic inhibition of arbuscular mycorrhiza development by root exudates of cucumber plants colonized by *Glomus mosseae*. Mycorrhiza 13:167–170
Weissenhorn I, Leyval C (1995) Root colonization in maize by a Cd-sensitive and a Cd-tolerant Glomus mosseae and Cadmium uptake in sand culture. Plant Soil 175:233–238
Weissenhorn I, Leyval C, Belgy G, Berthelin J (1995) Arbuscular mycorrhizal contribution to heavy metal uptake by maize (*Zea mays*) in pot culture with contaminated soil. Mycorrhiza 5:245–251

Whipps JM (1990) Carbon economy. In: Lynch JM (ed) The rhizosphere. Wiley, Chichester, pp 59–97

Williams C, Nascimento A (2006) Organic acids effects on desorption of heavy metals from a contaminated soil. Sci Agric 63:276–280

Wright SF, Miller PD (1994) Dynamic processes of vesicular-arbuscular mycorrhizae: A mycorrhizosystem within the agroecosystem. In: Hatfeld JL, Stewart BA (eds) Advances in soil science soil biology effects on soil quality. Lewis Publishers, Boca Raton, pp 29–59

Wright SF, Upadhyaya A (1996) Extraction of an abundant and unusual protein from soil and comparison with hyphal protein of arbuscular mycorrhizal fungi. Soil Sci 161:575–586

Wright SF, Upadhyaya A (1998) A survey of soils for aggregate stability and glomalin, a glycoprotein produced by hyphae of arbuscular mycorrhizal fungi. Plant Soil 198:97–107

Wright SF, Franke-Snyder M, Morton JB, Upadhyaya A (1996) Time-course study and partial characterization of a protein on hyphae of arbuscular mycorrhizal fungi during active colonization of roots. Plant Soil 181:193–203

Wright SF, Upadhyaya A, Buyer JS (1998) Comparison of N-linked oligosaccharides of glomalin from arbuscular mycorrhizal fungi and soils by capillary electrophoresis. Soil Biol Biochem 30:1853–1857

Xiong LM (1993) Vesicular-arbuscular mycorrhizae decrease cadmium uptake by plant. J Plant Res Environ 2:58–60

Yang MN, Wang J, Wang SH, Xu LL (2003) Salicylic acid induce aluminum tolerance by modulation of citrate efflux from roots of *Cassia tora*. Planta 217:168–174

Young CC, Chang CH, Chen LF, Chao CC (1998) Characterization of the nitrogen fixation and ferric phosphate solubilizing bacteria isolated from a Taiwan soil. J Chin Agric Chem Soc 36:201–210

Zak JC, Parkinson D (1982) Initial vesicular-arbuscular mycorrhizal development of the slender wheatgrass on two amended mine soils. Can J Bot 60:2241–2248

Zhang F, Romheld V, Marzchner H (1989) Effect of zinc deficiency in wheat on the release of zinc and iron mobilizing root exudates. Pflanzenernaehr Bodenkd 152:205–210

Zhou JL (1999) Zn biosorption by *Rhizopus arrhizus* and other fungi. Appl Microbiol Biotechnol 51(5):686–693

Zhu YG, Christie P, Laidlaw AS (2001) Uptake of Zn by Arbuscular mycorrhizal white clover from Zn-contaminated soil. Chemosphere 42:193–199

Role of Microorganisms in Alleviating Abiotic Stresses

6

Ajay M. Sorty, Utkarsh M. Bitla, Kamlesh Kumar Meena, and Narendra P. Singh

Abstract

Constantly changing agroclimatic scenario has raised serious threats to agricultural production and productivity nowadays. Current attempts toward attenuation of abiotic stressor(s) have met limited success. Among the several strategies proposed, microbial mitigation of abiotic stresses has gained rapid attention, particularly in light of its sustainable and green approach that utilizes the natural phenomenon of plant-microbe association and subsequent beneficial interactions. The role of phyllosphere, rhizosphere, and endophytic microorganisms in mitigating a variety of abiotic stressors is well known. However, limited information is available till date regarding the cumulative influence of abiotic stressor(s) on plant-microbe association and on the stress-mitigation potential of microorganisms as well. Microbial inoculation is frequently recommended under stress-prone environment; however, it appears quite crucial to understand the behavior of inoculants under stressed habitats, which could substantially reduce the failure encountered by microbial inocula. This chapter typically highlights the plant-microbial interactions under abiotic stresses, microbial adaptations, and the role of stress-resilient microbes in alleviating the same.

Keywords

Abiotic stress · Microbial community · Phyllosphere · Signaling · Biomolecules · Secondary metabolites

A. M. Sorty · U. M. Bitla · K. K. Meena (✉) · N. P. Singh
School of Edaphic Stress Management (SESM), ICAR-National Institute of Abiotic Stress Management, Baramati, Pune, Maharashtra, India
e-mail: kkmeenamicro@gmail.com

D. G. Panpatte et al. (eds.), *Microorganisms for Green Revolution*, Microorganisms for Sustainability 7,
https://doi.org/10.1007/978-981-10-7146-1_6

6.1 Introduction

Plants meet a variety of abiotic factors under constantly changing environmental conditions. Most of the extreme abiotic factors are detrimental to plants, which significantly influence the growth, development, and productivity. All such abiotic factors are cumulatively termed as abiotic stresses. Prominent abiotic stressors include salinity, drought, low/high light, heavy metals, acidity, sodicity, wind, heat, frost, flood (submergence), etc. (Wang et al. 2003; Agarwal and Grover 2006; Nakashima and Yamaguchi-Shinozaki 2006; Bailey-Serres and Voesenek 2008). Exposure of a stressor or a combination thereof may be either acute or chronic in nature. The influence of plant growth, however, depends on the intensity of the stressor, duration of the stressor, growth stage and physiology of the plant, and ultimately the susceptibility of the cultivar.

Plants execute intrinsic combat mechanisms to restrict the adverse influence of various stressor(s). These mechanisms include avoidance, early senescence, development of resistance, expressions of different heat shock proteins (HSPs), antioxidant enzymes, accumulation of osmolytes, etc. (Scarpeci et al. 2008; Atkinson et al. 2013; Prasch and Sonnewald 2013). However, these mechanisms often are overcome by the stressor, making the adverse influence visible in terms of retarded growth, development, and productivity of the plant.

The additive effect of plants' intrinsic stress-combat mechanisms has been observed in the presence of plant growth-promoting (PGP) microorganisms. Many PGP microbes have been shown to actively induce the synthesis and increase in the levels of antioxidant enzymes, accumulation of osmolytes, and expression of different stress-responsive genes. The ability of microbes to enhance germination and establishment of juvenile seedlings under the influence of different abiotic stressors is well known. A variety of PGP bacteria, fungi, and actinomycetes, either individually or in the consortium form, have been demonstrated for their plan-beneficial interactions with abiotically stressed environments (Braud et al. 2009; Hayat et al. 2010; Baltruschat et al. 2008; Wang et al. 2012). Besides executing direct interaction with the plant, the microbes also enhance sustenance of the plant under stress conditions by altering soil properties, e.g., mycorrhiza and other plant-beneficial fungi form a network of hyphae in root proximity, which promotes soil aggregation and thus the colonization (Calvet et al. 2004). This chapter typically covers the beneficial interactions between plant and the associated microflora under abiotically stressed habitats and the role of plant-associated microbes in mitigation of different abiotic stresses.

6.2 Plant-Microbe Interactions Under Abiotic Stress

Both the plant growth, development, and productivity and native microbial population from the soil are negatively influenced following exposure to abiotic stresses. The microbes being prokaryotic entities suffer drastically. However, a chronic exposure of the stressor can potentially induce adaptations in microbes to the altered environmental conditions. It is thus a fraction of soil flora managed to sustain under abiotic stress.

Besides availability of the soil microflora, plants suffering from abiotic stress fail to project colonizing signals in the form of root exudations (which gets altered following exposure to stress), thus leading to decreased colonization (Kamilova et al. 2006). This cumulatively influences the growth and development of the plant and worsens the adverse effects of the stressor.

6.2.1 Influence of Stress on Plants

Plants generally encounter a combination of abiotic stressors under in situ conditions. The influence of any number of stresses at a given point of time varies drastically with the determinant factors such as duration of exposure, growth stage, susceptibility, etc. Moreover, many times a combination of stresses helps reduce the adverse effects of each other (Iyer et al. 2013). Influence of temperature stress on an antioxidative defense system of wheat seedlings may be altered by water/salt stress in wheat seedlings (Keles and Oncel 2002). Mitigation of plants to a combination of drought and heat stress cytosolic APX1 plays a central role (Koussevitzky et al. 2008). Plants exposed to drought, heat stress or a combination of drought and heat stress have been shown to accrue sucrose other sugars such as maltose and glucose (Rizhsky et al. 2004).

At earlier stages, e.g., germination, abiotic stresses like drought, salinity, exposure of heavy metals, and acidic/sodic conditions exert prominent influence on seed germination that declines severalfold (McCue and Hanson 1990; Keles and Oncel 2002). Moreover, the seeds that managed to germinate successfully often show reduced vigor and thus exhibit poor establishment, which ultimately hinders the subsequent development of the plant and thus the productivity. Further, the plants suffering from abiotic stresses show altered root exudation, which hampers the rhizosphere signaling cascades that play crucial role in microbial colonization. Lack of important nodes in signaling cascades significantly reduces the microbial colonization, which further weakens the plants' combat mechanisms against the stress situation. Decreased resistance to the stressor significantly retards the growth and development of the plant, which ultimately result in declined yield.

6.2.2 Influence of Stress on Microbes and Microbial Colonization

As discussed earlier, microbes are relatively more susceptible to adverse conditions. They undergo rapid metabolic and physicochemical changes following the exposure to abiotic stress. This directly affects the qualitative and quantitative metabolism of the microorganisms from the stressed environment. Moreover, such stress-driven metabolic hindrances in microbes ultimately influence the important nodes of rhizosphere signaling cascades (Sadowasky 2005) and also the abundance of plant-beneficial biomolecules of microbial origin in the phyllosphere environment.

Another important factor to be considered while considering the influence of abiotic stress on microbes is the resilience of the native microbial community. The lesser the resilience of a microbial community to abiotic stressor(s), the lesser the chances of survival following the exposure. However, a chronic exposure to abiotic stressor(s) can induce adaptive changes in a fraction of native microbial population, which efficiently manage to perform routine metabolic reactions even under the influence of the stressor(s) (Meena et al. 2017). Thus sudden onset of stressor(s) can impose peak damage to microbial richness of the habitat, while a chronic exposure may potentially induce resilience among the candidate members of native microbial community. Abiotic stressors such as salinity, prolonged drought, acidity, sodicity, etc. often are chronic in nature; the habitats suffering from similar circumstances can potentially provide a plethora of adapted PGP microbial strains, which could probably be utilized as inoculants for alleviation of the particular stressor(s) at different sites (Omar et al. 2009; Tiwari et al. 2011; Sorty et al. 2016).

6.2.3 Interactive Effect of Stress on Microbial Colonization

Microbes colonize the phyllosphere by responding to different signals projected by plants. The colonizing microbes themselves also are known to communicate by projecting an array of signaling molecules that predominantly include microbial secondary metabolites. Stress-governed metabolic hindrance in microbes may potentially deviate the metabolic pathways and alter/completely cease the production of important metabolite mediators in signaling cascades (Paul 2012).

Similarly, the plants also suffer in terms of stress-induced metabolic hindrances, which affect the metabolic fluxes and thus the quality and quantity of root exudations, which fail to attract the microbes from the surrounding environment. Rhizosphere microorganisms, mainly beneficial bacteria and fungi, can enhance plant growth under stress environments and, consequently, enhance yield (Dimkpa et al. 2009), (Fig. 6.1). Absence of beneficial microbes in the phyllosphere region exerts negative influence on plants' growth and overall development. Thus ensuring optimum colonization under stress conditions represents an important task. Many observations have demonstrated the negative influence of abiotic stress conditions on microbial colonization in phyllosphere region. Under drought conditions, nodule initiation, formation, and development and nitrogen fixation are inhibited by lower water content (Serraj 2003); it inhibits nodule oxygen permeability and nitrogenase activity (Serraj and Sinclair 1996).

6.2.4 Microbes Mediated Alleviation of Abiotic Stress

Microbes carry out a variety of metabolic operations in phyllosphere region, which influence plant growth. Different biomolecules are produced by microbes that potentially act as plant growth regulators. Under stress-prone conditions, such molecules play vital role in plants' tolerance to the adverse circumstances. Most

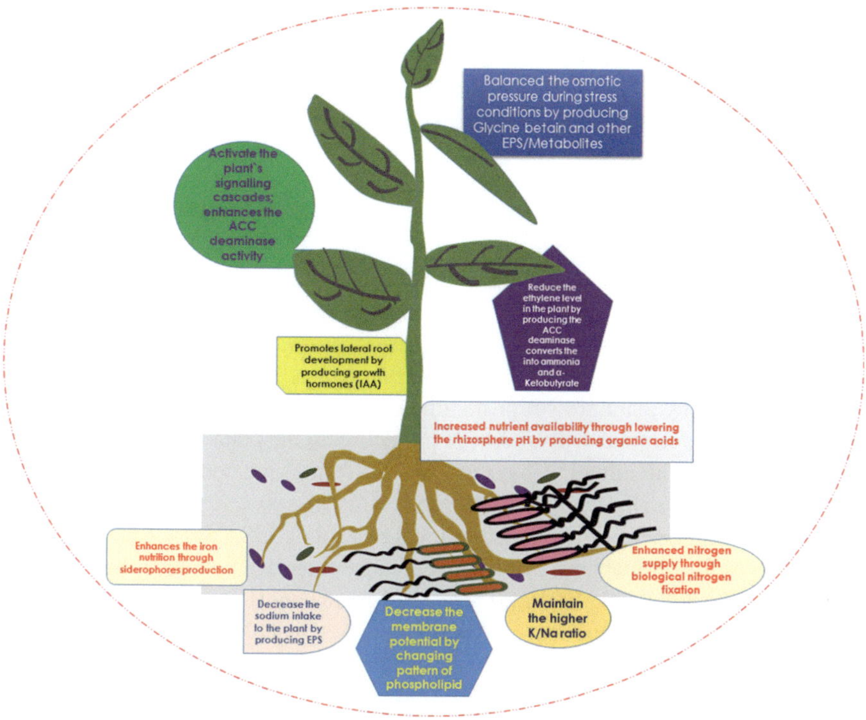

Fig. 6.1 The figure illustrates the mechanisms involved in microbe-mediated abiotic stress tolerance in crop plants. The application of nonpathogenic, plant growth-promoting bio-inoculants plays varying roles in the root zone or rhizosphere and enhances the stress tolerance. The presence of plant growth-promoting bacteria in the root zone alters the selectivity of the ions like Na^+, Ca^{++}, and K^+, thus enabling the plant to maintain higher K+/Na+ ratio. The members of PGP bio-inoculants make changes in components of the membrane particularly in phospholipids which ultimately alter the saturation patterns and reduce the membrane potential. Plant growth-promoting bacteria promote the lateral root development by producing the growth hormones like IAA which ultimately increases the total root surface area and enhances the plant tolerance toward different stressors. Bacteria help the plant accelerate the osmotic adjustment by producing variety of osmolytes like glycine, betaine, etc. Bio-inoculants activate the signaling cascades which help the host plant to maintain its defensive physiological state and also reduce the ethylene levels in the plant with the help of ACC deaminase. Bio-inoculants also enhance the nitrogen availability and other nutrients' availability through biological nitrogen fixation and lower the rhizosphere pH by producing the organic acids. Microorganisms also provide the fitness to the plants under abiotically stressed conditions by increasing the iron nutrition by producing siderophores. Further, microbes also reduce heavy metal toxicity by reducing their availability to the plants via production of diverse kinds of exopolysaccharides (EPS) and also through binding of the elements in biologically unavailable form

predominant molecules include microbial derivatives of plant growth hormones (PGHs), enzymes like ACC deaminase, siderophores, microbial exopolysaccharides (EPS), etc. (Fig. 6.1). All such direct and indirect biomolecular interactions between the associated microbes and plants are cumulatively studied under beneficial microbe-plant interactions under stress.

6.2.4.1 ACC Deaminase Production

Plant growth-promoting bacteria are able to produce ACC deaminase to lower plant ethylene level (Fig. 6.1). Ethylene is synthesized in higher plants, under a variety of abiotic as well as biotic stresses. Ethylene is a key modulator in the development and growth of the plant. Ethylene can affect all stages of plant development such as plant tissues like roots, stems, leaves, flowers, and fruits. Under abiotic stress conditions, the level of ethylene in higher plant is increased. ACC deaminase producing microbial strains facilitate plant growth, by lowering the levels of ethylene (Saleem et al. 2007). This enzyme is responsible for the cleavage of the plant ethylene precursor – 1-aminocyclopropane (ACC). Rhizobial strains that are able to produce ACC deaminase enzymes are 40% more efficient in producing nitrogen-fixing nodules than other strains that do not have this trait (Ma et al. 2003a, b). Production of ACC deaminase by plant growth-promoting bacterial strains also stimulates growth of plants in biotic stress that also provide protection to the plant by ethylene causing damage resulting in infection by a pathogen. Transgenic tomato (*Lycopersicon esculentum*) plants expressing the acdS gene from *Enterobacter cloacae* UW4 were distinctly positioned under the transcriptional control of 35S CaMV, rolD promoter, and prb-1b promoter that have been considered for their reaction due to submerging stress (Grichko and Glick 2001a, b). The transgenic canola plants were also developed expressing acdS gene from *Pseudomonas putida* UW4 under the control of root-specific rolD promoter from *Agrobacterium rhizogenes* to evaluate their response toward submergence tolerance (Farwell et al. 2007). Pea (*Pisum sativum* L.) seeds, coated with rhizobacterial strains capable of producing ACC deaminase, successfully eliminated the influence of drought stress on growth and ripening (Arshad et al. 2008). Soil inoculation with a natural root-associated bacterium *Variovorax paradoxus* 5C-2 was found effective in enhancing the yield of pea (*Pisum sativum*) plants grown in drying soil conditions (Belimov et al. 2009). Rhizobacteria isolated from rice rhizosphere containing ACC deaminase were also found effective in enhancing salt tolerance and consequently improving the growth and development of rice plants under salt-stress conditions (Bal et al. 2013).

6.2.4.2 Microbial Volatile Organic Compounds (MVOCs)

Microbial volatile organic compounds (MVOCs) are synthesized by numerous microorganisms ranging from bacteria to fungi. MVOCs have prospective as promising replacements to harmful insecticides, fungicides, and bactericides as well as genetic modification. MVOCs are complex mixture of low molecular weight compound and are vital infochemicals mediating essential communication systems in all kingdoms of life (Hare 2011; Dweck et al. 2015; Bitla et al. 2017). MVOCs are involved in plant rhizospheric processes like competence, pathogenesis, and symbiosis and also work as quorum-sensing signal molecules for both microbial growth and root development (Ortiz-Castro et al. 2008, 2011; Chernin et al. 2011). Rhizospheric bacterial strains release several VOCs that can potentially control both the plant growth promotion and root-system development (Gutierrez-Luna et al. 2010); some studies have demonstrated that PGPR volatile emission triggers the ISR in plants to tolerate abiotic stresses like salinity and drought (Yang et al. 2009). *Bacillus subtilis* strain GB03 and

Bacillus amyloliquefaciens strain IN937a emitted volatiles that considerably decrease the severity of disease on *Arabidopsis thaliana* affected by *Pectobacterium carotovorum* subsp. *carotovorum* and promote the growth of the plant (Ryu et al. 2003, 2004). *Fusarium oxysporum* MSA 35 produces VOC β-caryophyllene which induces shoot length, root length, and fresh weight of lettuce seedlings (Minerdi et al. 2011). 2R,3R-butanediol volatile compound produced by a *Pseudomonas chlororaphis* O6 is responsible for the development of induced systemic resistance *Arabidopsis thaliana* against drought (Cho et al. 2008). Rhizobacterial strain *B. subtilis* GB03 VOC emission controls the auxin production and cell development in *Arabidopsis thaliana* (Zhang et al. 2007). Roots of *A. thaliana* seedlings were inoculated with a suspension of *Bacillus subtilis*; acetoin emitted by *B. subtilis* triggers the ISR (Rudrappa et al. 2010). More than over 1000 microbial volatiles are described and documented in a distinct database for microbial VOCs called mVOC[1] (Lemfack et al. 2014). The numbers of volatiles reported are very low as compared to the microbial diversity present in soil; there is need of more study about MVOCs for application in agriculture against the abiotic stress.

6.2.4.3 Production of Microbial Derivatives of Plant Growth Hormones (PGHs)

Plant growth hormones synthesized by microbes are well studied. Hormones like auxins and cytokinins are known to play key role in the plant-microbe signaling as well as regulation of the plants' growth (Meena et al. 2012). Auxins and cytokinesis are also involved in the regulation of the symbiotic nitrogen fixation of rhizobium-plant interaction (Allen et al. 1953). The influence of cytokinins in the initiation of nodule organogenesis was also well studied in diverse legumes, where exogenous applications of cytokinins successfully induced amyloplast accumulation, cortical cell divisions, and the expression of initial nodulation indicators (Torrey 1961; Dehio and deBruijn 1992; Mathesius et al. 2000; Murray et al. 2007). Indole-3-acetic acid (IAA) is the most widely studied plant hormone – a carboxylic acid having a carboxyl group linked with a methylene group to the 3rd carbon of the indole ring (Thimann 1939). Commonly majority of the auxin-producing bacteria produce IAA, while some of the other bacteria produce GA3 and other derivatives. IAA also has the ability to influence the gene expression in microbes; thus IAA contributes central role in plant-microbe interaction (Fig. 6.1). Moreover, it is also involved in the adaptation of plant to abiotic stresses like salinity and drought (Spaepen and Vanderleyden 2011). IAA and GA3 induce the growth and are also involved in the development of plants' root system (Atzorn et al. 1988; Bottini et al. 2004). A recent study demonstrated that *Pseudomonas* sp., *Rhizobium* sp., *Enterobacter* sp., *Pantoea* sp., *Marinobacterium* sp., *Acinetobacter* sp., and *Sinorhizobium* sp., able to produce IAA, have significant influence on germination and seedling growth in wheat under saline condition. *Pseudomonas* sp. and *Acinetobacter* sp. were also described to enhance production of IAA in barley and oats, in salinity stress conditions (Chang et al. 2014). Cytokinins producing bacterial strains can also enhance the growth under drought stress (Arkhipova et al. 2007)

6.2.4.4 Nutrient Cycling

The plant depends on bioavailability of both macro- and micronutrients. Most of the times, abiotic stress situation alters the soil properties; for instance, conditions like salinity, acidity, alkalinity, metals contamination, etc. induce drastic changes in the physicochemical properties of soil. This badly affects the bioavailability of nutrients from the soil. The altered soil pH can potentially induce chemical changes among soil nutrients, thereby making them biologically unavailable. Such nutrients though are available in the soil, they become biologically unavailable, to which plants fail to absorb, which directly influence the plant growth. For instance, a change in pH toward the alkaline side drastically affects the Fe^{++} content from soil, which gets converted into Fe^{+++}, for uptake of which a strong chelating system is required, which constitutes a class of molecules called siderophores (Fig. 6.1). Though some monocots have been shown to produce siderophores, the iron supplemented by them appears insufficient for the plant. Under such circumstances, microbial siderophores can contribute significant role in Fe fulfillment to the plant (Kumar et al. 2008). The rhizosphere microbes produce relatively large quantities of siderophores which actively chelate Fe^{+++} and facilitate its bioavailability. Many siderophore-producing microbes have been shown to actively enhance iron fulfillment to the plant under iron-starved soils (Hao et al. 2012; Supanekar et al. 2013). Similarly, the bioavailability of other micronutrients is also known to facilitate the rhizosphere microbes under starved conditions.

Fixation of atmospheric nitrogen by rhizosphere microbes is another important task. Many microbial strains have been described in literature for their exceptional ability to fix atmospheric nitrogen under abiotic stress environment (Bianco and Defez 2009; Casanovas et al. 2002). Legume-rhizobium symbiosis has been found to be very sensitive to water-deficit condition (Kirda et al. 1989). Indigenous strains of *B. japonicum* were working inoculant than commercial strains of rhizobium once soybean grows under drought conditions (Hunt et al. 1988). The use of such resilient nitrogen fixers under the influence of stressors like salinity and altered pH conditions can ensure a sustainable supply of nitrogen, as addition of chemical fertilizers may worsen the situation over prolonged period. AM fungus *Glomus mosseae* was shown to contribute major role in the improvment in plant dry mass, nitrogen-fixing potential of nodules, and offering protection to the olive plant from harmful impact of salinity (Porras-Soriano et al. 2009).

Subsequently nitrogen and phosphate are the most necessary nutrients for the plant growth, but a soluble form of phosphate in soil is very low around 1 mg/kg or less (Goldstein 1994). Phosphate solubilization is another important trait of plant growth-promoting microbes. This trait is typically attributed to the organic acids produced by microbes as metabolic products (Fig. 6.1). The enzyme phytase has also been shown to have a significant role in phosphorus mobilization by microorganisms (Kohler et al. 2007). The ability of microbial strains to solubilize phosphate under abiotically stressed conditions like salinity is well known (Bianco and Defez 2010; Mishra et al. 2016); *P. oxalicum* and *P. expansum* indicated different stages of phosphate solubilization under saline stress (Hefnawy et al. 2014); bacterial strains from genera *Pseudomonas*, *Rhizobium*, and *Bacillus* and enterobacteria as well as

Penicillium and *Aspergillus* fungi are the most dominant P solubilizers (Whitelaw 2000). Such stress-resilient microbes demonstrate their metabolic efficiency and thus the adaptations under abiotic stress conditions. Similarly mobilization of potash is also achieved by rhizosphere microbes under stressed environment.

6.2.4.5 Microbial Exopolysaccharides

Microbial exopolysaccharides (EPS) are an important class of carbohydrate polymers that have a multifaceted role in the rhizosphere. They may be either homopolymers of heteropolymers, depending on the organism synthesizing. The unique property of microbial EPS to hold water (water holding capacity) further increases their importance under water-deficit conditions, where the presence of EPS can efficiently ensure the appropriate moisture supply in rhizosphere environment (Hepper 1975; Selvakumar et al. 2012; Alami et al. 2000). Additionally, microbial EPS have many reactive sites, where varieties of elements including micronutrients are trapped, which are then released slowly in rhizosphere environment. The affinity of many EPS toward Na^+ has been demonstrated in many studies; this highlights the potential use of this class of compounds in remediation of saline soils (Khan et al. 2016); further they can also potentially be applied for alleviation of salinity stress through point application.

EPS also have been shown to enhance microbial colonization by providing substratum for adherence to the microbes; this phenomenon is also helpful for the development of biofilm. Moreover microbial polysaccharides are also known to have a role in rhizosphere signaling, e.g., during nodulation, root colonization, etc. (Amellal et al. 1998). Considering the beneficial influence of microbial EPS in plant-microbe interactions, there is plenty of scope for their utilization in mitigation of abiotic stresses (Fig. 6.1). *Pseudomonas* sp. PMDzncd2003 on rice germination under salinity stress is verified; superior root colonizing ability of *Pseudomonas* sp. along with its capability to synthesize exopolysaccharides (EPS) indicates better tolerance toward salinity (Sen and Chandrasekhar 2014). *Pseudomonas putida* strain GAP-P45 mitigates drought stress in *Helianthus annuus* by producing exopolysaccharide (Sandhya et al. 2009).

6.3 Future Prospectus

A plethora of microbial strains have been described for their ability to promote growth by various mechanisms under abiotically stressed environments. However, the knowledge regarding their performance under field conditions appears quite limited. This leaves an excellent opportunity for the development of efficient inocula consisting of abiotic stress-resilient microbial strains, which could sustainably help mitigate the adverse effects of abiotic stressor(s) in agricultural crop. Moreover, there is also opportunity to explore novel diversity of resilient microorganisms from the habitat chronically exposed to abiotic stresses. Similarly the wild plants thriving luxuriantly under extreme environments may also be explored for candidate microbial strains which can open the new gateways in the area of microbial management of abiotic stress in agriculture. Similarly the current knowledge regarding microbial metabolism under the influence of abiotic stresses under field conditions is also in infancy. Detailed metabolomics account of stress-governed metabolic fluctuations in PGP microbes from rhizosphere habitat can potentially pave the way for the development of novel strategies relating to the use of microbial metabolic products for alleviation of abiotic stresses in stress-affected agricultural ecosystems.

6.4 Conclusions

Plant-associated microbes contribute a vital role in the sustenance and successful combat of the plant to adverse environmental circumstances generated by abiotic stressor(s). The habitats suffering prolonged exposure to abiotic stressor(s) can potentially act as reservoirs for stress-resilient PGP microbial strains that can serve the role as efficient stress alleviators in agriculture crop, thus promoting sustainable management of abiotic stresses under changing climate scenario.

Acknowledgments The authors are grateful to Indian Council of Agricultural Research for financial support.

References

Agarwal S, Grover A (2006) Molecular biology, biotechnology and genomics of flooding-associated low O2 stress response in plants. Crit Rev Plant Sci 25:1–21. doi:10.1080/07352680500365232

Alami Y, Achouak W, Marol C et al (2000) Rhizosphere soil aggregation and plant growth promotion of sunflowers by an exopolysaccharide-producing *Rhizobium* sp. strain isolated from sunflower roots. Appl Environ Microb 66:3393–3398

Allen EK, Allen ON, Newman AS et al (1953) Pseudonodulation of leguminous plants induced by 2-bromo-3,5-dichlorobenzoic acid. Am J Bot 40:429–435

Amellal N, Burtin G, Bartoli F et al (1998) Colonization of wheat rhizosphere by EPS producing *Pantoea agglomerans* and its effect on soil aggregation. Appl Environ Microbiol 64:3740–3747

Arkhipova TN, Prinsen E, Veselov S et al (2007) Cytokinin producing bacteria enhance plant growth in drying soil. Plant Soil 292:305–315. https://doi.org/10.1007/s11104-007-9233-5

Arshad M, Shaharoona B, Mahmood T et al (2008) Inoculation with Pseudomonas spp. containing ACC deaminase partially eliminates the effects of drought stress on growth, yield, and ripening of pea (*Pisum sativum* L.) Pedosphere 18:611–620

Atkinson NJ, Lilley CJ, Urwin PE et al (2013) Identification of genes involved in the response of arabidopsis to simultaneous biotic and abiotic stresses. Plant Physiol 162:2028–2041. https://doi.org/10.1104/pp.113.222372

Atzorn R, Crozier A, Wheeler CT et al (1988) Production of gibberellins and indole-3-acetic acid by Rhizobium phaseoli in relation to nodulation of *Phaseolus vulgaris* roots. Planta 175:532–538

Bailey-Serres J, Voesenek LACJ (2008) Flooding stress: acclimations and genetic diversity. Annu Rev Plant Biol 59(1):313–339

Bal HB, Nayak L, Das S et al (2013) Isolation of ACC deaminase PGPR from rice rhizosphere and evaluating their plant growth promoting activity under salt stress. Plant Soil 366:93–105

Baltruschat H, Fodor J, Harrach BD et al (2008) Salt tolerance of barley induced by the root endophyte *Piriformospora indica* is associated with a strong increase in anti-oxidants. New Phytol 180:501–510. https://doi.org/10.1111/j.1469-8137.2008.02583.x

Belimov AA, Dodd IC, Hontzeas N et al (2009) Rhizosphere bacteria containing 1-aminocyclopro pane-1-carboxylate deaminase increase yield of plants grown in drying soil via both local and systemic hormone signalling. New Phytol 181:413–423

Bianco C, Defez R (2009) Medicago truncatula improves salt tolerance when nodulated by an indole-3-acetic acid-overproducing Sinorhizobium meliloti strain. J Exp Bot. doi:10.1093/jxb/erp140

Bianco C, Defez R (2010) Improvement of phosphate solubilization and Medicago plant yield by an indole-3-acetic acid-overproducing strain of *Sinorhizobium meliloti*. Appl Environ Microbiol 76:4626–4632

Bitla UM, Sorty AM, Meena KK et al (2017) Rhizosphere signaling cascades: fundamentals and determinants. In: Singh DP, Singh HB, Prabha R (eds) Plant-microbe interactions in agro-ecological perspectives, vol I. Springer Nature, Singapore, pp 211–226

Bottini R, Cassán F, Piccoli P et al (2004) Gibberellin production by bacteria and its involvement in plant growth promotion and yield increase. Appl Microbiol Biotechnol 65:497–503

Braud A, Jezequel K, Bazot S et al (2009) Enhanced phytoextraction of an agricultural Cr-, Hg- and Pb-contaminated soil by bioaugmentation with siderophore producing bacteria. Chemosphere 74:280–286. https://doi.org/10.1016/j.chemosphere.2008.09.013

Calvet C, Estaún V, Camprubí A et al (2004) Aptitude for mycorrhizal root colonization in Prunus root stocks. Sci Hortic 100:39–49. https://doi.org/10.1016/j.scienta.2003.08.001

Casanovas EM, Barassi CA, Sueldo RJ et al (2002) Azospirillum inoculation mitigates water stress effects in maize seedlings. Cereal Res Commun 30:343–350

Chang P, Gerhardt KE, Huang XD et al (2014) Plant growth promoting bacteria facilitate the growth of barley and oats in salt-impacted soil: implications for phytoremediation of saline soils. Int J Phytorem 16:1133–1147. https://doi.org/10.1080/15226514.2013.821447

Chernin L, Toklikishvili N, Ovadis M et al (2011) Quorum-sensing quenching by rhizobacterial volatiles. Environ Microbiol Rep 3:698–704

Cho SM, Kang BR, Han SH et al (2008) 2R, 3R-butanediol, a bacterial volatile produced by *Pseudomonas chlororaphis* O6, is involved in induction of systemic tolerance to drought in *Arabidopsis thaliana*. Mol Plant Microbe Interact 21:1067–1075. https://doi.org/10.1094/MPMI-21-8-1067

Dehio C, deBruijn FJ (1992) The early nodulin gene SrEnod2 from *Sesbania rostrata* is inducible by cytokinin. Plant J 2:117–128. https://doi.org/10.1046/j.1365-313X.1992.t01-51-00999.x

Dimkpa C, Weinand T, Ash F et al (2009) Plant-rhizobacteria interactions alleviate abiotic stress conditions. Plant Cell Environ 32:1682–1694

Dweck HKM, Ebrahim SAM, Thoma M et al (2015) Pheromones mediating copulation and attraction in Drosophila. Proc Natl Acad Sci U S A 112:E2829–E2835. https://doi.org/10.1073/pnas.1504527112

Farwell AJ, Vesely S, Nero V et al (2007) Tolerance of transgenic canola plants (*Brassica napus*) amended with plant growth-promoting bacteria to flooding stress at a metal contaminated field site. Environ Pollut 147:540–545

Goldstein AH (1994) Involvement of the quinoprotein glucose dehydrogenase in the solubilization of exogenous phosphates by gram-negative bacteria. In: Torriani-Gorini A, Yagil E, Silver S (eds) Phosphate in microorganisms: cellular and molecular biology. ASM, Washington, DC, pp 197–203

Grichko VP, Glick BR (2001a) Flooding tolerance of transgenic tomato plants expressing the bacterial enzyme ACC deaminase controlled by the 35S, rolD or PRB-1b promoter. Plant Physiol Biochem 39:19–25

Grichko VP, Glick BR (2001b) Amelioration of flooding stress by ACC deaminase containing plant growth promoting bacteria. Can J Microbiol 47:77–80

Gutierrez-Luna FM, Lopez-Bucio J, tamirano-Hernandez J et al (2010) Plant growth-promoting rhizobacteria modulate root-system architecture in *Arabidopsis thaliana* through volatile organic compound emission. Symbiosis 51:75–83. https://doi.org/10.1007/s13199-010-0066-2

Hao L, Willis DK, Andrews-Polymenis H et al (2012) Requirement of siderophore biosynthesis for plant colonization by *Salmonella enterica*. Appl Environ Microbiol 78:4561–4570

Hare JD (2011) Ecological role of volatiles produced by plants in response to damage by herbivorous insects. Annu Rev Entomol 56:161–180. https://doi.org/10.1146/annurev-ento-120709-144753

Hayat R, Ali S, Amara U et al (2010) Soil beneficial bacteria and their role in plant growth promotion: a review. Ann Microbiol 60:579–598. https://doi.org/10.1007/s13213-010-0117-1

Hefnawy MA, Maisa MA, Shalaby GM et al (2014) Influence of salinity on phosphate solubilization by fungi. Middle East J Appl Sci 4:1080–1089

Hepper CM (1975) Extracellular polysaccharides of soil bacteria. In: Walker N (ed) Soil microbiology, a critical review. Wiley, New York, pp 93–111

Hunt PG, Matheny TA, Wollum AG et al (1988) Yield and N accumulation responses of late-season determinate soybean to irrigation and inoculation with various strains of *Bradyrhizobium japonicum*. Commun Soil Sci Plant Anal 19:1601–1612

Iyer NJ, Tang Y, Mahalingam R et al (2013) Physiological, biochemical and molecular responses to combination of drought and ozone in *Medicago truncatula*. Plant Cell Environ 36:706–720. https://doi.org/10.1111/pce.12008

Kamilova F, Kravchenko LV, Shaposhnikov A et al (2006) Effects of the tomato pathogen *Fusarium oxysporum* f. sp radicis-lycopersici and of the biocontrol bacterium *Pseudomonas fluorescens* WCS365 on the composition of organic acids and sugars in tomato root exudate. Mol Plant-Microbe Interact 19:1121–1126. https://doi.org/10.1094/MPMI-19-1121

Keles Y, Oncel I (2002) Response of antioxidative defense system to temperature and water stress combinations in wheat seedlings. Plant Sci 163:783–790

Khan A, Zhao XQ, Javed MT et al (2016) *Bacillus pumilus* enhances tolerance in rice (*Oryza sativa* L.) to combined stresses of NaCl and high boron due to limited uptake of NaCl. Environ Exp Bot 124:120–129. https://doi.org/10.1016/j.envexpbot.2015.12.011

Kirda C, Danso SKA, Zapata F et al (1989) Temporal water stress effects on nodulation, nitrogen accumulation and growth of soybean. Plant Soil 120:49–55

Kohler J, Caravaca F, Carrasco L et al (2007) Interactions between a plant growth-promoting rhizobacterium, an AM fungus and a phosphate solubilizing fungus in the rhizosphere of Lactuca sativa. Appl Soil Ecol 35:480–487

Koussevitzky S, Suzuki N, Huntington S et al (2008) Ascorbate peroxidase 1 plays a key role in the response of *Arabidopsis thaliana* to stress combination. J Biol Chem 283:34197–34203

Kumar KV, Singh N, Behl HM et al (2008) Influence of plant growth promoting bacteria and its mutant on heavy metal toxicity in Brassica juncea grown in fly ash amended soil. Chemosphere 72:678–683

Lemfack MC, Nickel J, Dunkel M et al (2014) mVOC: a database of microbial volatiles. Nucleic Acids Res 42:D744–D748. https://doi.org/10.1093/nar/gkt1250

Ma W, Guinel FC, Glick BR et al (2003a) The Rhizobium leguminosarum bv. viciae ACC deaminase protein promotes the nodulation of pea plants. Appl Environ Microbiol 69:4396–4402

Ma W, Sebestianova S, Sebestian J et al (2003b) Prevalence of 1- aminocyclopropaqne-1-carboxylate in deaminase in Rhizobia spp. Anton Leeuw 83:285–291

Mathesius U, Charon C, Rolfe BG et al (2000) Temporal and spatial order of events during the induction of cortical cell divisions in white clover by *Rhizobium leguminosarum* bv. trifolii inoculation or localized cytokinins addition. Mol Plant-Microbe Interact 13:617–628. https://doi.org/10.1094/MPMI.2000.13.6.617

McCue KF, Hanson AD (1990) Salt-inducible betaine aldehyde dehydrogenase from sugar beet: cDNA cloning and expression. Trends Biotechnol 8:358–362. https://doi.org/10.1016/0167-7799(90)90225-M

Meena KK, Kumar M, Kalyuzhnaya MG, Yandigeri MS, Singh DP, Saxena AK, Arora DK (2012) Epiphytic pink-pigmented methylotrophic bacteria enhance germination and seedling growth of wheat (Triticum aestivum) by producing phytohormone. Antonie Van Leeuwenhoek 101(4):777–786

Meena KK, Sorty AM, Bitla UM et al (2017) Abiotic stress responses and microbe-mediated mitigation in plants: the omics strategies. Front Plant Sci 8:172. https://doi.org/10.3389/fpls.2017.00172

Minerdi D, Bossi S, Maffei ME et al (2011) *Fusarium oxysporum* and its bacterial consortium promote lettuce growth and Expansin A5 gene expression through microbial volatile organic compound (MVOC) emission. FEMS Microbiol Ecol 76:342–351. https://doi.org/10.1111/j.1574-6941.2011.01051.x

Mishra BK, Meena KK, Dubey PN, Aishwath OP, Kant K, Sorty AM, Bitla U (2016) Influence on yield and quality of fennel (Foeniculum vulgare Mill.) grown under semi-arid saline soil, due to application of native phosphate solubilizing rhizobacterial isolates. Ecol Eng 97:327–333

Murray JD, Karas BJ, Sato S et al (2007) A cytokinins perception mutant colonized by *Rhizobium* in the absence of nodule organogenesis. Science 315:101–104. https://doi.org/10.1126/science.1132514

Nakashima K, Yamaguchi-Shinozaki K (2006) Regulons involved in osmotic stress-responsive and cold stress-responsive gene expression in plants. Physiol Plant 126:62–71. https://doi.org/10.1111/j.1399-3054.2005.00592.x

Omar MNA, Osman MEH, Kasim WA et al (2009) Improvement of salt tolerance mechanisms of barley cultivated under salt stress using *Azospirillum brasilense*. Tasks Veg Sci 44:133–147. https://doi.org/10.1007/978-1-4020-9065-3_15

Ortiz-Castro R, Martienz-Trujillo M et al (2008) N-acyl-L-homoserine lactones: a class of bacterial quorumsensing signals alter post-embryonic root development in Arabidopsis thaliana. Plant Cell Environ 31:1497–1509

Ortiz-Castro R, Diaz-Perez C, Martienz-Trujillo M et al (2011) Transkingdom signaling based on bacterial cyclodipeptides with auxin activity in plants. Proc Natl Acad Sci U S A 108:7253–7258

Paul D (2012) Osmotic stress adaptations in rhizobacteria. J Basic Microbiol 52:1–10

Porras-Soriano A, Soriano-Martin MS, Porras-Piedra A et al (2009) Arbuscular mycorrhizal fungi increased growth, nutrient uptake and tolerance to salinity in olive trees under nursery conditions. J Plant Physiol 166:1350–1359

Prasch CM, Sonnewald U (2013) Simultaneous application of heat, drought, and virus to Arabidopsis plants reveals significant shifts in signaling networks. Plant Physiol 162:1849–1866. doi:10.1104/pp.113.221044

Rizhsky L, Liang H, Shuman J et al (2004) When defense pathways collide: the response of Arabidopsis to a combination of drought and heat stress. Plant Physiol 134:1683–1696

Rudrappa T, Biedrzycki ML, Kunjeti SG et al (2010) The rhizobacterial elicitor acetoin induces systemic resistance in *Arabidopsis thaliana*. Commun Integr Biol 3:130–138. https://doi.org/10.4161/cib.3.2.10584

Ryu CM, Faraq MA, Hu CH et al (2003) Bacterial volatiles promote growth in Arabidopsis. Proc Natl Acad Sci U S A 100:4927–4932

Ryu CM, Faraq MA, Hu CH et al (2004) Bacterial volatiles induce systemic resistance in Arabidopsis. Plant Physiol 134:1017–1026

Sadowasky MJ (2005) Soil stress factors influencing symbiotic nitrogen fixation. In: Werner D, Newton WE (eds) Nitrogen fixation research in agriculture, forestry, ecology, and the environment. Springer, Dordrecht, pp 89–102

Saleem M, Arshad M, Hussain S et al (2007) Perspective of plant growth promoting Rhizobacteria (PGPR) containing ACC deaminase in stress agriculture. J Ind Microbiol Biotech 34:635–648

Sandhya V, Ali SKZ, Minakshi G et al (2009) Alleviation of drought stress effects in sunflower seedlings by the exopolysaccharides producing *Pseudomonas putida* strain GAP-P45. Biol Fertil Soils 46:17–26. https://doi.org/10.1007/s00374-009-0401-z

Scarpeci TE, Zanor MI, Valle EM et al (2008) Investigating the role of plant heat shock proteins during oxidative stress. Plant Signal Behav 10:856–857

Selvakumar G, Panneerselvam P, Ganeshamurthy AN et al (2012) Bacterial mediated alleviation of abiotic stress in crops. In: Maheshwari DK (ed) Bacteria in agrobiology: stress management. Springer, Berlin/Heidelberg, pp 205–224

Sen S, Chandrasekhar CN (2014) Effect of PGPR on growth promotion of rice (Oryza sativa L.) under salt stress. Asian J Plant Sci Res 4:62–67

Serraj R (2003) Effects of drought stress on legume symbiotic nitrogen fixation: physiological mechanisms. Indian J Exp Biol 41:1136–1141

Serraj R, Sinclair TR (1996) Inhibition of nitrogenase activity and nodule oxygen permeability by water deficit. J Exp Bot 47:1067–1073

Sorty AM, Meena KK, Choudhary K et al (2016) Effect of plant growth promoting bacteria associated with halophytic weed (*Psoralea corylifolia* L.) on germination and seedling growth of wheat under saline conditions. Appl Biochem Biotechnol 180:872–882. https://doi.org/10.1007/s12010-016-2139-z

Spaepen S, Vanderleyden J (2011) Auxin and plant-microbe interactions. Cold Spring Harb Perspect Biol 3(4), a001438-a001438

Supanekar S, Sorty A, Raut A (2013) Study of catechol siderophore froma newly isolated Azotobacter sp. SUP-III for its antimicrobial activity. J Microbiol Biotechnol Food Sci 3:270–273

Thimann KV (1939) Auxins and the inhibition of plant growth. Biol Rev 14:314–337

Tiwari S, Singh P, Tiwari R et al (2011) Salt-tolerant rhizobacteria-mediated induced tolerance in wheat (*Triticum aestivum*) and chemical diversity in rhizosphere enhance plant growth. Biol Fertil Soils 47:907–916. https://doi.org/10.1007/s00374-011-0598-5

Torrey JG (1961) Kinetin as trigger for mitosis in mature endomitotic plant cells. Exp Cell Res 23:281–299. https://doi.org/10.1016/0014-4827(61)90038-6

Wang W, Vinocur B, Altman A et al (2003) Plant responses to drought, salinity and extreme temperatures: towards genetic engineering for stress tolerance. Planta 218:1–14. https://doi.org/10.1007/s00425-003-1105-5

Wang C, Yang W, Wang C et al (2012) Induction of drought tolerance in cucumber plants by a consortium of three plant growth promoting rhizobacterium strains. PLoS One 7:e52565. https://doi.org/10.1371/journal.pone.0052565

Whitelaw MA (2000) Growth promotion of plants inoculated with phosphate solubilizing fungi. Adv Agron 69:99–151

Yang J, Kloepper JW, Ryu CM et al (2009) Rhizosphere bacteria help plants tolerate abiotic stress. Trends Plant Sci 14:1–4

Zhang H, Kim MS, Krishnamachari V et al (2007) Rhizobacterial volatile emissions regulate auxin homeostasis and cell expansion in Arabidopsis. Planta 226:839–851; PMID:17497164; https://doi.org/10.1007/s00425-007-0530-2

Role of Microbes in Climate Smart Agriculture

7

Sangita Mohanty and Chinmaya Kumar Swain

Abstract

Impact of climate change on agriculture and food production is already perceptible. In the present scenario of rising temperature, changing patterns of rain; increasing occurrence of extreme climatic events such as cyclone, drought and flood; the concept of climate smart agriculture (CSA) originated in order to make agriculture more resilient to climate change. Sustainably enhancing efficiency, adaptation to and mitigation of climate change are three pillars of CSA. Microorganisms are vital to several ecological processes in agroecosystem such as organic matter decomposition, nutrient cycling, N_2 fixation, phosphate solubilization, nutrient acquisition and recently discovered probiotics role. Appropriate management and exploitation of beneficial microbial functions such as use of biofertilizer, biopesticide, plant growth-promoting rhizobacteria, etc. help in achieving sustainable goal and alleviating adverse impact on environment. Microorganism can be used to facilitate adaptation to climate change by promoting growth and development and imparting resistance against several abiotic stresses. Soil microbes and their metabolic activity can influence land–atmosphere carbon exchanges in numerous ways, while these can be broadly divided into different groups as those that affect the ecosystem by methane and carbon dioxide uptake and that also control carbon loss from the soil through methane production and respiration. The role of microbe as a source and sink of greenhouse gas can be exploited to devise mitigation strategy for climate change.

Keywords

Climate smart agriculture · Rhizobacteria · Agroecosystem · Methane greenhouse gas · Mitigation

S. Mohanty (✉) · C. K. Swain
National Rice Research Institute, Cuttack, Odisha, India
e-mail: sangitamoha@gmail.com

D. G. Panpatte et al. (eds.), *Microorganisms for Green Revolution*,
Microorganisms for Sustainability 7,
https://doi.org/10.1007/978-981-10-7146-1_7

7.1 Introduction

Several scientific evidences show that the climate system is warming, and according to climate model projections during the twenty-first century, the global surface temperature is likely to rise further 0.3–1.7 °C in the lowest and 2.6–4.8 °C in highest emission scenario. Many of the unprecedented extreme climatic events taking place since 1950 point towards global climate change phenomenon. Intergovernmental Panel on Climate Change (IPCC) Fifth Assessment Report concluded that human influence has been the dominant cause of the prevailing phenomenon of global warming and climate change.

Worldwide, agriculture both contributes to and is threatened by climate change. The report of the IPCC indicates that agriculture contributes about 47 and 58% of total anthropogenic emissions of methane (CH_4) and nitrous oxide (N_2O), respectively. Climate change has already drastic impact on agriculture (Lobell et al. 2011) and is further going to impact directly and indirectly food production. Increase of mean temperature, changing patterns of rain and increasing frequency of extreme climatic events like cyclone, drought and flood will have profound impacts on agriculture (IPCC 2007).

The world's population is predicted to increase by one-third by 2050, and most of the additional two billion people will live in developing countries (FAO 2009). In order to meet the expected demand for food and feed, it has been estimated that the agricultural production will have to increase by 60%. Confronting food security challenges are huge because resources needed for sustaining or improving current production level already being threatened. Moreover, climate change is already negatively impacting on agricultural production worldwide and locally, and climate risks to cropping, livestock and fisheries are expected to increase in the coming decades. Transformation and reorientation of agricultural production under the new realities of climate change gave rise to concept of climate smart agriculture (CSA). According to the definition of Food and Agricultural Organization (FAO) of the United Nations, CSA is an integrative approach of agricultural production that aims at sustainably increasing efficiency, enhancing adaptation with mitigating GHG emission where possible and enhancing achievement of national food security goal.

Microorganisms are the most important components of agricultural ecosystem for which without them agriculture and food production simply could not exist. Microorganism plays essential functions in plant nutrient cycling in soil–plant–microbe–atmosphere continuum. Microbes are key components of carbon and nitrogen cycles and responsible for both the production and consumption of greenhouse gases such as carbon dioxide, methane and nitrous oxide. The large diversity of microbes provides an untapped opportunity of improving efficiency of agricultural production, adapting to and mitigating climate change effects, thus achieving goals of climate smart agriculture. Microorganisms that are disease causal organisms for weeds and insect pests can be used as biopesticides. Many beneficial plant–microbes are associated in the soil where they encourage for resistance or perform biological control functions. Soilborne free-living microorganisms contribute to the formation and building of the structure of soil, simultaneously the storage of nutrients and sequestration of carbon. However, some microbes acting in association with crop plants further regulate soil fertility and accessibility of nutrients. Soil microorganisms are also liable for bioremediation in the polluted sites by restoring soil fertility.

7.2 Microbial Functions in Agroecosystem

Microbial diversity in soil catalyse umpteen number of processes that are integral components of biogeochemical cycles and hence support several agroecosystem functions such as organic matter decomposition, availability of nutrient and overall soil and plant productivity. Microbes often form consortia and supply different limiting resources to plants. In addition to this presence of plant-specific microbe, e.g., N_2 fixers and mycorrhiza apart from enriching soil with specific nutrient also help the plants to acquire limiting nutrient required for their growth and production. As a result microbes promote agroecosystem functioning and sustainability and agricultural system.

7.2.1 Organic Matter Decomposition and Nutrient Cycling

The breakdown of complex organic molecules of dead material into simpler organic and inorganic molecules with biochemical transformation is known as biological process of decomposition (Juma 1998). In recycling energy and plant nutrients, microorganisms break down the organic matter, and any excess nutrients (N, P and S) are released into the soil in forms that plants can use properly; the release process is called mineralization. Therefore, microorganisms play an important role in nutrient cycling processes in soil with organic nutrients and some inorganic compounds that are supplied through exudation by plant roots and animal excreta and death of plants and animals.

7.2.2 N_2 Fixation (Nitrogen Fixation)

Nitrogen fixation is one of most crucial beneficial biological processes for the environmental sustainability of agriculture worldwide. Globally, the annual inputs of fixed nitrogen in form of crop legume–rhizobia symbioses are estimated to be 2.95 million tonnes and 18.5 million tonnes for pulses and oilseed legumes respectively (Howieson et al. 2005). However, legume plants and root-nodulating bacteria with their symbiotic associations belonging to the genera *Rhizobium*, *Mesorhizobium*, *Bradyrhizobium* and *Ensifer* produce around 80% of nitrogen in grains, with maximum profit each year. Some examples of free-living nitrogen-fixing organisms are as follows: *Azospirillum*, commonly connected with cereals in temperate zones; Beijerinckia, seems to be associated with sugar cane plantations in the tropical zones; and Azotobacter, plays an important role in nitrogen fixation in rice crops and also is used as biofertilizer for wheat, maize, barley, oat, sunflowers, line, beetroot, coconuts, coffee, tea and tobacco. However, some species belonging to the genera *Azospirillum*, *Herbaspirillum* and *Gluconacetobacter* are sugarcane endophytes and supply to its nitrogen fertilization. Nitrogen-fixing *Azorhizobium* strains have been isolated from wheat roots whereas *Rhizobium* and *Bradyrhizobium* from rice roots. Moreover, certain diazotrophic bacteria set up a truly mutualistic symbiosis with few plants through the formation of root nodules (García-Fraile et al. 2015).

7.2.3 Phosphate Solubilization

Phosphorus solubilizing microorganisms (PSMs) plays a vital role in phosphorus nutrition by enhancing the accessibility of P to plants through the release of organic and inorganic soil phosphorus pools by mineralization and solubilization (Walpola and MinHo 2012; Sharma et al. 2013). A diverse group of bacteria (*Pseudomonas*, *Bacillus*, actinomycetes) and fungi (*Aspergillus*, *Penicillium* species) are capable of solubilizing and mineralizing plant available forms of phosphorus in the soils, and its benefits from their use as inoculants are increasingly being recognized, especially in P-limited environments (Gyaneshwar et al. 2002). Bacteria are observed to be more effective than fungi in phosphorous solubilization (Sharma et al. 2013). *Penicillium bilaii* is another beneficial microflora which helps to unlock phosphate from the native soil. Fungus is thought to turn out an organic acid that dissolves the phosphate in soil so that plant roots can easily utilize it. Among soil bacterial groups, *Rhizobium*, ectorhizospheric strains of *Bacilli* and *Pseudomonas*, *Enterobacter* and endosymbiotic rhizobia are recorded as effective strains of phosphate solubilizers (Khan et al. 2009).

7.2.4 Nutrient Acquisition Through Mycorrhizal Symbiosis

Mycorrhizal symbiosis involves reciprocal transfer of carbon and nutrients between host plant and mycorrhizal fungus (arbuscular mycorrhizal fungi, AMF). The fungus provides the host plant with nutrients, i.e. nitrogen and phosphate, which increase resistance of the host to biotic and abiotic stress and in return plant transfers between 4–20% of the photosynthetically fixed carbon to mycorrhizal fungus. Although AMF symbiosis is widespread, its symbiotic functions differ from species to AMF isolates, host plants and soil properties. Arbuscular mycorrhizal fungi associations are generally considered to be diffuse and non-specific because of their multiple species colonization linking together two or more plants (Selosse Marc-Andre et al. 2006). Arbuscular mycorrhizal fungi symbiosis is one of the key beneficial biological processes for the treatment of contaminated soils and mining sites (Smith et al. 2008).

7.2.5 Probiotics Role of Microbial Community

Probiotics are the microorganisms that are thought to provide health benefits when consumed. Probiotics concept has been given by the Nobel laureate Élie Metchnikoff who suggested that dependence of the intestinal microbes on food makes it achievable to adopt measures to modify the flora in our bodies and to replace the harmful microbes by useful microbes. Soil probiotics are commonly considered as soil-based organisms (SBOs) because they are beneficial bacteria that live in soil. Plants have a limited capability to genetically adapt to rapid change in the environment, i.e. toxins, limited nutrients and drought, for which they might use microbes to have capacity to rapidly progress in the changing atmosphere at shorter life cycles. This is how plant mechanism shows a similar trend as of humans taking probiotics to improve health. Plant-specific stimulation of specific microbial groups in their rhizosphere zone suggests that plants might have evolved to tactically stimulate and support particular microbial groups which are capable of producing antibiotics as a defence against diseases caused by soilborne pathogens (Weller et al. 2002). Soil bacteria belonging to genus *Pseudomonas* are ubiquitous in most soils and have been linked to wide-ranging processes including nutrient cycling, plant growth promotion and inhibition, disease control, nitrogen fixation and bioremediation. Pseudomonads act as biocontrol potential against fungi and oomycete pathogens for more than two decades (deSouza 2002). Antibiosis is the most commonly suggested trait responsible for their activity against the plant pathogens, and a number of antimicrobial compounds have been identified, for example, 2,4-diacetylphloroglucinol (2,4-DAPG), phenazines, pyrrolnitrin, pyoluteorin, hydrogen cyanide and biosurfactant antibiotics (Picard et al. 2008). Their ability to respond quickly to changes in physical, chemical, carbon and nutritional conditions in the soil has been associated to their functional importance in agricultural ecosystems.

7.3 Role of Microbes in Sustainable Agriculture

Sustainable agriculture is that type of agriculture which focuses on producing long-term crops and livestock while having the minimal effects on environment. Such type of agriculture tries to find a good balance between the need for food production and preservation of the ecological system within the environment. One of the important criteria for ensuring sustainable agriculture is maintaining the quality of the soil of which microbes are integral part. The critical milestones of achieving sustainability are devising efficient nutrient recycling strategy, pest and disease control methods and minimizing negative effects of abiotic stress. Most of these activities are mediated by microbial services; hence appropriate management and exploitation of beneficial microbial functions can help in achieving sustainable goal and alleviating adverse impact on environment. Major impact of agricultural microbiology on sustainable agriculture would be to replacement and/or integration of agrochemicals (mineral fertilizers, pesticides) with the microbial preparations. Some of wide explanations on the use of microbes in sustainable agriculture are biofertilizer, biopesticide, and plant growth-promoting rhizobacteria (PGPR).

7.3.1 Biofertilizers

Biofertilizers are one of the best tools for sustainable agriculture which is a gift to the modern agricultural science. Biofertilizer is being applied in the agricultural field as a substitute to the organic manure. Organic manure-consisting compost, i.e. household wastes, a green manure and farm yard manure, may maintain the quality and sustainability of soil in the long run but not be able to meet the immediate requirement of crop. At the same time, chemical fertilizers have their own impact on environment in terms of fossil fuel burning, greenhouse gas emission and soil, water and air pollution. In addition to this, continuous use of chemical fertilizer alone causes nutrient imbalance in soil affecting its sustainability. Biofertilizer contains microorganisms which promote adequate supply of nutrients to the host plants and ensure proper improvement of growth and regulation in their physiology. Living microorganisms are used in the preparation of biofertilizers which have specific functions to enhance the plant growth and reproduction. Biofertilizer being a crucial component of organic farming plays a vital role in maintaining sustainability and long-term soil fertility. Some microorganisms which can absorb gaseous nitrogen directly from atmosphere and make it accessible to the plants can be identified and then multiplied in laboratories and introduced into the root zone of crop plants to supply nitrogen. Materials containing such organisms or plants are called biofertilizers. Some commonly used biofertilizer in agriculture are *Rhizobium*, *Azotobacter*, *Azospirillum* blue-green algae, *Azolla*, etc.

7.3.2 Biopesticide

Biopesticides are derived from natural materials such as plants and bacteria and contain minerals broadly used for controlling insects and disease-causing pathogens. They are broadly categorized as microbial pesticides, plant-included-protectants and biochemical pesticides that are produced through naturally occurring substances which control pests by harmless mechanisms. Microbial insecticide like *Bacillus thuringiensis* (*Bt*) produces toxin a which paralyse the mid gut of insect pest and prevent further feeding. Similarly spores of *Beauveria bassiana* and *Metarhizium anisopliae* penetrate the host cuticle and produce toxic metabolites known as beauvericin and destruxins, respectively, which cause death of the insects. Since biopesticides are inherently less toxic, affect only the target pest, are easily biodegradable and are effective in small quantities, they cause lower exposure and avoid pollution problems in the environment.

7.3.3 Plant Growth-Promoting Rhizobacteria (PGPR)

Plant growth-promoting rhizobacteria (PGPR) are some of the naturally occurring soil bacteria which have the ability to benefit plants by improving their immunity and productivity. However, these bacteria are associated with rhizosphere and some part of soil under the control of plant roots and their exudates. Due to their interactions with plants, PGPR can be separated into symbiotic bacteria (live inside the plants and exchange metabolites directly) and free-living rhizobacteria (live outside plant cells) (Gray and Smith 2005). Moreover, some of symbiotic bacteria have the ability to integrate their physiology with plant, resulting in formation of specific structures. In accordance with their mode of action, PGPR is being classified as biofertilizers, biopesticides and phytostimulators with certain bacteria having overlapping applications. The attachment of ACC (1-Aminocyclopropane-1-Carboxylate) deaminase gene and the presence of phytohormones like IAA (Indole Acetic Acid), siderophore, cytokinin, gibberellins, etc. seem to be responsible for enhancing the plant growth and yield and the nutrient uptake from various crop plants in different agroecosystems. According to research, the diversity of PGPR application can be used as a reliable component in the management of sustainable agricultural system.

However, selection of the best rhizobacterial strain for rhizosphere capability and making a good study on the ecology of introduced PGPR with the inhabitant PGPR and other microbial species in the plant rhizosphere will require a more widespread knowledge.

7.4 Microbial Contribution to Climate Change

Emission of some potent greenhouse gases responsible for global warming such as CH_4, N_2O and Carbon dioxide (CO_2) is essentially a by-product of microbial processes. Microorganisms are integral components of C–N cycle in soil; they break down organic matter and release carbon dioxide back into the atmosphere. Microbial respiration is an integral component of soil respiration which is a major pathway of carbon efflux from ecosystem and accounts for 25% of naturally emitted CO_2. Similarly a group of bacteria called methanogens are responsible for production of CH_4, another important greenhouse gas having global warming potential 25 times more than that of CO_2.

Natural emissions of N_2O primarily result from bacterial breakdown of nitrogen in soils and in the earth's oceans. In soils nitrous oxide is produced as a by-product of both denitrification and nitrification processes which are carried out by group of microorganisms known as denitrifiers and nitrifiers, respectively. Nitrification is the main source of N_2O under aerobic conditions, while denitrification dominates in anoxic environment. Denitrification is the microbial reduction of nitrate or nitrite form of N to dinitrogen or N oxides under anaerobic condition. Nitrification is the process of oxidation of ammonium form of N to nitrite or nitrate form and also responsible for the emission of N_2O from soil. In aerobic conditions, nitrifier nitrification involving ammonia oxidation by autotrophic ammonia-oxidizing bacteria is the major pathway of N_2O formation. Anaerobic ammonium oxidation (ANAMMOX) is an alternate microbial-mediated pathway responsible for N loss in the form of N_2 and N_2O in aquatic system. The global warming potential of nitrous oxide is 298 times more than the CO_2, and it accounts for about 19% of total global warming effect. Globally soils covered by natural vegetation are estimated to produce 6.6 Tg of N_2O annually, and oceans are thought to add around 3.8 Tg of N_2O annually to the atmosphere.

7.5 Microbial Response to Climate Change

Growth and activity of microbes and microbial processes are greatly affected by environmental factors such as temperature, moisture and substrate availability and all of which are likely to be affected by climate change. Effect of climate change on growth and activities of soil microbes may be direct and indirect. The direct effects include the influence of temperature, changing precipitation and extreme climatic events, whereas indirect effects are due to climate-driven changes which alter soil physicochemical conditions and plant productivity. Since soil matter decomposition is one of most widely contributions of soil microbes to climate change, global warming will accelerate rates of heterotrophic microbial activity, increasing the efflux of CO_2 to the atmosphere. The climate changes will have distinct indirect effects on soil microbial communities and their activity through its influence on vegetation composition and plant growth. Indirect effect of rising atmospheric concentrations of CO_2 on soil microbes is the first mechanism, and the second mechanism is the increased plant photosynthesis and transfer of photosynthate carbon to fine roots and mycorrhizal fungi (Zak et al. 1993; Bardgett Richard et al. 2008) and heterotrophic microbes (Hogberg and Read 2006). It is well-known that elevated/high CO_2 increases plant growth and photosynthesis under nutrient-rich conditions (Curtis and Wang 1998) which in turn increases the flux of carbon to roots and their symbionts and also heterotrophic microbes through root exudation of easily degradable sugars (Fig. 7.1). (Dıaz et al. 1993; Zak et al. 1993).

7.6 Role of Microbes in Combating Climate Change

7.6.1 Climate Change Adaptation

Microorganism can be used to facilitate adaptation to climate change by promoting growth and development and imparting resistance against several abiotic stresses. They offer an opportunity to rely on the biological processes rather than climate change-inducing synthetic chemicals. Soil microorganisms are involved in soil formation, regulate its characteristics, sustain its fertility, break down toxic compounds, enhance sustainable production and ultimately promote ecosystem resilience and sustainability. Microorganisms can be used to manage soil health and ecosystem resilience and to do away with the need for producing and transporting chemical fertilizers.

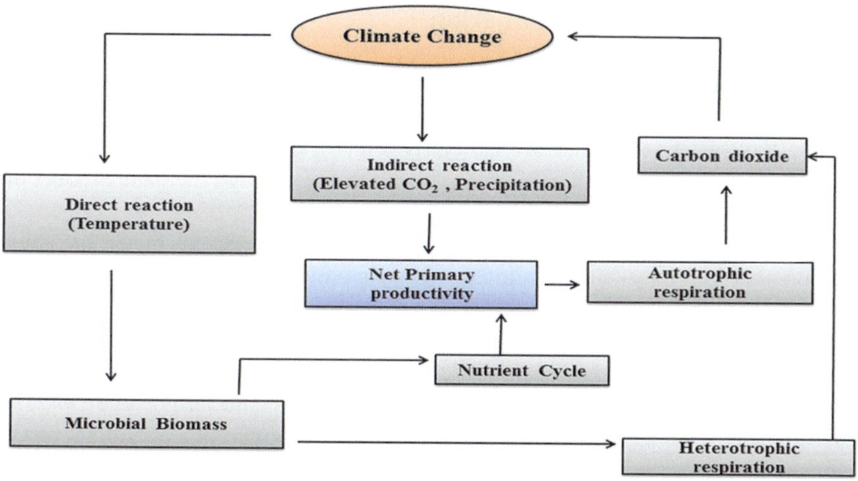

Fig. 7.1 Flow chart presentation of climate change on soil microbial communities and its direction of reaction to global warming through carbon dioxide production

Novel biological control agents can be used to limit the harmful impact of new and emerging pathogens and pests in a changing climate scenario. The most common biological control agents are *Neozygites fresenii* (parasitoid), used to control the cotton pest *Aphis gossypii*, and rust fungus *Maravalia cryptostegiae*, which is used in Australia to control the weed rubber vine. The bacterium *Bacillus thuringiensis*, which produces crystalline toxins that kill Lepidoptera and Diptera larvae, has been used on a large commercial. Both augmentative biological control, i.e. release of large number of previously mass-reared natural enemy in the environment, and conservation biological control, i.e. manipulation of environment to increase the fitness of natural enemies, can help adapt to climate change.

Beneficial plant–microbe interactions are known to stimulate plant growth and enhance their resistance to degenerative diseases and abiotic stresses. Beneficial bacteria when associated with plants improve nutrient acquisition, produce plant growth regulators and alter physiological and biochemical properties of the host plant and therefore help in protecting the plant roots against soilborne pathogens. Bacterial genera such as *Pseudomonas*, *Bacillus*, *Rhizobium*, *Azospirillum*, *Streptomyces* and *Serratia* come under this group. These plant-associated beneficial bacteria can be exploited for faster plant growth and enhanced disease resistance in changing climate conditions. Study shows C4 plant can cope with elevated CO_2 levels when combined with right strains of mycorrhizal fungi (Tang et al. 2009). Recently identified *Rhizobia* species associated with *Medicago sativa* has the potential to operate under various abiotic conditions such as low or high temperature or pH or low levels of organic matter in the soil. Thus the enormous unexplored reservoir of genetic and metabolic microbial diversity provides tremendous opportunity to identify novel genes for pest control, nitrogen fixation and biodegradation with help of recently developed tools such as metagenomics.

7.6.2 Climate Change Mitigation

Soil microbes and their metabolic activity can influence land–atmosphere carbon exchanges in numerous ways, while these can be broadly divided into different groups such as those that affect the ecosystem by methane and carbon dioxide uptake and that also control carbon loss from the soil through methane production and respiration (Fig. 7.1).

Methanotrophs or methane-oxidizing bacteria (MOB), present in aerobic soils, are potential biological sink to mitigate methane emission to the atmosphere. They utilize CH_4 as their sole source of carbon and energy. The methanotrophs can contribute up to 15% to the total global CH_4. However, they are sensitive to environmental perturbations, and they are difficult to isolate due to their slow growth rate and fixed attachment to soil particles. A recently discovered bacterium *Methylokorus infernorum*, found in geothermal areas in acidic and hot environment, utilizes methane gas. These bacteria can consume a huge quantity of methane, i.e. near about 11 kg/year, and also can be helpful in reducing methane emission from methane-producing factories and landfills. In addition to this, *Methylobacillus* bacteria are found to recycle carbon compounds such as methane, methanol and methylated amines. Apart from that there are some naturally occurring microbes that convert carbon dioxide into calcium carbonate. Some group of denitrifying bacteria are known to convert nitrous oxide into nitrogen gas. These microbes have great potential to mitigate GHG emissions.

In addition to this, use of beneficial microorganisms increased productivity, thus affecting GHG budget in terms of GHG emission per unit food production. Benefit accrued by use of beneficial microbes in terms of production can be considered as a contribution of microbe to climate change mitigation.

7.7 Future Prospects and Challenges

Microbial world is the largest uncultivated pool of biodiversity on earth, and till the last century, only a very small fraction, i.e. <10%, is known with respect to its nature and identification (Bhattacharyya and Jha 2012). Although microorganisms constitute the smallest forms of life, they play a vital role in every spectrum of activities within a living organism on earth. Therefore, study based on microbial ecology becomes a significant frontier in the present day of biological science. Soil bacteria sense the chemical-based messages and secrete chemicals of their own that can activate complex plant defences in the plant (Glick 2012). They contribute significantly to the utilization of greenhouse gases (Bardgett Richard et al. 2008), including CO_2, CH_4, N_2O and nitric oxide (NO). Microorganisms play a key role in the crop protection by enhancing disease resistance capacity of plants against the pathogens and exhibiting aggressive activities or substitute as biotic elicitors against different biotic and environmental factors. Among some microbes, fungi can colonize the upper parts of plants and provide protection against various biotic and abiotic like drought, heat, insects pest attack and pathogens (Singh et al. 2011). However, use of microbial-based fertilizer in agricultural practice is still not very popular. The main reasons are uncertainty of results and problems associated with identifying and tracking inoculated strains in the field. Now in the twenty-first century, microbial biotechnology and its application in sustainable development of agriculture are getting better consideration. Conservation of microbial diversity is crucial in the maintenance of species diversity of higher organisms and strategies for plant disease and nutrient management (Colwell and Munneke 1997). Climate change induces adaptation processes in the plants and microorganisms (Grover et al. 2011) and thus changes the ability of plant–microbe relations. An understanding of microbe–climate change feedback and potential positive and negative role of microbes in global climate change is essential to exploit them for climate change adaptation and mitigation.

7.8 Conclusion

Climate change phenomenon such as increase of mean temperature, changing patterns of rain and increasing frequency of extreme climatic events like cyclone, drought and flood will have profound impacts on agriculture. It is utmost necessary to reorient and transform agriculture to make it more resilient to climate change. Microorganisms carry out many vital ecosystem functions that are crucial for plant growth and development such as soil organic matter decomposition, nutrient cycling, N_2 fixation, PGPR activities and acquisition of nutrients through root–mycorrhiza association. Appropriate management and exploitation of beneficial microbial functions such as use of biofertilizer, biological control of disease and pest and plant growth-promoting rhizobacteria can contribute to both climate change adaptation and mitigation. Microorganism can be used to facilitate adaptation to climate change by promoting growth and development and imparting resistance against several abiotic stresses. Soil microbes and their metabolic activity influence land–atmosphere carbon exchanges in various ways. Group of bacteria that use CH_4 as their source of energy and food such as methanothrops can be exploited for climate change mitigation; however isolation, culture and inoculation of these bacteria are still a challenge to be addressed. Climate change affects not only growth and productivity of crop plants but also growth and activity of microbes in rhizosphere directly and indirectly. A proper understanding of microbial ecology and soil–plant–microbe interaction in a changing climate scenario is essential to use microbial technology for climate change adaptation and mitigation.

References

Bardgett Richard D, Freeman C, Ostle Nicholas J (2008) Microbial contributions to climate change through carbon cycle feedbacks. ISME J 2:805–814. https://doi.org/10.1038/ismej.2008.58

Bhattacharyya PN, Jha DK (2012) Plant growth-promoting rhizobacteria (PGPR): emergence in agriculture. World J Microbiol Biotechnol 28:1327–1350

Colwell PF, Munneke HJ (1997) The structure of urban land prices. J Urban Econ 41(3):321–336
Curtis PS, Wang X (1998) A meta-analysis of elevated CO2 effects on woody plant mass, form and physiology. Oecologia 113(3):299–313
deSouza JT (2002) Distribution, diversity and activity of antibiotic-producing pseudomonas spp. Wageningen University dissertation 3260
Dıaz S, Grime JP, Harris J, McPherson E (1993) Evidence of a feedback mechanism limiting plant-response to elevated carbon-dioxide. Nature 364:616–617
FAO (2009) The state of food and agriculture: livestock in the balance. FAO, Rome
García-Fraile P, Menéndez E, Rivas R (2015) Role of bacterial biofertilizers in agriculture and forestry. Bioengineering 2(3):183–205. https://doi.org/10.3934/bioeng.2015.3.183
Glick BR (2012) Plant growth-promoting bacteria: mechanisms and applications. Scientifica 2012:15
Gray EJ, Smith DL (2005) Intracellular and extracellular PGPR: commonalities and distinctions in the plant-bacterium signaling processes. Soil Biol Biochem 37:395–412
Grover M, Ali SZ, Sandhya V, Rasul A, Venkateswarlu B (2011) Role of microorganisms in adaptation of agriculture crops to abiotic stresses. World J Microbiol Biotechnol 27(5):1231–1240
Gyaneshwar P et al (2002) Role of soil microorganisms in improving P nutrition of plants. Plant Soil 245:83–83
Högberg P, Read DJ (2006) Towards a more plant physiological perspective on soil ecology. Trends Ecol Evol 21(10):548–554
Howieson JG, Yates RJ, O'Hara GW, Ryder M, Real D (2005) The interactions of Rhizobium leguminosarum biovar trifolii in nodulation of annual and perennial Trifolium spp. from diverse centres of origin. Aust J Exp Agr 45:199–207
Intergovernmental Panel on Climate Change (IPCC) (2007) Climate change 2007: mitigation of climate change. Contribution of Working Group III to the fourth assessment report of the IPCC
Juma NG (1998) The pedosphere and its dynamics: a systems approach to soil science, vol 1. Quality Color Press Inc., Edmonton. 315 pp
Khan A, Jilani G, Akhtar MS, Naqvi SM, Rasheed M (2009) Phosphorus solubilizing bacteria: occurrence, mechanisms and their role in crop production. J Agric Biol Sci 1:48–58
Lobell DB, Bänziger M, Magorokosho C, Vivek B (2011) Nonlinear heat effects on African maize as evidenced by historical yield trials. Nat Clim Chang 1(1):42–45
Picard C et al (2008) Genotypic and phenotypic diversity in populations of plant-probiotic Pseudomonas spp. colonizing roots. Naturwissenschaften 95:1–16
Selosse Marc-Andre et al (2006) Mycorrhizal networks: des liaisons dangereuses? Trends Ecol Evol 21:621–628
Sharma SB, Sayyed RZ, Trivedi MH, Gobi TA (2013) Phosphate solubilizing microbes: sustainable approach for managing phosphorus deficiency in agricultural soils. SpringerPlus 2:587
Singh LP, Gill SS, Tuteja N (2011) Unraveling the role of fungal symbionts in plant abiotic stress tolerance. Plant Signal Behav 6:175–191
Smith SE et al (2008) Mycorrhizal symbiosis. Academic, London
Tang J, Xu L, Chen X, Hu S (2009) Interaction between C4 barnyard grass and C3 upland rice under elevated CO2: impact of mycorrhizae. Acta Oecol 35:227–235
Walpola BC, Min-Ho Y (2012) Prospectus of phosphate solubilising microorganisms and phosphorus availability in agricultural soils: a review. Afr J Microbiol Res 6:6600–6605
Weller D et al (2002) Microbial populations responsible for specific soil suppressiveness to plant pathogens. Annu Rev Phytopathol 40:309–348
Zak DR, Pregitzer KS, Curtis PS, Teeri JA, Fogel R, Randlett DL (1993) Elevated atmospheric CO_2 and feedback between carbon and nitrogen cycles. Plant Soil 151(1):105–117

Stress-Tolerant Beneficial Microbes for Sustainable Agricultural Production

8

Reeta Goel, Deep Chandra Suyal, Vinay Kumar, Lata Jain, and Ravindra Soni

Abstract

Agriculture sector is a major contributor of national income in India, while ensuring food security and employment. Plant exposure to both biotic and abiotic stresses causes major losses to agricultural production worldwide. Biotic factors mainly include interaction with other pathogenic or parasitic microorganisms and insect pests, whereas abiotic factors include temperature, drought, water logging, and salinity. Microorganisms play an important role in the growth and development of plants. They confer several benefits to the plants and help them to alleviate the stress.

Keywords

Stress · Bacteria · Plant

R. Goel · D. C. Suyal
Department of Microbiology, CBSH, G.B. Pant University of Agriculture and Technology, Pantnagar, Uttarakhand, India

V. Kumar · L. Jain
ICAR-National Institute of Biotic Stress Management, Baronda Farm, Raipur, Chhattisgarh, India

R. Soni (✉)
Department of Agricultural Microbiology, College of Agriculture, Indira Gandhi Krishi VishvaVidyalaya, Raipur, Chhattisgarh, India
e-mail: rs31693@gmail.com

D. G. Panpatte et al. (eds.), *Microorganisms for Green Revolution*,
Microorganisms for Sustainability 7,
https://doi.org/10.1007/978-981-10-7146-1_8

8.1 Introduction

Microorganisms are ubiquitous in nature. However, their diversity, distribution, and community structure depend on various biotic and abiotic factors. As a consequence, they exhibit spatial and environmental variations and formed distinct communities under different ecosystems. Unfortunately, majority of microorganisms are unable to grow under in vitro conditions, and therefore, the accurate enumeration of their diversity is still lacking (Soni et al. 2010; Suyal et al. 2015a). However, an advance in the molecular biology led to the emergence of high-throughput next-generation sequencing (NGS) techniques which facilitate precise identification and characterizations and of microbial communities in specific environment.

Microorganisms play an important role in the growth and development of plants. However, they are often exposed to biotic stresses, viz., pathogens, foreign metabolites, as well as environmental stresses, viz., temperature, pH, moisture content, nutrient availability, etc. (Mendes et al. 2013). Therefore, they exhibit a wide range of adaptive features to sense and respond the stresses. Moreover, they can modify their signal transduction networks to cope with stress conditions. Plant growth-promoting rhizobacteria (PGPR) can be classified as bio-fertilizers and is one of the most cost-effective and sustainable way to increase nutrient uptake, plant productivity, and immunity. They confer several benefits to the plants and help them to alleviate the stress. Therefore, exploration of stress tolerant bacteria can help in sustainable agricultural plans in environmentally unfavorable conditions.

8.2 Plant Stress and Microbes

8.2.1 Abiotic Stress

8.2.1.1 Drought Tolerance

Water deficit is a serious issue in many areas of the world and is usually countered by extensive irrigation. It is expected to cause serious plant growth problems by drought for more than 50% of the arable lands by 2050 (Vinocur and Altman 2005). Drought stress confines crop growth and yield, especially in arid and semiarid regions, and can influence chemical composition of plant, often resulting in decreased carbohydrates (Emerson et al. 2014; Nichols et al. 2014; Vander Weijde et al. 2016) and accumulation of compounds including soluble sugars and amino acids that protect against osmotic stresses (Chaves et al. 2003; Ings et al. 2013). It also directly or indirectly affects the soil microbiota by creating osmotic stress, which leads to microbial death and cell lysis (Turner et al. 2003). However, plants that developed under drought tended to support higher bacterial and fungal population and increased fungal/bacterial ratios in the soil (terHorst et al. 2014). Therefore, inoculation of plants with native beneficial microorganisms may increase drought tolerance of plants growing in arid or semiarid areas (Marulanda et al. 2007). Priming with isolates of rhizospheric bacteria from harsh environments is a promising, novel way to improve plant water use efficiency (Timmusk et al. 2014). The

ACC deaminase activity of *Achromobacter piechaudii* was shown to confer tolerance to water deficit in tomato and pepper (Mayak et al. 2004a, b). Similarly, Arshad et al. (2008) obtained that a strain of *Pseudomonas* spp. with ACC deaminase activity partially eliminated the effect of drought stress on the growth of peas (*Pisum sativum L.*). Furthermore, inoculation of *Azospirillum* with wheat under drought stress conditions resulted in a significant increase in water content (Creus et al. 2004; Pereyra et al. 2012; Saghafi et al. 2013; Kasim et al. 2013). Several information on *Bacillus* and *Pseudomonas* strains imparting drought tolerance in plants are already available in public domain (Arkhipova et al. 2007, Vardharajula et al. 2011; Wolter and Schroeder 2012; Lim and Kim 2013; Kumar et al. 2014; Gagné-Bourque et al. 2016). It is also evident that microbial associates of roots can significantly influence the response of tree seedlings to drought (Rincón et al. 2008). Not only bacteria and fungi, even virus can also improve the drought tolerance in plant (Xu et al. 2008). In recent years, many experiments have been conducted on the role of microbial association like *Arbuscular mycorrhizal* (AM) symbiosis and PGPR in drought-stressed crop plants (Wu et al. 2008). AM symbiosis protects host plants against the detrimental effects of drought stress through mechanisms of drought avoidance (Augé et al. 2001; Ruiz-Lozano and Aroca 2010; Marulanda et al. 2009).

8.2.1.2 Heavy Metal Tolerance

All the metals having density > 5 g/cm^3 are known as heavy metals (Pan et al. 2016). They have long-term persistence in nature and can accumulate in food chains. Some heavy metals are important for metabolism, viz., iron, nickel, zinc, and copper. Others like cadmium, silver, and mercury have no biological role and are harmful to the organisms, even at very low concentrations (Khan and Goel 2014). Heavy metals are difficult to degrade and/or eliminate from the environment. However, various bacteria are reported to have resistance to toxic metals. It is well accepted that chromium (Cr) is a toxic agent for the growth and development of plants (Mohanty and Patra 2013). Similarly, aluminum (Al) is known as an inhibitory element for the growth of plants, especially in acidic soils where the most toxic form of aluminum (Al_3^+) is prevalent (Liu et al. 2014). Heavy metals at higher levels have the capability to interact with several imperative cellular biomolecules such as nuclear proteins and DNA. This would inflict serious morphological, metabolic, and physiological irregularities in plants (Emamverdian et al. 2015).

Under heavy metal stress conditions, bacteria have adopted many strategies to overcome the metal toxicity, viz., metal adsorption, bioaccumulation, expulsion of metal outside the cell, and biotransformation. Moreover, microbial traits, viz., release of chelating agents, acidification of adjacent environment, and ability to change in redox potential, can reduce heavy metal toxicity in soil. Therefore, metal-resistant bacterial strains can be used as bio-inoculants in contaminated soils. They convert bioavailable forms of metal to the non-bioavailable forms and protect the plants from the phytotoxicity of excessive metals. Several plant growth-promoting metal-resistant microorganisms, viz., *Pseudomonas* (Khan and Goel 2014), *Commamonas* (Rani et al. 2013), *Proteus* and *Bacillus* (Saluja et al. 2011, 2012), *Staphylococcus*, *Planococcus*, and *Vibrio* (Zampieri et al. 2016), *Mycorrhiza*

(Dhawia et al. 2016), and *Actinomycetes* (Taj and Rajkumar 2016), are reported. Recently, two cadmium (Cd)-resistant strains, i.e., *Bacillus megaterium* H3 and *Neorhizobium huautlense* T1-17, were investigated for metal immobilization and provide a safe way to produce rice in Cd-contaminated soils (Li et al. 2017). Moreover, complete genome sequence of the heavy metal-resistant bacterium *Agromyces aureus* AR33 has been submitted recently (Corretto et al. 2017).

8.2.1.3 Cold Tolerance

Cold-adapted microorganisms constitute a major fraction of Earth's biomass and perform crucial roles in biogeochemical cycles (Siddiqui and Cavicchioli 2006). These bacteria synthesize cold-shock proteins (Csps) and cold acclimation proteins (Caps) in response to low temperature. In addition to this, they are able to synthesize housekeeping proteins even under cold conditions.

Several earlier reports reveal successful implementation of the cold-tolerant plant growth-promoting rhizobacteria (Rani et al. 2013; Suyal et al. 2014a). Earlier, Katiyar and Goel (2003) evaluated the effect of cold-tolerant mutant of *P. fluorescens* on growth and development of mung bean and observed that inoculated plants were better than respective control. Moreover, Trivedi and Sa (2008) observed that two *Pseudomonas corrugata* mutants were more efficient than wild-type strain within the temperature range of 4–28 °C. Selvakumar et al. (2013) revealed the solubilization of rock phosphate using psychrotolerant *Pseudomonas* spp. isolated from Indian Himalayas. Furthermore, Majeed et al. (2015) analyzed the effect of plant growth promontory rhizobacteria isolated from wheat rhizosphere of the Himalayan region of Kashmir. Bergottini et al. (2015) studied the bio-inoculation effect of yerba maté seedlings (*Ilex paraguariensis* St. Hill.) with native plant growth-promoting rhizobacteria. Furthermore, seven diazotrophs were isolated from the Himalayan rhizospheric soil, and recently, their proteome was documented (Suyal et al. 2014b; Soni et al. 2015).

8.2.1.4 Salt Tolerance

It has been estimated that worldwide 20% of total cultivated and 33% of irrigated agricultural lands are afflicted by high salinity (Jamil et al. 2011). The impacts of salinity include less agricultural productivity, less economic returns, and soil erosions (Hu and Schmidhalter 2002; Vaishnav et al. 2016). During salt stress, sodium ion ($Na+$) is excluded from shoots, and potassium ($K+$) is accumulated in shoots, thereby stabilizing the high cytosolic $K+/Na+$ ratio especially in leaves (Serrano and Rodríguez-Navarro 2001; Shi et al. 2002; Ren et al. 2005; Sunarpi et al. 2005).

Effective existence of bacteria in saline environment due to excessive accumulation of secondary metabolites may result in better root colonization and plant growth (Hirsch 2010; Karlidag et al. 2011). Bacteria-induced salt tolerance in plants has been observed for several PGPR strains (Mayak et al. 2004a, b; Barriuso et al. 2008; Zhang et al. 2009). Salt-enhanced salt tolerance of *Zea mays* upon co-inoculation with *Rhizobium* and *Pseudomonas* is correlated with decreased electrolyte leakage and maintenance of leaf water contents (Bano and Fatima 2009). Paul and Nair (2008) reported that *Pseudomonas fluorescens* MSP-393, a PGPR strain, as a means of salt tolerance, de novo synthesized the osmolytes, alanine, glycine, glutamic acid, serine, threonine, and aspartic acid in their cytosol. Two PGPR *B. megaterium* and *Enterobacter* sp. induce salt tolerance and consequent improvement in the growth of okra plants under salt stress (Habib et al. 2016). *Bacillus* and other members of this genus were dominant salt-tolerant plant growth-promoting rhizobacteria, which could serve as a suitable bio-inoculant for crops grown under saline soils (Sapsirisopa et al. 2009; Upadhyay et al. 2012; Han et al. 2014; Islam et al. 2016). Further, most other *Azospirillum* species can tolerate only 2% NaCl (Upadhyay et al. 2009). Ramadoss et al. (2013) reported that halotolerant bacteria *Hallobacillus* sp. SL3 and *Bacillus halodenitrificans* PU62 isolated from saline environments have potential to enhance plant growth under saline stress. Further, soil bacteria also conferred a positive relationship and improved salt stress tolerance in transgenic pea (Ali et al. 2015a, b). Recently four endophytic bacteria from genera *Sphingomonas*, *Pantoea*, *Bacillus*, and *Enterobacter* were reported with some unique properties that are very valuable for exploiting bio-inoculants aiding in the efforts to establish a sustainable and large-scale feedstock production system for hybrid *Pennisetum*, particularly, on the saline marginal lands (Li et al. 2016). Similarly some other endophytes were also reported for improving plant growth in different crops like sugarcane (Pirhadi et al. 2016) and caliph medic legume (Yaish et al. 2016).

8.2.1.5 Heat Tolerance

The increasing threat of climate change is already having a substantial impact on agricultural production worldwide as heat waves cause significantly yield losses with great risks for future global food security (Christensen and Christensen 2007). High soil temperature is a major constraint for biological nitrogen fixation in legume crops. Heat stress tolerance can be conferred to crop plants by genetic improvement (Singh and Grover 2008), but the use of bacteria and fungi serving as biological control agents can be more appropriate eco-friendly approach to overcome this problem. Recent reports suggest that PGPR also enhance the tolerance of plants to temperature stress (Ali et al. 2009; Abd El-Daim et al. 2014; Nehra et al. 2007). Seed treatment with PGPR strains that colonize roots significantly improved heat stress tolerance of wheat seedlings (Abd El-Daim et al. 2014). Nehra et al. (2007) reported that heat-resistant/heat-tolerant mutants of *Rhizobium* sp. (*Cajanus*) can tolerate thermal stress and can fix atmospheric N2 more efficiently than the parent strain under natural high-temperature conditions. Further, inoculation with thermotolerant PGP *Pseudomonas putida* strain AKMP7 alleviated heat stress and consequently improved the growth of wheat plant in the presence of heat stress (Ali et al. 2011). The bacterial heat-shock response is not limited to changes in temperature and is a general stress response, as many of the heat-shock proteins are induced by other environmental changes, such as the addition of ethanol, heavy metals, high osmolarity, pollutants, starvation, or interaction with eukaryotic hosts (Ron et al. 2000).

8.2.1.6 Nutrient Deficiency

Rhizospheric microorganisms are able to influence the status of nutrients in the plants (Mendes et al. 2013), viz., diazotrophs (Suyal et al. 2014a; Soni et al. 2015), mycorrhiza (Dhawia et al. 2016), and P-solubilizing bacteria (Rani et al. 2013; Singh et al. 2013). Among diazotrophs, besides *Rhizobium* and *Bradyrhizobium*, several other genera are also reported (Gaby and Buckley 2011; Suyal et al. 2014b). Microorganisms can also help in the uptake of few trace elements, viz., iron uptake by secreting the siderophores (Hider and Kong 2010). Moreover, microbial communities are also found to release the cations from minerals and thus contribute in plant health under nutrient-limiting soils (Mapelli et al. 2012; Mendes et al. 2013). Bacterial endophytes are also involved in plant growth promotion directly through the production of plant growth regulators or interactions that alter/induce endogenous plant hormone production or activity that increases accessibility of nutrients such as nitrogen and phosphorus (Glick 2012).

8.2.2 Biotic Stresses

8.2.2.1 Bacterial Diseases

Plant pathogenic bacteria cause many serious diseases of plants throughout the world (Vidhyasekaran 2002). Bacterial blight is a destructive disease of domesticated rice (*Oryza sativa*) caused by the pathogen *Xanthomonas oryzae* pv. *oryzae* (Grennan 2008). *Burkholderia solanacearum* causes a vascular wilt disease and has been ranked as the second most important bacterial pathogen (Yuliar et al. 2015). Furthermore, the use of bacterial and fungal antagonists has already been applied to control several plant diseases including citrus canker caused by *Xanthomonas axonopodis* pv. *citri* (Shahriari et al. 2005). Bacterial panicle blight, which is caused by the Gram-negative bacterial pathogens *Burkholderia glumae* and *B. gladioli*, is another important rice disease in many rice-growing regions around the world (Nandakumar et al. 2009; Ham and Groth 2011). Recently biological control activities of rice-associated *Bacillus* (RAB) sp. strains has been observed against bacterial panicle blight of rice (Shrestha et al. 2016). Sadik et al. (2013) found that *P. agglomerans* 2066-7 strain has an antagonist effect against onion bacterial disease. Furthermore, *Bacillus subtilis* has been reported to be effective in the management of bacterial wilt disease in tomato (Sinha et al. 2012). Avirulent strains of *R. solanacearum* and *Pseudomonas* spp., followed by *Bacillus* spp. and *Streptomyces* spp., were reported for their potential value in controlling bacterial wilt (Ramesh and Phadke 2012). Several uncommon BCAs have also been reported to control bacterial wilt (Xue et al. 2009; Nion and Toyota 2015; Huang et al. 2013).

8.2.2.2 Fungal Diseases

Fungal pathogens are considered the most important microbial agents causing serious losses in crop productivity. A widely used fungal biocontrol agent in the soil is *Trichoderma*, such as *Trichoderma harzianum*, which is an antagonist of *Rhizoctonia*, *Pythium*, *Fusarium*, and other fungal pathogens observed that three species of

Trichoderma, i.e., *T. viride*, *T. harzianum*, and *Trichoderma* sp., suppressed effectively the growth of pathogenic fungi, namely, *Aspergillus niger*, *A. flavus*, *Phytophthora* sp., *Fusarium oxysporum*, *Rhizoctonia solani*, *Penicillium notatum*, and *Alternaria solani*. Anthracnose disease caused by fungal pathogen, *Colletotrichum* species, is detrimental to numerous plant species. This disease is usually controlled by application of fungicides which has implications on human health as well as environmental safety. The root rot fungi which pose serious threat to forest nurseries include the species of *Fusarium*, *Phytophthora*, *Pythium*, *Rhizoctonia*, *Macrophomina*, and *Cylindrocladium* (Haggag 2002). These pathogens are significantly inhibited by mycorrhizal fungi, *Pisolithus tinctorius* and *Laccaria laccata* (Dar et al. 2011). Further, *Paxillus involutus* also effectively controlled root rot caused by *Fusarium* species in red pine (Pal and Gardener 2006). Bacteria signify an important group of biocontrol agents, and several commercial products are nowadays available to control fungus/fungal diseases. Recently, a group of antagonistic microorganisms include *Bacillus* spp., *Pseudomonas* spp., *Paenibacillus* spp., and *Streptomyces* spp., which are mainly isolated to control the pepper anthracnose from the rhizosphere and plant microflora (Lee et al. 2011; Lamsal et al. 2012; Yoo et al. 2012). Rhizobacterial strains of *Brevibacillus brevis*, *Stenotrophomonas maltophilia*, and *Brevibacillus formosus* are found to be highly effective in suppression of wilt and root rot diseases of common sage, *Salvia officinalis*, caused by different isolates of *Fusarium oxysporum* and *F. solani* (Omar and Ahmed 2014). *Piriformospora indica* confers disease resistance systemically by colonizing the roots of many plant species and stimulates growth, biomass, and seed production of the hosts. *P. indica* confers resistance against biotic stress, and colonization by the fungus stimulates the host to synthesize the phosphatidic acid. Viruses of the genus *Totivirus* infect smut fungus, whereas those belonging to the genus *Victorivirus* infect filamentous fungi (Ghabrial and Nibert 2009). Many other microbial antagonists have also been reported to possess antagonistic activities against plant fungal pathogens, such as *Pseudomonas fluorescens*, *Agrobacterium radiobacter*, *Bacillus subtilis*, *B. cereus*, *B. amyloliquefaciens*, *Trichoderma virens*, *Burkholderia cepacia*, *Saccharomyces* sp., and *Gliocadium* sp. (Suprapta 2012).

8.2.2.3 Viral Diseases

Several plant viruses have been identified which cause various diseases and significant crop losses. Due to the lack of proper viricides, virus diseases are conventionally controlled using certified virus-free planting material, eradicating infected plants and spraying chemicals against virus vectors. Viral diseases can be checked by cross-protection phenomenon in which prior infection with one virus can protect plants against economic damage caused by another isolate of the same or a closely virus-related virus (Zhou and Zhou 2012). This method was initially developed to further study the mechanism of mild strain cross-protection against *Citrus tristeza* virus. Pepino mosaic virus (PepMV) is currently considered as one of the most dangerous pathogens infecting tomato crops worldwide. Hasiów-Jaroszewska et al. (2014) examined differences in cross-protection between pathotypes of Pepino mosaic virus and obtained efficient cross-protection. To cross-protect cucumber plants from zucchini yellow mosaic virus (ZYMV), Kosaka et al. (2006) used cold treatment to obtain an attenuated isolate of ZYMV. Recently, Ogai and co-worker (2013) reported that the attenuated isolate effectively protected commercially cultivated pepper plants against PMMoV infection. Ali et al. (2015a, b) report on a new strategy toward improving plant resistance to geminiviruses using the bacterial CRISPR/Cas system.

8.2.2.4 Insect Diseases

Over 100 bacteria have been identified as insect pathogens, among which Bacillus thuringiensis has got the maximum importance as microbial control agent. It is a Gram-positive bacterium that produces proteinaceous crystalline (Cry) inclusion bodies during sporulation. Cry proteins once ingested by the insect are solubilized in the midgut and are then cleaved by digestive proteases. Some of the resulting polypeptides are able to bind to midgut epithelial cell receptors resulting in cell lysis and finally insect death (Gahan et al. 2010). Entomopathogenic fungi are also important natural regulators of insect populations and have potential as mycoinsecticide agents against diverse insect pests in agriculture. The use of the insect pathogenic fungus *Metarhizium anisopliae* against adult *Aedes aegypti* and *Aedes albopictus* mosquitoes has also been reported. Further, baculovirus biopesticide has many advantages as a tool in the insect pest management program, including the highly specificity, no adverse effect on vertebrates and plants, and ease of genetic manipulation (Inceoglu et al. 2001). *Cydia pomonella* granulovirus (CpGV) provided a highly effective pathogen for control of important insect pests worldwide that are responsible for huge economic loss in apple and pear orchards every year. This virus has been used for several years as a bioinsecticide in codling moth (*Cydia pomonella*) control (Jehle et al. 2017).

Some important biocontrol agents are summarized in Table 8.1.

8.3 Microbial Mechanisms for Controlling Biotic and Abiotic Stresses

In response to biotic and abiotic stresses, there is often a significant increase in the endogenous ethylene production that has adverse effects on plant growth. The role of exopolysaccharides (EPSs) from a plant growth-promoting rhizobacterium, on elicitation of plant resistance toward some biotic and a biotic stress, was investigated by several investigators (Wafaa et al. 2014). According to Grover et al. (2010), certain microbial types may mitigate the impact of soil drought through production of exopolysaccharates, induction of resistance genes, increased circulation of water in the plant, and the synthesis of ACC deaminase, indoleacetic acid, and proline. The enzyme 1-aminocyclopropane-1-carboxylate deaminase (ACC deaminase) catalyzes the degradation of 1-aminocyclopropane-1-carobylic acid (ACC), the immediate precursor of the plant hormone ethylene into α-ketobutyrate and ammonia. The degradation of ACC by bacterial ACC deaminase releases plant stress and

Table 8.1 Biological control by microorganism against biotic stresses caused by virus, bacteria, fungi, and insect in different crops/plants/trees

Diseases	Name of bio-agents	Pathogen	Crop	References
Bacterial				
Bacterial wilt	*Pseudomonas aeruginosa* T1	*Ralstonia solanacearum*	Tomato	Maji and Chakrabartty (2014)
	Pseudomonas putida R6			
	Bacillus thuringiensis, *Bacillus cereus*	*Ralstonia solanacearum*	Eucalyptus	Santiagoa et al. (2015)
Bacterial blight	*P. fluorescens*	*Xanthomonas oryzae* pv. *oryzae* (*Xoo*)	Rice	Velusamy et al. (2013)
	Pantoea vagans, *Peudomonas fluorescens*	*Xanthomonas arboricola* pv. *juglandis* (*Xaj*)	Walnut	Ozaktan et al. (2012)
Bacterial panicle blight	*Bacillus* sp.	*Burkholderia glumae* and *B. gladioli*	Rice	Shrestha et al. (2016)
Fungal diseases				
Collar rot	*Pseudomonas fluorescens*	*Aspergillus niger* *Aspergillus flavus*	Peanut (*Arachis hypogaea*)	Dey et al. (2004)
Stem rot		*Sclerotium rolfsii*		
Various fungal diseases in sesame	*Paenibacillus Polymyxa* E681	*Fusarium* and other fungi	Sesame	Ryu et al. (2004)
White rot	*Mesorhizobium loti* MP6	*Sclerotinia sclerotiorum*	Mustard (*Brassica campestris*)	Chandra et al. (2007)
Fusarium wilt	*Bacillus cereus* BS 03	*Fusarium udum*	Pigeon pea (*Cajanus cajan*)	Dutta et al. (2008)
	Pseudomonas aeruginosa RRLJ04			
Charcoal rot	*Pseudomonas* sp.	*Rhizoctonia bataticola*	Groundnut (*Arachis hypogaea*)	Gupta et al. (2002)
Fungal biocontrol	*Azotobacter* sp., *Pseudomonas* sp.		Wheat (*Triticum aestivum*)	Wachowska et al. (2004)
Crown rot	*Bacillus subtilis*, *Pseudomonas fluorescens*	*Phytophthora cactorum*, *Phytophthora fragariae*	Strawberry	Vestberg et al. (2004)
Red steel				
Anthracnose	*Pseudomonas fluorescens* strain 89B-61	*Colletotrichum lagenarium*	Cucumber	Wei et al. (1991)

(continued)

Table 8.1 (continued)

Diseases	Name of bio-agents	Pathogen	Crop	References
Root rot, root knot	*Pseudomonas aeruginosa, Bacillus subtilis*	*Phytophthora*	Mung bean	Siddiqui et al. (2001)
Insect				
Aphid	*Bacillus licheniformis, Bacillus amyloliquefaciens Bacillus cereus* MJ-1	*Myzus persicae*	Tomato (*Lycopersicon esculentum*), pepper,red pepper	Lucas et al. (2004), Kokalis-Burelle et al. (2002), and Joo et al. (2005)
Green peach aphids	*Bacillus amyloliquefaciens, Bacillus subtilis*	*Myzus persicae* (Sluzer)	Bell pepper (*Capsicum annuum*)	Herman et al. (2008)
Virus				
Virus	*Bacillus amyloliquefaciens* strain1N 937[a]	Tomato mottle virus	Tomato	Murphy et al. (2000)
	Bacillus subtilis 1N 937b			
Viral disease	*Pseudomonas fluorescens*	Tobacco necrosis virus	Tobacco (*Nicotiana tobacum*)	Park and Kloepper (2000)

rescues normal plant growth (Glick et al. 2007). Strains of rhizobacteria, namely, *Rhizobium leguminosarum* bv. *viciae*, *R. japonicum*, *R. hedysari*, *B. japonicum*, *R. gallicum*, *B. elkani*, *M. loti*, and *S. meliloti*, had been known to produce ACC deaminase (Okazaki et al. 2004; Madhaiyan et al. 2006). However, it has also been found that drought-tolerant plants increase their polyamine levels to a much greater extent than sensitive ones. A significant decrease was observed on polyamines accumulation when PGPRs were applied on stressed soybean plants (Zahedi and Samira 2015; Zhou et al. 2016). Proline, an amino acid, accumulates in many plant species in response to environmental stress. An osmoprotective function of proline was discovered first in bacteria, where a causal relationship between proline accumulation and salt tolerance has been reported (Szabados and Savoure 2009). A feedback-insensitive bacterial proBA genes were overexpressed in transgenic *Arabidopsis*, leading to proline hyperaccumulation and enhanced osmotolerance (Chen et al. 2007). Trehalose metabolism in rhizobia plays a crucial role in signaling plant growth, yield, and adaptation to abiotic stress, and its manipulation has a major agronomical impact on leguminous plants. Suárez et al. (2008) overexpressed trehalose-6-phosphate synthase gene in the symbiotic bacterium *Rhizobium etli* in order to increase drought tolerance and yield in legumes. Glycine betaine accumulates in cells of a number of halophytes and bacteria as an adaptive response to high salt. Bacterial Trk and Ktr, fungal Trk, and plant HKT form a family of membrane transporters permeable to K(+) and/or Na(+) involved in adaptation to osmotic or salt stress in plants (Hauser and Horie 2010; Corratgé-Faillie et al. 2010). Salt

Overly Sensitive 1 (SOS1) gene, which has been identified in a genetic screen (Wu et al. 1996), encodes a putative membrane protein that is homologous to plasma membrane Na+/H+ antiporters from bacteria and fungi (Shi et al. 2000). The activity of endophyte *Burkholderia phytofirmans* PsJN is affected by plant drought stress; it senses plant stress signals and adjusts its gene expression accordingly to cope with and adapt to the altered conditions (Sheibani-Tezerji et al. 2015).

Plants have the ability to acquire induced systemic resistance (ISR) to pathogens with plant growth-promoting bacteria (PGPBs) and PGPBs in association with plant roots that can prime the plant innate immune system and confer resistance to a wide range of pathogens with a minimal impact on yield and growth (Van Hulten et al. 2006). Pieterse et al. (2014) reported that most of the bacterial endophytes have the ability to stimulate a latent disease defense mechanism that confers an enhanced level of protection to a broad spectrum of pathogens. Plants recognize pathogens either directly or indirectly. In direct recognition, plants can recognize extracellular molecules known as microbe-associated molecular patterns or pathogen-associated molecular patterns like bacterial flagellin, lipopolysaccharides, and peptidoglycans (Boller and Felix 2009) and/or intracellular effector proteins (e.g., Avr3a, Avrk, and Avra10 proteins) or tissue damage using pattern recognition receptor (PRR) proteins located on the cell surface or intracellularly (Allen et al. 2004; Rivas and Thomas 2005; Boller and Felix 2009). SA acts as an endogenous signal involved in systemic acquired resistance (SAR), an inducible resistance against a broad spectrum of pathogens including bacteria, fungi, and viruses which cause necrosis through programmed cell death of infected cells, known as the hypersensitive response (Durrant and Dong 2004).

8.4 Our Leads

Our group under the energetic supervision of Dr Reeta Goel covered different aspects of plant stress and role of microbes to overcome these stresses. Saluja et al. (2011) highlighted identical mechanisms of arsenic resistance in two bacterial strains, namely, AG24 and AGM13, which were taken from two different ecosources and states of India. Gupta et al. (2005) did in situ characterization of mercury-resistant plant growth-promoting fluorescent pseudomonads. Nag (2015) in his master's research also reported several *Azotobacter* isolates having abilities to tolerate different types of stresses along with plant growth-promoting activities. Besides this our group efficiently did several important researches on metal bioremediation by microbes (Gupta et al. 2002, Tripathi et al. 2005, Rani and Goel (2009), Rani et al. 2008, 2013, Saluja et al. 2012).

Besides metal toxicity, cold stress tolerance was also studied by us (see Sect. 8.2.1.3). Previously, we have confirmed the presence of diversified microbial assemblage in rhizospheric soil of Himalayan red kidney bean (RKB) (Suyal et al. 2015a, b). Moreover, seven cold-adapted diazotrophs were isolated from the Western Indian Himalayan (WIH) RKB rhizosphere, and their proteome was documented (Suyal et al. 2014b). Furthermore, psychrophilic *Pseudomonas migulae* S10724

(JX173286), which was originally isolated from WIH RKB rhizosphere, was reported to promote the growth of *Vigna radiata* (L.) Wilczek (Suyal et al. 2014a).

References

Abd El-Daim IA, Bejai S, Meijer J (2014) Improved heat stress tolerance of wheat seedlings by bacterial seed treatment. Plant Soil 379:337–350

Ali SKZ, Sandhya V, Grover M, Kishore N, Venkateswar Rao L, Venkateswarlu B (2009) *Pseudomonas sp.* strain AKM-P6 enhances tolerance of sorghum seedlings to elevated temperatures. Biol Fertil Soils 46:4555

Ali SKZ, Sandhya V, Grover M, Linga VR, Bandi V (2011) Effect of inoculation with a thermo tolerant plant growth promoting *Pseudomonas putida* strain AKMP7 on growth of wheat (*Triticum spp.*) under heat stress. J Plant Interact 6(4):239–246

Ali Z, Abulfaraj A, Idris A, Ali S, Tashkandi M, Mahfouz MM (2015a) CRISPR/Cas9-mediated viral interference in plants. Genome Biol 16:238

Ali S, Ehetshamul-Haque S, Shahid Shaukat S (2001) Use of rhizobacteria in the control of root rot-root knot disease complex of mungbean. J Phytopathol 149(6):337–346

Ali Z, Ullah N, Naseem S, Inam-Ulhuaq M, Jacobsen HJ (2015b) Soil bacteria conferred a positive relationship and improved salt stress tolerance in transgenic pea (*Pisum sativum* L.) harboring Na+/H+ antiporter. Turk J Bot 39:962–972

Allen TD, Eby LT, Poteet ML, Lentz E, Lima L (2004) Career benefits associated with mentoring for proteges: a meta-analysis. J Appl Psychol 89(1):127–136

Arkhipova T, Prinsen E, Veselov S, Martinenko E, Melentiev A, Kudoyarova G (2007) Cytokinin producing bacteria enhance plant growth in drying soil. Plant Soil 292:305–315. https://doi.org/10.1007/s11104-007-9233

Arshad M, Shaharoona B, Mahmood T (2008) Inoculation with Pseudomonas spp. containing ACC-deaminase partially eliminates the effects of drought stress on growth, yield, and ripening of pea (*Pisum sativum* L.) Pedosphere 18:611–620

Augé RM, Stodola AJW, Tims JE, Saxton AM (2001) Plant Soil 230(1):87–97

Bano A, Fatima M (2009) Salt tolerance in *Zea mays* (L.) following inoculation with *Rhizobium* and *Pseudomonas*. Biol Fertil Soils 45:405–413

Barriuso J, Solano BR, Gutierrez Manero FJ (2008) Protection against pathogen and salt stress by four plant growth-promoting rhizobacteria isolated from *Pinus sp.* on *Arabidopsis thaliana*. Phytopathology 98:666–672. https://doi.org/10.1094/PHYTO-98-6-0666

Bergottini VM, Otegui MB, Sosa DA, Zapata PD, Mulot M, Rebord M, Junier P (2015) Bio-inoculation of yerba mate seedlings (Ilex paraguariensis St. Hill.) with native plant growth-promoting rhizobacteria: a sustainable alternative to improve crop yield. Biol Fertil Soils 51(6):749–755

Boller T, Felix G (2009) A renaissance of elicitors: perception of microbe-associated molecular patterns and danger signals by pattern- recognition receptors. Annu Rev Plant Biol 60(1):379–406

Chandra S, Choure K, Dubey RC, Maheshwari DK (2007) Rhizosphere competent Mesorhizobium loti MP6 induces root hair curling, inhibits Sclerotinia sclerotiorum and enhances growth of Indian mustard (Brassica campestris). Braz J Microbiol 38:124–130

Chaves MM, Maroco JP, Pereira JS (2003) Understanding plant responses to drought- from genes to the whole plant. Funct Plant Biol 30:239–264

Chen M et al (2007) Expression of Bacillus subtilis proBA genes and reduction of feedback inhibition of proline synthesis increases proline production and confers osmotolerance in transgenic Arabidopsis. J Biochem Mol Biol 40:396–403

Christensen JH, Christensen OB (2007) A summary of the PRUDENCE model projections of changes in European climate by the end of this century. Clim Chang 81:7–30. https://doi.org/10.1007/s10584-006- 9210-7

Corratgé-Faillie C, Jabnoune M, Zimmermann S, Véry A-A, Fizames C, Sentenac H (2010) Potassium and sodium transport in non-animal cells: the Trk/Ktr/HKT transporter family. Cell Mol Life Sci 67:2511

Corretto E, Antonielli L, Sessitsch A, Compant S, Höfer C, Puschenreiter M, Brader G (2017) Complete genome sequence of the heavy metal resistant bacterium Agromyces aureus AR33T and comparison with related Actinobacteria. Stand Gen Sci. https://doi.org/10.1186/s40793-016-0217-z

Creus CM, Sueldo RJ, Barassi CA (2004) Water relations and yield in Azospirillum-inoculated wheat exposed to drought in the field. Can J Bot 82:228–273

Dar GH, Beig MA, Ahanger FA, Ganai NA, Ashraf Ahangar M (2011) Management of root rot caused by *Rhizoctonia solani* and *Fusarium oxysporum* in blue pine (*Pinus wallichiana*) through use of fungal antagonists. Asian J Plant Pathol *5:62–74*

Dey R, Pal KK, Bhatt DM, Chauhan SM (2004) Growth promotion and yield enhancement of peanut (Arachis hypogaea L) by application of plant growth promoting rhizobacteria. Microbiol Res 159:371–394

Dhawia F, Dattaa R, Ramakrishna W (2016) Mycorrhiza and heavy metal resistant bacteria enhance growth, nutrient uptake and alter metabolic profile of sorghum grown in marginal soil. Chemosphere 157:33–41

Durrant WE, Dong X (2004) Systemic acquired resistance. Annu Rev Phytopathol 42(1):185–209

Dutta S, Mishra AK, Kuma BSD (2008) Induction of systemic resistance against fusarial wilt in pigeon pea through interaction of plant growth promoting rhizobacteria and rhizobia. Soil Biol Biochem 40:452–461

Emamverdian A, Ding Y, Mokhberdoran F, Xie Y (2015) Heavy metal stress and some mechanisms of plant defense response. Sci World J 756120:18

Emerson R, Hoover A, Ray A, Lacey J, Cortez M, Payne C et al (2014) Drought effects on composition and yield for corn stover, mixed grasses, and Miscanthus as bioenergy feedstocks. Biofuels 5(3):275–291

Gaby JC, Buckley DH (2011) A global census of nitrogenase diversity. Environ Microbiol 13:1790–1799

Gagné-Bourque F, Bertrand A, Claessens A, Aliferis KA, Jabaji S (2016) Alleviation of drought stress and metabolic changes in timothy (*Phleum pratense* L.) colonized with *Bacillus subtilis* B26. Front Plant Sci 7:584

Gahan LJ, Pauchet Y, Vogel H, Heckel DG, Mauricio R (2010) An ABC transporter mutation is correlated with insect resistance to bacillus thuringiensis Cry1Ac Toxin. PLoS Genet 6(12):e1001248

Ghabrial SA, Nibert ML (2009) Victorivirus, a new genus of fungal viruses in the family Totiviridae. Arch Virol 154:373–379

Glick BR (2012) Plant growth-promoting bacteria: mechanisms and applications, Scientifica, vol. 2012, Article ID 963401, 1–15

Glick BR et al (2007) Promotion of plant growth by bacterial ACC deaminase. Crit Rev Plant Sci 26:227–242

Grennan AK (2008) Ethylene response factors in jasmonate signaling and defense response. Plant Physiol 146(4):1457–1458

Grover M, Ali SZ, Sandhya V, Rasul A, Venkateswarlu B (2010) Role of microorganisms in adaptation of agriculture crops to abiotic stresses. World J Microbiol Biotechnol 27:1231–1240

Gupta CP, Dubey RC, Maheshwari DK (2002) Plant growth causing charcoal rot of peanut by fluorescent Pseudomonas. Biol Fertl Soil 35:295–301

Gupta A, Rai V, Bagdwal N, Goel R (2005) In situ characterization of mercury-resistant growth-promoting fluorescent pseudomonads. Microbiol Res 160(4):385–388

Habib SH, Kausar H, Md Saud H (2016) Plant growth-promoting rhizobacteria enhance salinity stress tolerance in okra through ROS-scavenging enzymes. Biomed Res Int 6284547:10

Haggag WM (2002) Sustainable agriculture management of plant diseases. J Biol Sci 2:280–284

Ham JH, Groth D (2011) Bacterial panicle blight, an emerging rice disease. Louisiana Agric 2011:16–17

Han Q-Q, Lü XP, Bai JP, Qiao Y, Paré PW, Wang SM et al (2014) Beneficial soil bacterium *Bacillus subtilis* (GB03) augments salt tolerance of white clover. Front Plant Sci 5:525

Hasiów-Jaroszewska B, Minicka J, Pospieszny H (2014) Cross-protection between different Pathotypes of Pepino mosaic virus representing Chilean 2 genotype. Acta Sci Pol, Hortorum Cultus 13(5):177–185

Hauser F, Horie T (2010) A conserved primary salt tolerance mechanism mediated by HKT transporters: a mechanism for sodium exclusion and maintenance of high K+/Na+ ratio in leaves during salinity stress. Plant Cell Environ 33:552–565

Herman MAB, Nault BA, Smart CD (2008) Effects of plant growth promoting rhizobacteria on bell pepper production and green peach aphid infestation in New York. Crop Prot 27:996–1002

Hider RC, Kong X (2010) Chemistry and biology of siderophores. Nat Prod Rep 27:637–657

Hirsch AM (2010) How rhizobia survive in the absence of a legume host, a stressful world indeed. In: Symbiosis and stress: joint ventures in biology (Eds.) J. Seckbach, M. Grube. cellular origin, life in extreme habitats and astrobiology. 17(4):375–391Dordrecht: Springer

Hu Y, Schmidhalter U (2002) Limitation of salt stress to plant growth. In: Hock B, Elstner CF (eds) Plant toxicology. Marcel Dekker Inc., New York, pp 91–224

Huang J, Wei Z, Tan S, Mei X, Yin S, Shen Q, Xu Y (2013) The rhizosphere soil of diseased tomato plants as a source for novel microorganisms to control bacterial wilt. Appl Soil Ecol 72:79–84

Inceoglu AB, Kamita SG, Hinton AC, Huang Q, Severson TF, Kang K, Hammock BD (2001) Recombinant baculoviruses for insect control. Pest Manag Sci 57:981–987

Ings J, Mur LAJ, Robson PRH, Bosch M (2013) Physiological and growth responses to water deficit in the bioenergy crop *Miscanthus x giganteus*. Front Plant Sci 4:468

Islam F, Yasmeen T, Arif MS, Ali S, BAli HS, Zhou W (2016) Plant growth promoting bacteria confer salt tolerance in *Vigna radiata* by up-regulating antioxidant defense and biological soil fertility. Plant Growth Regul 80:23–36

Jamil A, Riaz S, Ashraf M, Foolad MR (2011) Gene expression profiling of plants under salt stress. Crit Rev Plant Sci 30(5):435–458

Jehle JA, Schulze-Bopp S, Undorf-Spahn K, Fritsch E (2017) Evidence for a second type of resistance against cydia pomonella granulovirus in field populations of codling moths. Appl Environ Microbiol 83(2):e02330

Joo GJ, Kin YM, Kim JT, Rhee IK, Kim JH, Lee IJ (2005) Gibberellins producing rhizobacteria increase endogenous gibberellins content and promote growth of red peppers. J Microbiol 43(6):510–515

Karlidag H, Esitken A, Yildirim EMF, Turan MD (2011) Effects of plant growth promoting bacteria on yield, growth, leaf water content, membrane permeability and ionic composition of strawberry under saline conditions. J Plant Nutr 34:34–45

Kasim WA, Osman ME, Omar MN, El-Daim IAA, Bejai S, Meijer J (2013) Control of drought stress in wheat using plant growth promoting bacteria. J Plant Growth Regul 32:122–130

Katiyar V, Goel R (2003) Solubilization of inorganic phosphate and plant growth promotion by cold tolerant mutants of Pseudomonas fluorescens. Microbiol Res 158:1–6

Khan M, Goel R (2014) Expression, purification and in silico studies of cold resistant protein from plant growth promoting Ps fluorescens mutant CRPF1. Curr Biotechnol 3(3):266–272. https://doi.org/10.2174/ 22115501030314091812011 3

Kokalis-Burelle N, Vavrina CS, Rosskopf EN, Shelby RA (2002) Field evaluation of plant growth-promoting rhizobacteria amended transplant mixes and soil solarization for tomato and pepper production in Florida. Plant Soil 238:257–266

Kosaka Y, Ryang B-S, Kobori T, Shiomi H, Yasuhara H, Kataoka M (2006) Effectiveness of an attenuated Zucchini yellow mosaic virus isolate for cross-protecting cucumber. Plant Dis 90:67–72

Kumar GP, Mir Hassan Ahmed SK, Desai S, Leo Daniel Amalraj E, Rasul A (2014) *In Vitro* screening for sbioticd tress Tolerance in Potent Biocontrol and Plant Growth Promoting Strains of Pseudomonas and Bacillus spp. Int J Bacteriol 2014:195946

Lamsal K, Kim SW, Kim YS, Lee YS (2012) Application of rhizobacteria for plant growth promotion effect and biocontrol of anthracnose caused by on pepper. Mycobiology 40(4):244

Lee G-W, Kim M-J, Park J-S, Chae J-C, Soh B-Y, Jae-Eun J, Lee K-J (2011) Biological control of phytophthora blight and anthracnose disease in red-pepper using bacillus subtilis S54. Res Plant Dis 17(1):86–89

Li X, Geng X, Xie R, Fu L, Jiang J, Gao L, Sun J (2016) The endophytic bacteria isolated from elephant grass (*Pennisetum purpureum* Schumach) promote plant growth and enhance salt tolerance of Hybrid *Pennisetum*. Biotechnol Biofuels 9:190

Li Y, Pang HD, He LY, Wang Q, Sheng XF (2017) Cd immobilization and reduced tissue Cd accumulation of rice (Oryza sativa wuyun-23) in the presence of heavy metal-resistant bacteria. Ecotoxicol Environ Saf 138:56–63

Lim J-H, Kim SD (2013) Induction of drought stress resistance by multi-functional PGPR *Bacillus licheni formis* K11 in pepper. Plant Pathol J 29(2):201–208

Liu J, Piñeros MA, Kochian LV (2014) The role of aluminum sensing and signaling in plant aluminum resistance. J Integr Plant Biol 56(3):221–230

Lucas GJA, Probanza A, Ramos B, Palomino MR, Gutierrez Mañero FJ (2004) Effect of inoculation of Bacillus licheniformis ontomato and pepper. Agronomie 24:169–176

Madhaiyan M, Poonguzhali S, Ryu J, Sa T (2006) Regulation of ethylene levels in canola (Brassica campestris) by 1-aminocyclopropane-1- carboxylate deaminase-containing Methylobacterium fujisawaense. Planta 224:268–278

Majeed A, Abbasi MK, Hameed S, Imran A, Rahim N (2015) Isolation and characterization of plant growth-promoting rhizobacteria from wheat rhizosphere and their effect on plant growth promotion. Front Microbiol 6:198

Maji S, Chakrabartty PK (2014) Biocontrol of bacterial wilt of tomato caused by Ralstonia solanacearum by isolates of plant growth promoting rhizobacteria. AJCS 8(2):208–214

Mapelli F, Marasco R, Balloi A, Rolli E, Cappitelli F, Daffonchio D, Borin S (2012) Mineral-microbe interactions: biotechnological potential of bioweathering. J Biotechnol 157:473–481

Marulanda A, Porcel R, Barea JM, Azcón R (2007) Drought tolerance and antioxidant activities in lavender plants colonized by native drought-tolerant or drought-sensitive Glomus species. Microb Ecol 54:543–552

Marulanda A, Barea J-M, Azcón R (2009) Stimulation of plant growth and drought tolerance by native microorganisms (AM Fungi and Bacteria) from dry environments: mechanisms related to bacterial effectiveness. J Plant Growth Regul 28:115–124

Mayak S, Tirosh T, Glick BR (2004a) Plant growth-promoting bacteria that confer resistance to water stress in tomatoes and peppers. Plant Sci 166:525–530

Mayak S, Tirosh T, Glick BR (2004b) Plant growth-promoting bacteria confer resistance in tomato plants to salt stress. Plant Physiol Biochem 42:565–572. https://doi.org/10.1016/j.plaphy.2004.05.009

Mendes R, Garbeva P, Raaijmakers JM (2013) The rhizosphere microbiome: significance of plant beneficial, plant pathogenic, and human pathogenic microorganisms. FEMS Microbiol Rev 37(5):634–663

Mohanty M, Patra HK (2013) Effect of ionic and chelate assisted hexavalent chromium on mung bean seedlings (Vigna radiata L. wilczek. var k-851) during seedling growth. J Stress Physiol Biochem 9(2):232–241

Murphy JF, Zender GW, Schuster DJ, Sikora EJ, Polston JE, Kloepper JW (2000) Plant growth promoting rhizobacterial mediated protection in tomato against tomato mottle virus. Plant Dis 84:779–784

Nag NK (2015) Selection of stress tolerant effective azotobacter isolates for climatic conditions of Chhattisgarh. M.Sc. (Ag) thesis submitted in Indiragandhi Krishi Vishwavidyalaya Raipur

Nandakumar R, Shahjahan AKM, Yuan XL, Dickstein ER, Groth DE, Clark CA et al (2009) Burkholderia glumae and B. gladioli cause bacterial panicle blight in rice in the southern United States. Plant Dis 93(9):896–905

Nehra K, Yadav AS, Sehrawat AR, Vashishat RK (2007) Characterization of heat resistant mutant strains of *Rhizobium sp.* [Cajanus] for growth, survival and symbiotic properties. Indian J Microbiol 47:329–335

Nichols VA, Miguez FE, Jarchow ME, Liebman MZ, Dien BS (2014) Comparison of cellulosic ethanol yields from midwestern maize and reconstructed tallgrass prairie systems managed for bioenergy. Bioenergy Res 7(4):1550–1560
Nion YA, Toyota K (2015) Recent trends in control methods for bacterial wilt diseases caused by ralstonia solanacearum. Microbes Environ 30(1):1–11
Ogai R, Kanda-Hojo A, Tsuda S (2013) An attenuated isolate of Pepper mild mottle virus for cross protection of cultivated green pepper (Capsicum annuum L.) carrying the L3 resistance gene. Crop Prot 54:29–34
Omar AM, AhmedI. S. Ahmed (2014) Antagonistic and inhibitory effect of some plant rhizobacteria against different fusarium isolates on salvia officinalis. Am Eurasian J Agric Environ Sci 14(12):1437–1446
Ozaktan H, Erdal M, Akkopru A, Aslan E (2012) Biological control of bacterial blight of walnut by antagonistic bacteria. J Plant Pathol 94(1, Supplement):S1.53–S1.56
Okazaki Y, Ishihara A, Nishioka T, Iwamura H (2004) Identification of a dehydrodimer of avenanthramide phytoalexin in oats. Tetrahedron 60(22):4765–4771
Pal KP, Gardener BM (2006) Biological control of plant pathogens. Plant Health Instr. https://doi.org/10.1094/PHI-2006-117-02
Pan XD, Wu PG, Jiang XJ (2016) Levels and potential health risk of heavy metals in marketed vegetables in Zhejiang, China. Sci Rep 6:20317
Park KS, Kloepper JW (2000) Activation of PR-1a promoter by rhizobacteria which induces systemic resistance in tobacco against Pseudomonas syringae pv. Tabaco Biol Cont 18:2–9
Paul D, Nair S (2008) Stress adaptations in a plant growth promoting rhizobacterium (PGPR) with increasing salinity in the coastal agricultural soils. J Basic Microbiol 48(5):378–384. https://doi.org/10.1002/jobm. 200700365
Pereyra MA, García P, Colabelli MN, Barassi CA, Creus CM (2012) A better water status in wheat seedlings induced by Azospirillum under osmotic stress is related to morphological changes in xylem vessels of the coleoptile. Appl Soil Ecol 53:94–97
Pieterse CMJ, Zamioudis C, Berendsen RL, Weller DM, Van Wees SCM, Bakker PAHM (2014) Induced systemic resistance by beneficial microbes. Annu Rev Phytopathol 52:347–375
Pirhadi M, Enayatizamir N, Motamedi H, Sorkheh K (2016) Screening of salt tolerant sugarcane endophytic bacteria with potassium and zinc for their solubilizing and antifungal activity. Biosci Biotech Res Comm 9(3):530–538
Ramadoss D, Lakkineni VK, Bose P, Ali S, Annapurna K (2013) Mitigation of salt stress in wheat seedlings by halotolerant bacteria isolated from saline habitat. SpringerPlus 2:6
Ramesh R, Phadke GS (2012) Rhizosphere and endophytic bacteria for the suppression of eggplant wilt caused by Ralstonia solanacearum. Crop Prot 37:35–41
Rani A, Goel R (2009) Strategies for crop improvement in contaminated soils using metal-tolerant bioinoculants. In: Khan MS, Zaidi A, Musarrat J (eds) Microbial strategies for crop improvement. Springer, Berlin, pp 85–104
Rani A, Shouche YS, Goel R (2008) Declination of copper toxicity in pigeon pea and soil system by growth-promoting Proteus vulgaris KNP3 strain. Curr Microbiol 57(1):78
Rani A, Souche Y, Goel R (2013) Comparative in situ remediation potential of pseudomonas putida 710A and commamonas aquatica 710B using plant (Vigna radiata (L.) wilczek) assay. Ann Microbiol 63(3):923–928
Ren ZH, Gao JP, Li LG, Cai XL, Huang W, Chao DY, Zhu MZ, Wang ZY, Luan S, Lin HX (2005) A rice quantitative trait locus for salt tolerance encodes a sodium transporter. Nat Genet 37:1141–1146
Rincon A, Valladares F, Gimeno TE, Pueyo JJ (2008) Water stress responses of two Mediterranean tree species influenced by native soil microorganisms and inoculation with a plant growth promoting rhizobacterium. Tree Physiol 28(11):1693–1701
Rivas S, Thomas CM (2005) Molecular interactions between tomato and the leaf mold pathogen. Annu Rev Phytopathol 43(1):395–436
Ron EZ, Segal G, Sirkis R, Robinson M, Graur D (2000) Regulation of heat-shock response in bacteria. Microbial biosystems: new frontiers. In: Bell CR, Brylinsky M, Johnson-Green P (eds) Proceedings of the 8th international symposium on microbial Ecology. Atlantic Canada Society for Microbial Ecology, Halifax, Canada

Ruiz-Lozano JM, Aroca R (2010) Modulation of aquaporin genes by the arbuscular mycorrhizal symbiosis in relation to osmotic stress tolerance. In: Seckbach J, Grube M (eds) Symbioses and stress: joint ventures in biology, cellular origin, life in extreme habitats and astrobiology. Springer, Dordrecht, pp 359–374

Ryu CM, Farag MA, CH H, Reddy MS, Kloepper JW, Paré PW (2004) Bacterial volatiles induce systemic resistance in Arabi-dopsis. Plant Physiol 134:1017–1026

Sadik S, Mazouz H, Bouaichi A, Benbouazza A, Achbani EH (2013) Biological control of bacterial onion diseases using a bacterium, *Pantoea Agglomerans* 2066-7. Int J Sci Res (IJSR) 4(1):2319–7064

Saghafi K, Ahmadi J, Asgharzadeh A, Bakhtiari S (2013) The effect of microbial inoculants on physiological responses of two wheat cultivars under salt stress. Int J Adv Biol Biom Res 1(4):421–431

Saluja B, Gupta A, Goel R (2011) Mechanism of arsenic resistance prevalent in Bacillus species isolated from soil and ground water sources of India. Ekologija 57(41):55–161

Saluja B, Tripathi M, Goel R (2012) Molecular and functional characterization of cadmium resistant Proteus vulgaris strain KNP3 to unravel its resistance mechanistic. Chem Ecol 8(1):17–23

Santiagoa TR, Grabowskib C, Rossatoa M, Romeiroa RS, Mizubuti ESG (2015) Biological control of eucalyptus bacterial wilt with rhizobacteria. Biol Control 80:14–22

Sapsirisopa S, Chookietwattana K, Maneewan K, Khaengkhan P (2009) Effect of salt-tolerant *Bacillus* inoculum on rice KDML 105 cultivated in saline soil. As J Food Ag-Ind, special issue S69–S74

Selvakumar G, Joshi P, Suyal P, Mishra PK, Joshi GK, Venugopalan R, Gupta HS (2013) Rock phosphate solubilization by psychrotolerant Pseudomonas spp. and their effect on lentil growth and nutrient uptake under polyhouse conditions. Ann Microbiol 63(4):1353–1362

Serrano R, Rodríguez-Navarro A (2001) Ion homeostasis during salt stress in plants. Curr Opin Cell Biol 13:399–404

Shahriari F, Khodakaramian G, Heydari A (2005) Assessment of antagonistic activity of Pseudomonas fluorescens biovars toward Pectobacterium carotovorum subsp. atrosepticum. J Sci Technol Agric Nat Resour 8:201–211

Sheibani-Tezerji R, Rattei T, Sessitsch A, Trognitz F, Mitter B (2015) Transcriptome profiling of the endophyte *Burkholderia phytofirmans* PsJN indicates sensing of the plant environment and drought stress. MBio 6(5):e00621-15. https://doi.org/10.1128/mBio.00621-15

Shi H, Ishitani M, Kim C, Zhu J-K (2000) The Arabidopsis thaliana salt tolerance gene SOS1 encodes a putative Na+/H+ antiporter. Proc Natl Acad Sci 97(12):6896–6901

Shi H, Quintero FJ, Prado JM, Zhu JK (2002) The putative plasma membrane Na+-H+ antiporter SOS1 controls longdistance Na+ transport in plants. Plant Cell 14:465–477

Shrestha BK, Karki HS, Groth DE, Jungkhun N, Ham JH (2016) Biological control activities of rice-associated bacillus sp. strains against sheath blight and bacterial panicle blight of rice. PLoS One 11(1):e0146764

Siddiqui KS, Cavicchioli R (2006) Cold-Adapted Enzymes. Annu Rev Biochem 75(1):403–433

Singh A, Grover A (2008) Genetic engineering for heat tolerance in plants. Physiol Mol Biol Plants 14(1-2):155–166

Singh AV, Chandra R, Goel R (2013) Phosphate solubilization by Chryseobacterium sp. and their combined effect with N and P fertilizers on plant growth promotion. Arch Agron Soil Sci 59(5):641–651

Sinha S, Singh D, Yadav DK, Upadhyay BK (2012) Utilization of plant growth promoting Bacillus subtilis isolates for the management of bacterial wilt incidence in tomato caused by Ralstoniasolanacearum race 1 biovar 3. Indian Phytopathol 65(1):18–24

Soni R, Saluja B, Goel R (2010) Bacterial community analysis using temporal temperature gradient gel electrophoresis (TTGE) of 16S rDNA PCR products of soil metagenome. Ekologija 56(3–4):94–98; Saluja B, Gupta A, Goel R (2011) Mechanism of arsenic resistance prevalent in Bacillus species isolated from soil and ground water sources of India. Ekologija 57(4):155–161

Soni R, Suyal DC, Agrawal K, Yadav A, Souche Y, Goel R (2015) Differential proteomic analysis of Himalayan psychrotolerant diazotroph Pseudomonas palleroniana N26 strain under low temperature diazotrophic conditions. CryoLetters 36:74–82

Suárez R, Wong A, Ramírez M, Barraza A, Orozco M d C, Cevallos MA, Lara M, Hernández G, Iturriaga G (2008) Improvement of drought tolerance and grain yield in common bean by overexpressing trehalose-6-phosphate synthase in rhizobia. Mol Plant-Microbe Interact 21(7):958–966

Sunarpi HT, Motoda J, Kubo M et al (2005) Enhanced salt tolerance mediated byAtHKT1 transporter-induced Na+ unloading from xylem vessels to xylem parenchyma cells. Plant J 44:928–938

Suprapta DN (2012) Potential of Microbial antagonists as biocontrol agents against plant fungal pathogens. J ISSAAS 18(2):1–8

Suyal DC, Shukla A, Goel R (2014a) Growth promotory potential of the psychrophilic diazotroph Pseudmonas migulae S10724 against Native Vigna radiata (L.) Wilczek. 3Biotech 4:665–668. https://doi.org/10.1007/s13205-014-0259-0

Suyal DC, Yadav A, Shouche Y, Goel R (2014b) Differential proteomics in response to low temperature diazotrophy of Himalayan psychrophilic nitrogen fixing Pseudomonas migulae S10724 strain. Curr Microbiol 68(4):543–550. https://doi.org/10.1007/s00284-013-0508-1

Suyal DC, Yadav A, Shouche Y, Goel R (2015a) Bacterial diversity and community structure of Western Indian Himalayan red kidney bean (Phaseolus vulgaris) rhizosphere as revealed by 16S rRNA gene sequences. Biologia 70(3):305–313. https://doi.org/10.1515/biolog-2015-0048

Suyal DC, Yadav A, Shouche Y, Goel R (2015b) Diversified diazotrophs associated with the rhizosphere of Western Indian Himalayan native red kidney beans (Phaseolus vulgaris L.) 3Biotech 5:433–441. https://doi.org/10.1007/s13205-014-0238-5

Szabados L, Savoure A (2009) Proline: a multifunctional amino acid. Trends Plant Sci 15(2):89–97

Taj ZZ, Rajkumar M (2016) Perspectives of plant growth-promoting actinomycetes in heavy metal phytoremediation. In: Subramaniam G, Arumugam S, Rajendran V (eds) Plant growth promoting actinobacteria. Springer, Singapore, pp 213–231

terHorst CP, Lennon JT, Lau JA (2014) The relative importance of rapid evolution for plant-microbe interactions depends on ecological context. Proc R Soc B 281:20140028

Timmusk S, Abd El-Daim IA, Copolovici L, Tanilas T, Kannaste A, Behers L, Nevo E, Seisenbaeva G, Elna S¨m, lo Niinemets Ü (2014) drought-tolerance of wheat Improved by rhizosphere bacteria from harsh environments: enhanced biomass production and reduced emissions of stress volatiles. PLoS One 9(5):e96086

Tripathi M, Munot HP, Shouche Y, Meyer JM, Goel R (2005) Isolation and functional characterization of siderophore-producing lead- and cadmium-resistant pseudomonas putida KNP9. Curr Microbiol 50(5):233–237

Trivedi P, Sa T (2008) Pseudomonas corrugata (NRRL B-30409) mutants increased phosphate solubilization, organic acid production, and plant growth at lower temperatures. Curr Microbiol 56(2):140–144

Turnera BL, Driessena JP, Haygarthb PM, Mckelvi ID (2003) Potential contribution of lysed bacterial cells to phosphorus solubilisation in two rewetted Australian pasture soils. Soil Biol Biochem 35:187–189

Upadhyay SK, Singh DP, Saikia R (2009) Genetic diversity of plant growth promoting rhizobacvteria isolated from rhizosphere soil of wheat under saline condition. Curr Microbiol 59:489–496

Upadhyay SK, Maurya SK, Singh DP (2012) Salinity tolerance in free living plant growth promoting Rhizobacteria. Ind J Sci Res 3(2):73–78

Vaishnav A, Kumari S, Jain S, Choudhary DK, Sharma KP (2016) Molecular characterization of potential salt tolerant bacteria for soybean growth promotion. Int J Bioassays 5(12):5118–5123

van Hulten M, Pelser M, van Loon LC, Pieterse CMJ, Ton J (2006) Costs and benefits of priming for defense in Arabidopsis. Proc Natl Acad Sci 103(14):5602–5607

Vander Weijde T, Huxley LM, Hawkins S, Sembiring EH, Farrar K, Dolstra O et al (2016) Impact of drought stress on growth and quality of miscanthus for biofuel production. GCB Bioenergy. https://doi.org/10.1111/gcbb.12382

Vardharajula S, Zulfikar Ali S, Grover M, Reddy G, Bandi V (2011) Drought-tolerant plant growth promoting Bacillus spp.: effect on growth, osmolytes, and antioxidant status of maize under drought stress. J Plant Interact 6:1–14. https://doi.org/10.1080/17429145.2010.535178

Velusamy P, Immanue JE, Gnanamanickam SS (2013) Rhizosphere bacteria for biocontrol of bacterial blight and growth promotion of rice. Rice Sci 20(5):356–362

Vestberg M, Kukkonen S, Saari K, Parikka P, Huttunen J, Tainino L, Devos N, Weekers F, Kevers C, Thonart P, Lemoine MC, Cordier C, Alabouvette C, Gianinazzi S (2004) Microbial inoculation for improving the growth and health of micropropagated strawberry. Appl Soil Ecol 27:243–258

Vidhyasekaran P (2002) Bacterial disease resistance in plants. Molecular biology and biotechnological applications. The Haworth Press, Binghamton. 452 pp

Vinocur B, Altman A (2005) Recent advances in engineering plant tolerance to abiotic stress: achievements and limitations. Curr Opin Biotechnol 16:123–132

Wachowska U, Majchrzak B, Borawska M, Karpinska Z (2004) Biological control of winter wheat pathogens by bacteria. Acta fytotech zootech 7, 2004, special number. In: Proceedings of the XVI. Slovak and Czech plant protection conference organized at Slovak Agricultural University in Nitra, Slovakia

Wafaa MHW, Hussein MM, Mehanna HM, El-Moneim D (2014) Bacteria polysaccharides elicit resistance of wheat against some biotic and abiotic stress. Int J Pharm Sci Rev Res 29(2):292–298

Wei G, Kloepper JW, Tuzun S (1991) Induction of systemic resistance of cucumber to Colletotrichum orbiculare by select strains of plant growth-promoting rhizobacteria. Phytopathology 81:1508–1512

Wolter A, Schroeder F-G (2012) Effect of drought stress on the productivity of ivy treated with rhizobacterium Bacillus subtilis. In: Proceedings of the international symposium on soilless cultivation 1004, Shanghai, pp 107–113

Wu SJ, Lei D, Zhu JK (1996) SOS1, a genetic locus essential for salt tolerance and potassium acquisition. Plant Cell 8:617–627

Wu QS, Xi RX, Zou YN (2008) Improved soil structure and citrus growth after inoculation with three Arbuscular mycorrhizal fungi under drought stress. Eur J Soil Biol 44:122–128

Xu P, Chen F, Mannas JP, Feldman T, Sumner LW, Roossinck MJ (2008) Virus infection improves drought tolerance. New Phytol 180:911–921. https://doi.org/10.1111/j.1469-8137.2008.02627

Xue Q-Y, Chen Y, Li S-M, Chen L-F, Ding G-C, Guo D-W, Guo J-H (2009) Evaluation of the strains of Acinetobacter and Enterobacter as potential biocontrol agents against Ralstonia wilt of tomato. Biol Control 48:252–258

Yaish MW, Al-Lawati A, Jana GA, Vishwas Patankar H, Glick BR (2016) Impact of soil salinity on the structure of the bacterial endophytic community identified from the roots of caliph medic (*Medicago truncatula*). PLoS One 11(7):e0159007. https://doi.org/10.1371/journal.pone.0159007

Yoo JH, Park IC, Kim WG (2012) Biocontrol of Anthracnose of Chili Pepper by Bacillus sp. NAAS-1. Kor J Mycol 40(4):277–281

Zahedi H, Samira A (2015) Effect of plant growth promoting rhizobacteria (PGPR) and water stress on phytohormones and polyamines of soybean. Indian J Agric Res 49(5):427–431

Zampieri BDB, Pinto AB, Schultz L, de Oliveira MA, de Oliveira AJFC (2016) Diversity and distribution of heavy metal-resistant bacteria in polluted sediments of the araca bay, sao sebastiao (sp), and the relationship between heavy metals and organic matter concentrations. Microb Ecol 72:582

Zhang H, Sun Y, Xie X, Kim M-S, Dowd SE, Paré PW (2009) A soil bacterium regulates plant acquisition of iron via deficiency-inducible mechanisms. Plant J 58(4):568–577

Zhou C, Zhou Y (2012) Strategies for viral cross protection in plants. Methods Mol Biol 894:69–81

Zhou C, Ma Z, Zhu L, Xiao X, Xie Y, Zhu J, Wang J (2016) Rhizobacterial strain *Bacillus megaterium* BOFC15 induces cellular polyamine changes that improve Plant growth and drought resistance. Int J Mol Sci 17:976

ACC-Deaminase Producing Rhizobacteria: Prospects and Application as Stress Busters for Stressed Agriculture

9

Ruchi Soni, Sarita K. Yadav, and Ajay Singh Rajput

Abstract

The plant hormone ethylene is responsible for a multiple biological processes in plants and also regulates plants' response to various biotic-abiotic stress conditions. Ethylene with low concentration enhances seedling emergence, root hair development, and root extension, while its higher concentration resulted into inhibition of root elongation. The significant function of 1-aminocyclopropane-1-carboxylic acid (ACC)-deaminase is reported in alleviating environmental biotic-abiotic stress conditions, and ACC serves as the precursor of plant hormone ethylene synthesized in plant tissues during stressful conditions. The enzyme ACC-deaminase produced by several rhizobacteria catalyzes the deprivation of ACC, consequently reducing deleterious ethylene level in the stressed plant. ACC-deaminase enzyme basically cleaves the ethylene precursor ACC to α-ketobutyrate and ammonia, which is further metabolized by rhizobacteria for their growth and development. ACC-deaminase producing rhizobacteria are responsible in facilitating the resistance to plants against stressful environmental conditions including salinity, drought, flooding, metal and organic contamination, and phytopathogens. ACC-deaminase is an inducible enzyme and its synthesis is stimulated in the presence of substrate ACC. ACC-deaminase is encoded by gene AcdS and regulated in a discrepancy way under different stressed environmental conditions. Application of rhizobacterial AcdS gene has been employed in developing transgenic plants conferring ACC-deaminase gene which showed enhanced tolerance against biotic-abiotic stresses in plants. The rhizobacteria acquiring AcdS gene encoding enzyme ACC-deaminase are known to exhibit productivity in a variety of crops under stressful conditions; thus, they

R. Soni (✉) · S. K. Yadav · A. S. Rajput
Regional Centre of Organic Farming, Department of Agriculture and Cooperation, Ministry of Agriculture and Farmers Welfare, Government of India, Nagpur, Maharashtra, India
e-mail: ruchihpau1987@gmail.com

D. G. Panpatte et al. (eds.), *Microorganisms for Green Revolution*, Microorganisms for Sustainability 7,
https://doi.org/10.1007/978-981-10-7146-1_9

can be considered as potential candidates for development of microbial inoculants and stress busters in stressed agriculture.

Keywords
ACC-deaminase activity · AcdS gene · Biotic-abiotic stress · Ethylene level · Plant growth

9.1 Introduction

The crop productivity may be affected by a wide range of biotic-abiotic environmental stresses in the agriculture. These stressful regimes include extreme conditions of drought, nutrient deficiency, salinity, radiation, temperature, toxic metals and organic contaminants, water logging, and various phytopathogens. As a result, plant growth and productivity is detrimentally affected by these diverse environmental stress factors. It has been reported that the loss in the productivity incurred with these stress factors is estimated to be >50% for the major growing crops in the agriculture. Thus, these stress regimes are the principal adverse factors affecting the productivity of agriculture crops worldwide. Problems acquired due to nutritional limitation can be conquered using synthetic inputs such as chemical fertilizers, insecticides and pesticides, and biological approaches like biofertilizers, but adverse effects of calamities due to biotic-abiotic factors by non-biological processes is extremely challenging for the farmers and researchers also. Plants respond these stress regimes by mediating the level of phytohormone ethylene which results into regulating the expression of stress-encoding proteins responsible for protection from the harmful effects of biotic-abiotic stresses. Ethylene in low concentration enhances seedling emergence, root hair development, and root extension; however, when produced higher than its threshold level, it acts as stress ethylene and inhibits plant growth and productivity by reducing plant root-shoot elongation and proliferation and declining other growth parameters.

The plant growth-promoting rhizobacteria (PGPR) exist in the vicinity of the root zone due to the presence of a number of nutrients provided through root exudates which are metabolized by these bacteria for their growth and development. The rhizobacteria belonging to the genera include *Acetobacter*, *Arthrobacter*, *Azoarcus*, *Azospirillum*, *Azotobacter*, *Bacillus*, *Burkholderia*, *Enterobacter*, *Gluconacetobacter*, *Herbaspirillum*, *Klebsiella*, *Paenibacillus*, *Pseudomonas*, *Serratia*, *Stenotrophomonas*, etc. The use of these rhizobacteria has been emphasized worldwide, and now there is also a replenish interest of the researchers in their application in major growing crops in modern agriculture. Some of these plant beneficial rhizobacteria possess an enzyme 1-aminocyclopropane-1-carboxylate (ACC)-deaminase, which reduces detrimental effect of ethylene in the plants. ACC an immediate precursor of ethylene, is hydrolysed by enzyme ACC-deaminase to α-ketobutyrate and ammonia and consequently reducing ethylene level in the stressed plant. PGPR exhibiting ACC-deaminase activity to provide tolerance against biotic-abiotic stresses revolutionizes stress regimes confer induced systemic tolerance in the host plants. Thus, PGPR acquiring ACC-deaminase activity is of immense importance in agriculture in inhibiting the adverse effect of environmental stress factors.

The selection of ACC-deaminase producing rhizobacteria with established mechanisms of plant growth promotion has been emphasized for realizing their potential for improving crop productivity in the stressed agriculture. Application of ACC-deaminase-encoding AcdS gene also has been employed in the development of transgenic plants which exhibit enhanced tolerance against biotic-abiotic stresses in plants. Thus, they can be considered as potential candidates in the development of microbial inoculants for improving crop productivity and implemented as stress busters in stressed agriculture. Therefore, keeping in view the above factors, the present chapter addresses the correlation of ethylene biosynthesis and plant stress, strategies adopted by ACC-deaminase producing rhizobacteria in stressed agriculture, and details of biochemical properties, genetics, and regulation of ACC-deaminase enzyme which plays a promising role in reducing the detrimental level of ethylene in stressed plants and rescuing them from the adverse effects of environmental stress conditions.

9.2 Correlation Between Ethylene Biosynthesis and Plant Stress

Phytohormone ethylene with optimal threshold value is important in growth and development of the plants and plays a key role in adventitious root extension, root hair elongation, seed germination enhancement, seed dormancy break down, etc., while its higher value results into defoliation, root elongation inhibition, nodulation reduction in legumes, leaf senescence, abscission, and chlorosis (Fig. 9.1). Therefore, regulation of ethylene production in plant roots for their improved

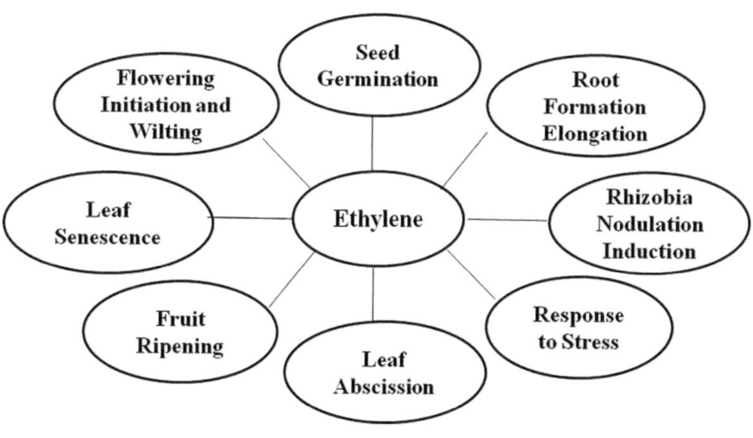

Fig. 9.1 The phytohormone ethylene affecting different physiological conditions responsible for plant growth and development

growth and productivity is an imperative factor. The production of ethylene in the plants entirely depends on various biotic-abiotic environmental stress conditions. Ethylene biosynthesis is significantly affected when the plants are subjected to stressful regimes of salinity, drought, flooding, heavy metal, and organic contamination (Glick 2007). Acceleration in ethylene synthesis is also reported in the pathogen-infected plants resulting into tissue damage, necrosis, and low endogenous resistance (Frankenberger and Arshad 1995). The higher ethylene production has also known to be exerted by tissue wounding, most preferably by the stimulation of ACC-synthase activity (Arshad and Frankenberger 2002). Maintaining ethylene level to its threshold value by chemical approaches or biological processes can significantly narrow down its magnitude and in turn reduce stress-induced tissue damage to the plants (VanLoon 2007). This section clearly indicates that with the increment in concentration, phytohormone ethylene plays a role of negative regulator and reduces plant growth and development. However under stress conditions, surveillance on ethylene evolution achieved by the chemical or biological inhibitors will be beneficial for improving agricultural crops. Thus, plant growth and productivity in the adverse environmental conditions confronted with various biotic-abiotic stresses and ultimately release of stress hormone ethylene may be improved by employing some promising technology in stressed agriculture.

9.3 Various Biotic-Abiotic Stress Conditions

9.3.1 High-Salinity and Drought Conditions

Soil salinity and limited rainfall result into the excessive use of alternative sources including groundwater pump by the tube wells, and sewage for irrigation is a serious problem in agriculture. In hot and dry areas, the soils are saline causing reduction in plant growth and crop productivity worldwide. Saline- and drought-induced stress in the plants results into enhanced stress ethylene production. It has been reported that transgenic plants and ACC-deaminase producing rhizobacteria may protect plant from these stressful environmental conditions (Glick 2005).

9.3.2 Water-Logged Conditions

During flooding, the plant root system frequently becomes anaerobic because of water logging conditions that result in inducing the expression of ACC-synthase and simultaneously stimulating the ACC accumulation in root tissues (Else and Jackson 1998). In this case, ACC does not convert into ethylene as it happens in the other stress regimes because enzyme ACC-oxidase requires oxygen to catalyze this reaction which is not possible in water-logged conditions. Besides that, a significant proportion of accumulated ACC is transported to the shoots from the roots where ACC-oxidase acquires an aerobic environment to produce ethylene and causes chlorosis epinasty and necrosis and ultimately limits crop productivity.

9.3.3 Heavy Metal and Organic Contamination

Plants subjected to high level of heavy metals stress resulting in increased synthesis of stress hormone ethylene also become iron deficient (Burd et al. 1998; Farwell et al. 2007). In such conditions, soil can be immediately remedied by applying rhizobacteria exhibiting ACC-deaminase activity and siderophore-producing attribute. Moreover, many heavy metals and organic contaminants are present at high threshold levels in the soil causing low bioavailability; as a result only a small proportion is taken up by the plants for their growth and development. Many plants in association with versatile rhizobacteria can easily degrade many toxic heavy metals and organic pollutants which are inhibitory to plant growth and productivity.

9.3.4 Phytopathogens

When plant is infected with various phytopathogens like bacteria, fungi, and viruses or nematodes, it induces increased ethylene biosynthesis. In this regard, a severe damage to the plant occurs as a response to increased level of stress ethylene, and many chemical inhibitors have been reported to reduce sternness of exogenous ethylene toward phytopathogens infection (Abeles et al. 1992; Husen et al. 2011). In recent years, transgenic plants that are resistant to a wide range of pathogens and some rhizobacteria possessing ACC-deaminase activity which protect plants form deleterious ethylene damage from pathogens have become popular approaches in agriculture.

9.4 Chemical and Biological Inhibitors Lowering Stress Ethylene Level

Many chemicals including aminoethoxyvinyl glycine (AVG), a very specific and effective ACC-synthase inhibitor, 1-methylcyclopropene (1-MCP), and silver thiosulfate ethylene biosynthesis inhibitors are extremely toxic to the food and also destroy chemical and physical characteristics of the soil (Yuhashi et al. 2000). Moreover, 1-MCP is commercially used to control wilting in flowers and post-harvest disease occurrence and spoilage of fruits. Cobalt (Co^{2+}) with concentration 10–100μM has also been reported to inhibit the conversion of ACC into ethylene and decline the level of ACC-oxidase (McKeon et al. 1995). The chemical inhibitors are not safe and frequent approaches to use in the fields to eliminate the deleterious effects of biotic-abiotic environmental stresses. Therefore, an eco-friendly and environmentally sustainable mechanism like ACC-deaminase producing rhizobacteria which decrease the harmful stress ethylene level along with plant growth-promoting attributes should be preferred in stressed agriculture.

9.5 ACC-deaminase Producing Rhizobacteria and Their Ecological Significance

PGPR associated with plant roots possess multiple plant growth-promoting attributes conferring capability to improve crop biomass and productivity (Soni et al. 2017). ACC-deaminase enzyme activity is considered as one of the most important attributes among these plant beneficial rhizobacteria (Nascimento et al. 2014). A large number of PGPR exhibiting ACC-deaminase activities are reported in the literature showing inhibition to ethylene concentration, thereby alleviating the harmful effects exerted by ethylene and promoting plant growth and development (Table 9.1). ACC-deaminase producing bacteria exert beneficial effects on plant by rescuing them from the detrimental effects of environmental stress conditions, hindering senescence, providing biocontrol efficiency against a wide range of phytopathogens, and inducing nodulation in leguminous crops (Ali et al. 2012). Shaharoona et al. (2007) have also demonstrated the decrease in ACC level in etiolated pea seedlings inoculated with ACC-deaminase rhizobacteria in the same way as achieved with chemical inhibitor Co^{2+}, which resulted into significant increment in seedling length and root elongation in comparison to control.

At extreme environmental conditions, variation in ACC-deaminase activity in diverse microbial species may also exert beneficial effects in phytoremediation at different environmental sites and stressful conditions. ACC-deaminase-containing rhizobacteria assist the host plants coupled with phytoremediation by biotransforming noxious elements, mediating rhizodegradation through root exudates, detoxificating heavy metals and organic-inorganic contaminants, and regulating rhizoremediation by root elongation along with promoting their survival under adverse environmental conditions (Gamalero and Glick 2012; Singh et al. 2015; Singh and Jha 2016). A positive correlation was investigated between the increase in bacterial ACC-deaminase activity and reduction in heavy metal accumulation (cadmium and nickel) in the plant tissue and significant increment in root length (Belimov et al. 2005; Rodriguez et al. 2008). The occurrence of ACC-deaminase in human opportunistic pathogenic bacteria *Burkholderia cenocepacia* J2315, plant pathogenic fungi *Aspergillus* spp. and *Myceliophthora thermophila*, and endophytic fungi *P. citrinum* suggests that the respective enzyme may have a significant role in ecological fitness (Jia et al. 2000). Thus, ACC-deaminase bacteria in association with various bacterial and fungal species help in their colonization and survival in the extreme environmental conditions and also possess additional advantage to the plant by providing extra nutrients after degradation of ACC into α-ketobutyrate and ammonia.

Table 9.1 ACC-deaminase producing rhizobacteria exhibiting tolerance to the host plants under biotic-abiotic stresses and normal environmental conditions

Name of rhizobacteria	Host plant	Stress tolerance	References
Achromobacter piechaudii ARV8	*Lycopersicon esculentum*	Salinity	Mayak et al. (2004)
Acidovorax facilis	*Brassica juncea*	Cadmium	Belimov et al. (2005)
Alcaligenes sp. AF288728 and *Alcaligenes xylosoxidans* AF302096	*Brassica napus*	Polluted soil	Belimov et al. (2001)
Bacillus circulans DUC1	*Brassica campestris*	Normal conditions	Ghosh et al. (2003)
Burkholderia phytofirmans PsjN and *Enterobacter* sp. FD17	*Zea Mays*	Drought	Naveed et al. (2014)
Burkholderia phytofirmans PsJN	*Vitis vinifera*	Low temperature	Ait Bakra et al. (2006)
Kluyvera ascorbata SUD165	*Brassica campestris*	Nickel	Burd et al. (1998)
Enterobacter aerogenes NBRIK24	*Brassica juncea*	Fly-ash soil	Kumar et al. (2008)
Enterobacter cloacae and *Bacillus drentensis*	*Vigna radiata*	Salinity	Mahmood et al. (2016)
Escherichia coli DH5α/p4U2	*Brassica campestris*		Shah et al. (1998)
Klebsiella oxytoca 10MKR7 and *E. sakazakii* 8MR5	*Zea mays*	Striga-infested soils	Babalola et al. (2003)
Klebsiellaoxytoca EF508705	*Gossypium*	Salinity	Yue et al. (2007)
Many rhizobacteria	*Triticum aestivum*	Salinity	Nadeem et al. (2010)
Methylobacterium fujisawaense and *Methylobacterium oryzae* sp.	*Brassica campestris*	Normal conditions	Madhaiyan et al. (2006)
P. fluorescens ACC-5	*Pisum sativum*	Drought	Zahir et al. (2008)
P. fluorescens YsS6	*Lycopersicon esculentum*	Salinity	Ali et al. (2014)
Pseudomonas putida UW4	*Lycopersicon esculentum*	Flooding	Grichko and Glick (2001)
Pseudomonas putida	*Vigna radiate*	Salinity	Mayak et al. (1999)
P. putida UW4	*Lycopersicon esculentum*	Salinity	Yan et al. (2014)
P. putida UW4	*Pinus sabiniana*	Plant pathogen	Nascimento et al. (2013)
Pseudomonas putida (WPTe)	*Papaver somniferum* L.	Downy mildew	Barnawal et al. (2017)
Pseudomonas sp.	*Pisum sativum*	Drought	Arshad et al. (2008)
Rhodococcus sp. and *Variovorax paradoxus*	*Brassica juncea*	Cadmium	Belimov et al. (2005)
Viridibacillus arenosi strain IHB B 7171	*Camellia sinensis*	Acidic soil	Thakur et al. (2017)

(continued)

9.6 Role of ACC-deaminase Producing Rhizobacteria in Reducing Stress Ethylene Level

Tryptophan and many other micro-macro molecules in the root exudates act as precursor for synthesis of indole-3-acetic acid (IAA) by rhizobacteria, where some of the IAA concentration is also acquired by the plant for its growth. This IAA along with endogenous plant IAA can induce plant cell division and proliferation and root cell elongation and stimulate the transcription of the enzyme ACC-synthase responsible for the synthesis of ACC (Poel and Straeten 2014). Furthermore, the ethylene biosynthesis is induced by ACC-synthase enzyme that transforms S-adenosyl methionine (SAM) into ACC and 5′methylthioadenosine (MTA) which is further recycled to L-methionine. In the next step, ACC-oxidase facilitates the conversion of ACC to ethylene and uncharacterized ACC-N-malonyl transferase assists the formation of 1-malonyl-ACC (MACC) simultaneously (Fig. 9.2). A large number of ACC-oxidase isoforms have been known to exhibit differential activity under varying physiological conditions. Some amount of the ACC is consumed by the rhizobacteria and subsequently degraded by ACC-deaminase enzyme into α-ketobutyrate and ammonia in turn reducing the IAA-induced ACC level along with the ethylene concentration in the plant. Plant beneficial rhizobacteria exhibiting ACC-deaminase activity reduces stress ethylene level followed by increment in the plant growth and crop productivity under various biotic-abiotic stresses.

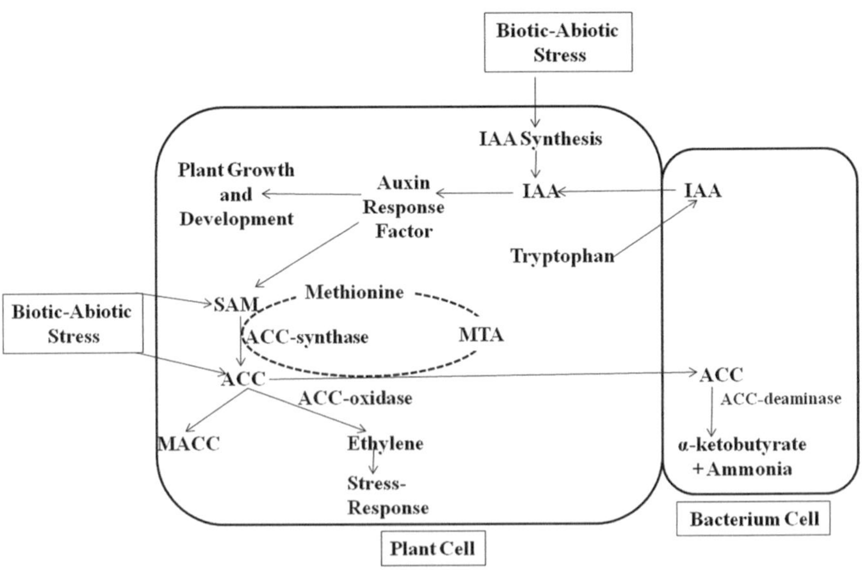

Fig. 9.2 A schematic diagram showing mechanism of ACC-deaminase producing rhizobacteria facilitating plant growth in correlation with IAA. Here, *ACC* 1-aminocyclopropane-1-carboxylate, *IAA* indole-3-acetic acid, *MACC* 1-malonyl-ACC, *MTA* 5-methylthioadenosine, *SAM* S-adenosyl methionine

Thus, the plants associated with these bacteria in their vicinity should possess longer root-shoot lengths and are expected to be more resistant to various ethylene-inducing environmental stresses. The enzyme ACC-deaminase is more often present in rhizobacteria at a low concentration, and its induction is comparatively a sluggish and multifarious process until the plant is subjected to any environmental stress. Irrespective of the moderate induction of a low ACC-oxidase level in the plant, it has relatively higher affinity for ACC in comparison to ACC-deaminase produced by rhizobacteria. Thus, when ACC-deaminase producing rhizobacteria are present in the rhizosphere, levels of ethylene totally depend upon the ACC-oxidase to ACC-deaminase ratio (Glick et al. 1998). At last, this section concludes that there is a direct correlation between IAA and ACC-deaminase activity in reducing the stress ethylene level and ACC-deaminase also facilitates the induction of plant growth and development in association with IAA.

9.7 Biochemical Properties and Mode of Action of ACC-Deaminase Enzyme

ACC-deaminase enzyme (EC: 4.1.99.4) catalyzes the transformation of ACC to α-ketobutyrate, and ammonia was first discovered by Honma and Shimomura in 1978. The ACC-deaminase is pyridoxal phosphate dependent, and cofactor of the enzyme is tightly associated with coenzyme pyridoxal phosphate with approximately one mole of pyridoxal phosphate per trimeric subunit. ACC-deaminase enzyme is among a large group of enzymes and the cofactor pyridoxal phosphate necessitates the enzymatic activity. On the basis of its three-dimensional structure, these enzymes have been grouped into four folding domains including alanine racemase, aspartate aminotransferase, D-amino acid aminotransferase, and tryptophan synthase, where ACC-deaminase enzyme belongs to domain tryptophan synthase (Jansonius 1998). ACC-deaminase enzyme was commonly present in diverse bacterial and fungal genera belonging to relatively soil microorganisms. The enzyme was first purified from a rhizobacterial strain *Pseudomonas* sp. ACP and exhibits the same molecular weight and isoforms as that of partially purified enzyme obtained from strains *Pseudomonas chlororaphis* 6G5 (Klee et al. 1991) and *Pseudomonas putida* GR12-2 (Jacobson et al. 1994). The size of the enzyme was reported 110–112 KDa and105 KDa from strains *Pseudomonas* sp. ACP and *P. putida* GR12-2, respectively. The enzyme exhibit trimeric form with subunit mass of approximately 36,500 daltons. The absorption spectra of enzyme purified from bacteria showed a maximum absorption of 416 and 326nm at pH 6 and 9, respectively. The Km value of ACC-deaminase enzyme extracted at pH 8 from different microbes including bacteria and fungi was recorded in the range of 1.5–17.4mM indicating its low affinity toward ACC substrate. The optimum Km value of the enzyme extracts from different microbes for ACC was reported at pH 8–8.5 and temperature 30°C (Glick et al. 1998).

ACC-deaminase is an inducible enzyme and substrate ACC is required to stimulate its synthesis in adverse conditions. The induction of the enzyme is a complex and sluggish process because ACC requires few hours for induction, and once the process initiates the activity of the enzyme gradually decreases. Moreover, the enzyme induction is directly correlated with the concentration of substrate ACC. The enzyme activity is induced immediately after transferring the bacteria to minimal medium supplemented with ACC as sole nitrogen source from a nutrient-rich media, whereas a low activation in enzyme was achieved in minimal media supplemented with ammonium sulfate as a nitrogen source. Amino acids such as L-alanine, DL-alanine, and D-serine also have role in enzyme induction and expression, while L-isomers of other amino acids including L-alanine, L-serine, L-homoserine, and L-α aminobutyric acid inhibit the enzyme activity.

9.8 Genetics, Expression, and Regulation of ACC-Deaminase Gene

ACC-deaminase enzyme encoding ACC-deaminase structural gene (AcdS) has been reported in a wide range of bacterial and fungal genera as discussed above. It can be easily detected in many genera belonging to gram-positive bacteria, gram-negative bacteria, endophytes, rhizobia, and fungi (Jia et al. 2000; Hontzeas et al. 2005; Rashid et al. 2012; Glick 2014; Singh and Jha 2017), although the expression of putative AcdS gene may vary from one microbe to another. Several universal primer pairs have been reported to analyze the presence of AcdS gene encoding ACC-deaminase activity in bacteria; however, its complete genetic composition and function has been well defined in very few bacterial species. The nucleotide sequences of AcdS gene showed similar nucleotide sequences of other genes, namely, dcyD and yedO, encoding for PLP-dependent enzyme D-cysteine desulfhydrase (Singh et al. 2015). AcdS gene is present in the bacterial chromosomal DNA. Regulation and expression of AcdS subsequently depends upon the presence, absence, and amount of oxygen, ACC substrate concentration, and product accumulation in the cell.

The general regulatory mechanism of AcdS gene in a bacterial cell is described in Fig. 9.3 (Singh et al. 2015). Regulatory elements regulate the expression of ACC-deaminase gene consist AcdR, a regulatory gene positioned on 5′ upstream of AcdS, and promoter regions including Lrp box, AcdB box, FNR box, and CRP box as binding sites of regulatory proteins Lrp, AcdB, fumarate, and nitrate reductase protein and cAMP receptor protein, respectively (Cheng et al. 2008). In the presence of substrate ACC, Lrp emphasizes the formation of an octamer that actively binds to the complex of ACC substrate and AcdB protein that encodes for the glycerophosphoryl diester phosphodiesterase. This complex binds to the promoter region and activates transcription of AcdS gene. During the ACC-deaminase enzyme-catalyzed reaction, α-ketobutyrate is produced as a product of ACC cleavage which further synthesizes leucine responsible for negative regulation of ACC-deaminase gene activity. The increase in leucine concentration initiates the formation of inactive LRP dimer form which promotes the inhibition of AcdS gene transcription. Data obtained from IMG (The Integrated Microbial Genomes) database analysis showed the occurrence of AcdR encoding LRP and its homologous nucleotide sequences in

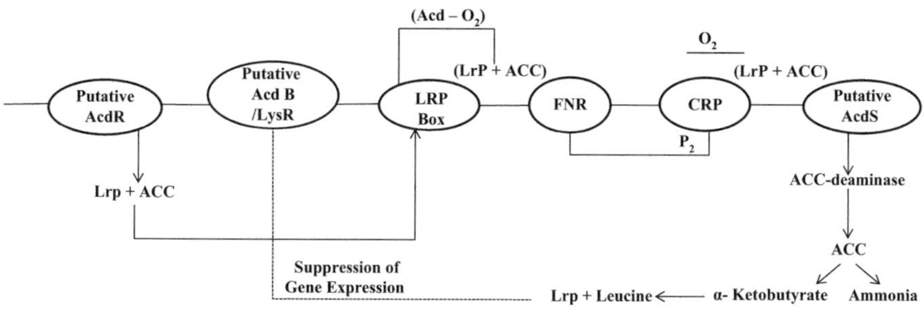

Fig. 9.3 The regulation and expression pathway of AcdS gene. Here, *AcdB* encoding for glycerophosphoryl diester phosphodiesterase, *AcdR* regulatory gene for ACC-deaminase, *AcdS* gene for encoding ACC-deaminase, *CRP* c-AMP receptor protein, *FNR* fumarate nitrate reductase protein, *LRP* leucine responsive protein

the major groups of bacteria. The phylogenetic evaluation of AcdR gene sequences revealed that AcdS and AcdR have been emerged in a similar fashion.

9.9 Transgenic Plants with ACC-Deaminase Activity

ACC-deaminase rhizobacteria exhibiting reduced ethylene level and ultimately significant increment in plant growth and productivity motivated researchers to study transgenic plants conferring gene encoding ACC-deaminase activity exposed to environmental stress conditions in vitro and in vivo as a biotechnological approach. A wide range of transgenic plants expressing bacterial AcdS gene have been employed to reduce the deleterious stress ethylene levels in plants through genetic engineering (Grichko and Glick 2001; Farwell et al. 2007; Zhang et al. 2008). The transgenic canola plants (*Brassica napus*) containing AcdS gene exhibited significant higher growth under salinity stress in comparison to non-transgenic plant (Sergeeva et al. 2006). Two tomato cultivars were transformed with AcdS gene of *Pseudomonas chlororaphis* which led to the extension of fruit ripening duration along with statistical higher reduction in ethylene concentration over the parental line (Reed et al. 1995). Klee and Kishore (1992) also reported 77% reduction in ethylene level and prolonged senescence in transgenic tomato and tobacco plants expressing bacterial AcdS gene. Apse et al. (1999) investigated that water-logged plants inoculated with ACC-deaminase producing rhizobacteria and the transgenic plants expressed in this enzyme accumulate less ACC in the roots than uninoculated and non-transgenic plants. Similarly, plants transformed with AcdS gene are more resistant to the deleterious effects of metals as compared to the non-transformed plants. Furthermore, transgenic tomato plants expressing AcdS gene significantly restricted the damage occurred from *Verticillium* wilt over the control non-transgenic plants (Robison et al. 2001). There is a limited report in literature regarding the evaluation of performance of transgenic plant exhibiting bacterial AcdS gene in vivo. However, future prospects should be focused on application of transgenic plants under field conditions, their crop productivity, and survival under adverse environmental conditions followed by genetic modification of the target gene for the concerned gene identification using modern biotechnological tools.

9.10 Conclusion and Future Prospects

In the present scenario, the continuous use of chemical and synthetic inputs created environmental pollution which have significantly affected the plant growth confronted with persuade of biotic-abiotic stresses in the agriculture. Rhizobacteria possessing ACC-deaminase enzyme activity has a capability to regulate the resistance mechanisms against various environmental stress regimes and simultaneously enhanced tolerance in the associated plants. These rhizobacteria exhibit the potential to promote crop growth and productivity in stressed agriculture as stress busters by lowering down phytohormone ethylene level. ACC-deaminase is an inducible enzyme, and its regulation and expression is subsequently based on the availability of ACC substrate and various environmental and physiological factors. However, lack of information regarding the mechanisms of regulation and expression of AcdS gene is a major constraint for the application of ACC-deaminase producing rhizobacteria. The molecular mechanism based on ACC signaling and its other associated putative signaling modules involved in ACC pathway are also yet to be discovered. The knowledge about regulatory pathways of AcdS gene will provide an insight into the exploitation of ACC-deaminase producing bacteria for improving crop productivity under stress agriculture. Therefore, further investigation is needed to explore AcdS gene in detail and its role in the developing transgenic plants so that they can be widely used under the field conditions.

References

Abeles FB, Morgan PW, Saltveit JME (1992) Ethylene in plant biology, 2nd edn. Academic, New York

Ait Bakra E, Nowak J, Clement C (2006) Enhancement of chilling resistance of inoculated grapevine plantlets with a plant growth promoting rhizobacterium *Burkholderia phytofirmans* strain PsJN. Appl Environ Microbiol 72:7246–7252. https://doi.org/10.1128/AEM.01047-06

Ali S, Charles T, Glick BR (2012) Delay of flower senescence by bacterial endophytes expressing 1-aminocyclo propane-1-carboxylate deaminase. J Appl Microbiol 113:1139–1144. https://doi.org/10.1111/j.1365-2672.2012.05409.x

Ali S, Charles TC, Glick BR (2014) Amelioration of high salinity stress damage by plant growth-promoting bacterial endophytes that contain ACC deaminase. Plant Physiol Biochem 80:160–167. https://doi.org/10.1016/j.plaphy.2014.04.003

Apse MP, Aharon GS, Snedden WA, Blumwald E (1999) Salt tolerance conferred by over expression of a vacuolar Na+/H+ antiport in Arabidopsis. Science 285:1256–1258

Arshad M, Frankenberger WTJ (2002) Ethylene: agricultural sources and applications. Kluwer Academic/Plenum Publishers, Dordrecht

Arshad M, Shaharoona B, Mahmood T (2008) Inoculation with *Pseudomonas* spp. containing ACC-deaminase partially eliminates the effects of drought stress on growth, yield, and ripening of pea (*Pisum sativum* L.) Pedosphere 18:611–620. https://doi.org/10.1016/S1002-0160(08)60055-7

Babalola OO, Osir EO, Sanni AI, Odhiambo GD, Bulimo WD (2003) Amplification of 1-amino-cyclopropane-1-carboxylic (ACC) deaminase from plant growth promoting bacteria in Striga infested soil. Afr J Biotechnol 2:157–160. https://doi.org/10.5897/AJB2003.000-1032

Barnawal D, Pandey SS, Bharti N, Pandey A, Ray T, Singh S, Chanotiya CS, Kalra A (2017) ACC deaminase-containing plant growth-promoting rhizobacteria protect *Papaver somniferum* from downy mildew. J Appl Microbiol 122:1286–1298. https://doi.org/10.1111/jam.13417

Belimov AA, Safronova VI, Sergeyeva TA, Egorova TN, Matveyeva VA, Tsyganov VE, Borisov AY, Tikhonovich IA, Kluge C, Preisfeld A, Dietz KJ, Stepanok VV (2001) Characterization of plant growth promoting rhizobacteria isolated from polluted soils and containing 1-aminocycl opropane-1-carboxylate deaminase. Can J Microbiol 47:642–652

Belimov AA, Hontzeas N, Safronova VI, Demchinskaya SV, Piluzza G, Bullitta S, Glick BR (2005) Cadmium tolerant plant growth promoting bacteria associated with roots of Indian mustard (*Brassica juncea* L Czern.) Soil Biol Biochem 37:241–250. https://doi.org/10.1016/j.soilbio.2004.07.033

Burd GI, Dixon DG, Glick BR (1998) A plant growth-promoting bacterium that decreases nickel toxicity in seedlings. Appl Environ Microbiol 64:3663–3668

Cheng Z, Duncker BP, McConkey BJ, Glick BR (2008) Transcriptional regulation of ACC deaminase gene expression in *Pseudomonas putida* UW4. Can J Microbiol 54:128–136. https://doi.org/10.1139/W07-128

Else MA, Jackson MB (1998) Transport of 1-aminocyclopropane-1-carboxylic acid (ACC) in the transpiration stream of tomato (*Lycopersicon esculentum*) in relation to foliar ethylene production and petiole epinasty. Aust J Plant Physiol 25:453–458. https://doi.org/10.1071/PP97105

Farwell AJ, Vesely S, Nero V, Rodriguez H, McCormack K, Shah S, Dixon DG, Glick BR (2007) Tolerance of transgenic canola plants (*Brassica napus*) amended with plant growth-promoting bacteria to flooding stress at a metal contaminated field site. Environ Pollut 147:540–545. https://doi.org/10.1016/j.envpol.2006.10.014

Frankenberger WT, Arshad M (1995) Phytohormones in soils: microbial production and function. Marcel Dekker Inc., New York

Gamalero E, Glick BR (2012) Plant growth-promoting bacteria and metal phytoremediation. In: Anjum NA, Pereira ME, Ahmad I, Duarte AC, Umar S, Khan NA (eds) Phytotechnologies: remediation of environmental contaminants. CRC Press, Boca Raton, pp 361–376

Ghosh S, Penterman JN, Little RD, Chavez R, Glick BR (2003) Three newly isolated plant growth-promoting bacilli facilitate the seedling growth of canola, *Brassica campestris*. Plant Physiol Biochem 41:277–281. https://doi.org/10.1016/S0981-9428(03)00019-6

Glick BR (2005) Modulation of plant ethylene levels by the bacterial enzyme ACC deaminase. FEMS Microbiol Lett 251:1–7. https://doi.org/10.1016/j.femsle.2005.07.030

Glick BR (2007) Promotion of plant growth by bacterial ACC deaminase. Crit Rev Plant Sci 26:227–242. https://doi.org/10.1080/07352680701572966

Glick BR (2014) Bacteria with ACC deaminase can promote plant growth and help to feed the world. Microbiol Res 169:30–39. https://doi.org/10.1016/j.micres.2013.09.009

Glick BR, Penrose DM, Li J (1998) A model for lowering of plant ethylene concentrations by plant growth promoting bacteria. J Theoret Biol 190:62–68. https://doi.org/10.1006/jtbi.1997.0532

Grichko VP, Glick BR (2001) Amelioration of flooding stress by ACC deaminase-containing plant growth-promoting bacteria. Plant Physiol Biochem 39:11–17. https://doi.org/10.1016/S0981-9428(00)01212-2

Honma M, Shimomura T (1978) Metabolism of 1-aminocyclopropane1-carboxylic acid. Agric Biol Chem 42:1825–1831. https://doi.org/10.1271/bbb1961.42.1825

Hontzeas N, Richardson AO, Belimov AA, Safranova VI, Abu-Omar MM, Glick BR (2005) Evidence for horizontal gene transfer (HGT) of ACC deaminase genes. Appl Environ Microbiol 71:7556–7558. https://doi.org/10.1128/AEM.71.11.7556-7558.2005

Husen E, Wahyudi AT, Suwanto A, Giyanto (2011) Growth enhancement and disease reduction of soybean by 1-aminocyclopropane-1-carboxylate deaminase-producing Pseudomonas. Am J Appl Sci 8:1073–1080. https://doi.org/10.3844/ajassp.2011.1073.1080

Jacobson CB, Pasternak JJ, Glick BR (1994) Partial purification and characterization of 1-aminocyclopropane1-carboxylate deaminase from the plant growth promoting rhizobacterium *Pseudomonas putida* GR12-2. Can J Microbiol 40:1019–1025. https://doi.org/10.1139/m94-162

Jansonius NJ (1998) Structure, evolution and action of vitamin B6-dependent enzymes. Curr Opin Struct Biol 8:759–769

Jia YJ, Ito H, Matsui H, Honma M (2000) 1-Aminocyclopropane-1carboxylate (ACC) deaminase induced by ACC synthesized and accumulated in *Penicillium citrinum* intracellular spaces. Biosci Biotechnol Biochem 64:299–305. https://doi.org/10.1271/bbb.64.299

Klee HJ, Kishore GM (1992) Control of fruit ripening and senescence in plants, U.S. Patent Number 5702933. Monsanto, St. Louis

Klee HJ, Hayford MB, Kretzmer KA, Barry GF, Kishore GM (1991) Control of ethylene synthesis by expression of a bacterial enzyme in transgenic tomato plants. Plant Cell 3:1187–1193. https://doi.org/10.1105/tpc.3.11.1187

Kumar KV, Singh N, Behl HM, Srivastava S (2008) Influence of plant growth promoting bacteria and its mutant on heavy metal toxicity in *Brassica juncea* grown in fly ash amended soil. Chemosphere 72:678–683. https://doi.org/10.1016/j.chemosphere.2008.03.025

Madhaiyan M, Poonguzhali S, Ryu J, Sa T (2006) Regulation of ethylene levels in canola (*Brassica campestris*) by 1-aminocyclopropane 1-carboxylate deaminase-containing *Methylobacterium fujisawaense*. Planta 224:268–278. https://doi.org/10.1007/s00425-005-0211-y

Mahmood S, Daur I, Al-Solaimani SG, Ahmad S, Madkour MH, Yasir M, Hirt H, Ali S, Ali Z (2016) Plant growth promoting rhizobacteria and silicon synergistically enhance salinity tolerance of mung bean. Front Plant Sci 7:876. https://doi.org/10.3389/fpls.2016.00876

Mayak S, Tirosh T, Glick BR (1999) Effect of wild-type and mutant plant growth-promoting rhizobacteria on the rooting of mung bean cuttings. J Plant Growth Regul 18:49–53. https://doi.org/10.1007/PL00007047

Mayak S, Tirosh T, Glick BR (2004) Plant growth-promoting bacteria confer resistance in tomato plants to salt stress. Plant Physiol Biochem 42:565–572. https://doi.org/10.1016/j.plaphy.2004.05.009

McKeon TA, Fernandez-Maculet JC, Yang SF (1995) Biosynthesis and metabolism of ethylene. In: Davies PJ (ed) Plant hormones physiology, biochemistry and molecular biology. Kluwer Academic Publishers, Dordrecht, pp 118–139

Nadeem SM, Zahir ZA, Naveed M, Asghar HN, Arshad M (2010) Rhizobacteria capable of producing ACC-deaminase may mitigate salt stress in wheat. Soil Sci Soc Am J 74:533–542. https://doi.org/10.2136/sssaj2008.0240

Nascimento FX, Vicente CSL, Barbosa P, Espada M, Glick BR, Oliveira S (2013) The use of the ACC-deaminase producing bacterium *Pseudomonas putida* UW4 as a biocontrol agent for pine wilt disease. Biocontrol 58:427–433. https://doi.org/10.1007/s10526-012-9500-0

Nascimento FX, Rossi MJ, Soares CRFS, McConkey BJ, Glick BR (2014) New insights into 1-aminocyclopropane-1-carboxylate (ACC) deaminase phylogeny, evolution and ecological significance. PLoS ONE 9:e99168. https://doi.org/10.1371/journal.pone.0099168

Naveed M, Hussain MB, Zahir ZA, Mitter B, Sessitsch A (2014) Drought stress amelioration in wheat through inoculation with *Burkholderia phytofirmans* strain PsJN. Plant Growth Regul 73:121–131. https://doi.org/10.1007/s10725-013-9874-8

Rashid S, Charles TC, Glick BR (2012) Isolation and characterization of new plant growth-promoting bacterial endophytes. Agric Ecosyst Environ 61:217–224. https://doi.org/10.1016/j.apsoil.2011.09.011

Reed AJ, Magin KM, Anderson JS, Austin GD, Rangwala T, Linde DC, Love JN, Rogers SG, Fuchs RL (1995) Delayed ripening in tomato plants expressing the enzyme 1aminocyclopropane-1-carboxylic acid deaminase: molecular characterization, enzyme expression, and fruit ripening traits. J Agric Food Chem 43:1954–1962. https://doi.org/10.1021/jf00055a036

Robison MM, Griffith M, Pauls KP, Glick BR (2001) Dual role of ethylene in susceptibility of tomato to *Verticillium* wilt. J Phytopathol 149:385–388. https://doi.org/10.1046/j.1439-0434.2001.00639.x

Rodriguez H, Vessely S, Shah S, Glick BR (2008) Effect of a Nickel-tolerant ACC deaminase-producing *Pseudomonas* strain on growth of non transformed and transgenic canola plants. Curr Microbiol 57:170–174. https://doi.org/10.1007/s00284-008-9181-1

Sergeeva E, Shah S, Glick BR (2006) Growth of transgenic canola (*Brassica napus* cv. westar) expressing a bacterial 1-aminocyclopropane-1-carboxylate (ACC) deaminase gene on high concentrations of salt. World J Microbiol Biotech 22:277–282. https://doi.org/10.1007/s11274-005-9032-1

Shah S, Li J, Moffatt BA, Glick BR (1998) Isolation and characterization of ACC deaminase genes from two different plant growth-promoting rhizobacteria. Can J Microbiol 44:833–843

Shaharoona B, Jamro GM, Zahir ZA, Arshad M, Memon KS (2007) Effectiveness of various *Pseudomonas* spp., and *Burkholderia caryophylli* containing ACC-deaminase for improving growth and yield of wheat (*Triticum aestivum* L.) J Microbiol Biotechnol 17:1300–1307

Singh RP, Jha PN (2016) A halotolerant bacterium *Bacillus licheniformis* HSW-16 augments induced systemic tolerance to salt stress in wheat plant (*Triticum aestivum*). Front Plant Sci 7. https://doi.org/10.3389/fpls.2016.01890

Singh RP, Jha PN (2017) The draft genome sequence of the plant growth promoting rhizospheric bacterium *Enterobacter cloacae* SBP-8. Genomics Data 12:81–83. https://doi.org/10.1016/j.gdata.2017.03.006

Singh RP, Shelke GM, Kumar A, Jha PN (2015) Biochemistry and genetics of ACC deaminase: a weapon to "stress ethylene" produced in plants. Front Microbiol 6:937. https://doi.org/10.3389/fmicb.2015.00937

Soni R, Kumar A, Kanwar SS, Pabbi S (2017) Efficacy of liquid formulation of versatile rhizobacteria isolated from soils of the North-Western Himalayas on *Solanum lycopersicum*. Indian J Tradit Know 16:660–668

Thakur R, Sharma KC, Gulati A, Sud RK, Gulati A (2017) Stress-tolerant *Viridibacillus arenosi* strain IHB B 7171 from tea rhizosphere as a potential broad-spectrum microbial inoculants. Indian J Microbiol 57:195–200. https://doi.org/10.1007/s12088-017-0642-8

Van de Poel B, Van der Straeten D (2014) 1-aminocyclopropane-1-carboxylic (ACC) in plants: more than just the precursor of ethylene. Front Plant Sci 5:640. https://doi.org/10.3389/fpls.2014.00640

VanLoon LC (2007) Plant responses to plant growth-promoting rhizobacteria. Eur J Plant Pathol 119:243–254

Yan J, Smith MD, Glick BR, Liang Y (2014) Effects of ACC deaminase containing rhizobacteria on plant growth and expression of Toc GTPases in tomato (*Solanum lycopersicum*) under salt stress. Botany 92:775–781. https://doi.org/10.1139/cjb-2014-0038

Yue H, Mo W, Li C, Zheng Y, Li H (2007) The salt stress relief and growth promotion effect of RS-5 on cotton. Plant Soil 297:139–145. https://doi.org/10.1007/s10658-007-9165-1

Yuhashi KI, Ichikawa N, Ezura H, Akao S, Minakawa Y, Nukui N, Yasuta T, Minamisawa K (2000) Rhizobitoxine production by *Bradyrhizobium elkanii* enhances nodulation and competitiveness on *Macroptilium atropurpureum*. Appl Environ Microbiol 66:2658–2663

Zahir ZA, Munir A, Asghar HN, Shaharoona B, Arshad M (2008) Effectiveness of rhizobacteria containing ACC deaminase for growth promotion of peas (*Pisum sativum*) under drought conditions. J Microbiol Biotechnol 18:958–963

Zhang Y, Zhao L, Wang Y, Yang B, Chen S (2008) Enhancement of heavy metal accumulation by tissue specific coexpression of iaaM and ACC deaminase genes in plants. Chemosphere 72:564–571. https://doi.org/10.1016/j.chemosphere.2008.03.043

Algae-Based Biofertilizers: A Biorefinery Approach

10

Durga Madhab Mahapatra, H. N. Chanakya, N. V. Joshi, T. V. Ramachandra, and G. S. Murthy

Abstract

Current perturbations in the agrarian economies and the agro-environment have sparked concerns regarding the future food security and a dire need for sustainable agricultural practices without jeopardizing the environmental assets. One such major requirement for an agrarian society is a resilient nutrient source for agriculture. In this regard algal cells that are cosmopolitan in nature with unparalleled characteristics of high biomass productivity, high photosynthetic efficiency and ability to grow in barren and non-arable lands are attractive. They grow in a wide range of water systems especially the ones that are highly enriched

D. M. Mahapatra (✉)
Department of Biological and Ecological Engineering [BEE], College of Engineering, Oregon State University, Corvallis, OR 97331, USA

Centre for Sustainable Technologies (formerly known as ASTRA) [CST], Centre for infrastructure Sustainable Transportation and Urban Planning [CiSTUP], Indian Institute of Science, Bangalore, India

Energy & Wetlands Research Group, Centre for Ecological Sciences [CES], Indian Institute of Science, Bangalore, India
e-mail: mahapatd@oregonstate.edu; durgam@iisc.ac.in; durgamadhab.mahapatra@gmail.com

H. N. Chanakya
Centre for Sustainable Technologies (formerly known as ASTRA) [CST], Centre for infrastructure Sustainable Transportation and Urban Planning [CiSTUP], Indian Institute of Science, Bangalore, India

N. V. Joshi
Energy & Wetlands Research Group, Centre for Ecological Sciences [CES], Indian Institute of Science, Bangalore, India

Centre for infrastructure Sustainable Transportation and Urban Planning [CiSTUP], Indian Institute of Science, Bangalore, India

D. G. Panpatte et al. (eds.), *Microorganisms for Green Revolution*, Microorganisms for Sustainability 7,
https://doi.org/10.1007/978-981-10-7146-1_10

with salts (saline) and nutrients (eutrophied) and in numerous contaminated and polluted systems as urban wastewaters. Apart from these primary benefits, the algal route also offers important byproducts that can be further used as value-added products in industries and as C neutral commodities that help to evade climate change by negating greenhouse gas emissions. The most important aspects are its efficiency as a biofertilizer that dynamically improves the soil health and its physicochemical behaviour. The wastewater-grown algal microflora is exceptional in imparting the appropriate mineral nutrient mix with essential vitamins and plant growth promoters together with increasing the water holding capacity of the soil. Algal communities from wastewaters with optimal NPK ratio and secondary nutrients are therefore model biofertilizers. They can be substitutes of the conventional chemical fertilizers due to its ubiquity, enhanced metabolic flux, short generation time and inherent capabilities to transform inert N into plant-available N (N fixation). All the above-mentioned characteristics, techno-economic feasibility and environmental benefits make algae the most beneficial and demanding bioresource of the twenty-first century.

Keywords
Biofertilizer · Algae · *Spirulina* · Wastewater · Nitrogen · Nutrients

10.1 Introduction

Agricultural food security is essential for development and globalization. Existing global population is 7.5 billion and is expected to reach ~10 billion by 2050. Ensuring adequate food supply globally would require an annual production of cereal grains by 50% that indicates an increase from 2 to 3 billion tons/annum. This mammoth goal presses serious concerns on improving present-day strategies and developing technologies to increase the agricultural productivity and thus exerts a great burden on the present agricultural sector to accomplish the towering food demands. This giant upsurge in anticipated food production can be only possible by

T. V. Ramachandra
Centre for Sustainable Technologies (formerly known as ASTRA) [CST], Centre for infrastructure Sustainable Transportation and Urban Planning [CiSTUP], Indian Institute of Science, Bangalore, India

Energy & Wetlands Research Group, Centre for Ecological Sciences [CES], Indian Institute of Science, Bangalore, India

Centre for infrastructure Sustainable Transportation and Urban Planning [CiSTUP], Indian Institute of Science, Bangalore, India

G. S. Murthy
Department of Biological and Ecological Engineering [BEE], College of Engineering, Oregon State University, Corvallis, OR 97331, USA

either (a) increasing the area under cultivation or (b) increasing the efficiency/productivity of the existing cultivable lands. The possibility of procuring more land and reclamations into agricultural lands is presently rampant (Ramachandra et al. 2009, 2013) and has higher environmental risks and externalities (Ramachandra et al. 2015). Higher land use and land cover changes have altered the physical, chemical and biological integrity of the soil (pedosphere) and have rendered them infertile. Thus, there are low possibilities of the first option although presently there are a lot of research on salt-tolerant species possibly grown in saline soils, but such research are still within laboratories and are presently under progress that needs more field trails and analysis. However, the second option though challenging is in fact practical, and various research outcomes show revealing results in improvement of soil fertility and agricultural productivity using environmental-friendly irrigation and agricultural processes that ensures food security sustainably.

Conventional cropping practices are presently overdependent on man-made fertilizers, weedicides and insecticides, tilling practices and intensive irrigation, which have surged a reasonably high quantum of food grains across the last five decades. This rapid surge in production of food grains termed as green revolution, although has created surplus food grains in developing nations, but has intensely costed the quality of the environmental assets and raised concerns for human health and hygiene. Very apparent among them are the air, water and soil pollution, loss in soil fertility, alterations in soil property, excessive use of natural resources, increased nutrient enrichments and accumulation with consequent algal blooms, and moreover all this has added to increased cost in agricultural production. Thus, it becomes a challenge to identify resilient solutions to address these problems in figuring out sustainable nutrition sources, efficient, less-energy and resource-intensive agricultural practices to cope up with the towering food and resource demands in the future. Integrating the traditional agricultural practices like crop mulching, transplantation and crop rotation with ubiquitously found algal biomass can serve as a sustainable solution for higher agricultural productivity with no environmental externalities. Traditional agricultural practices were driven mostly by naturally fixed nutrient sources as green plant-based fertilizers and organic manures and using local water resources as man-made lakes and ponds with efficient bio-based yard manures that taps nutrients from domestic wastewaters. Due to globalization and higher concentrations of population in urban conglomerates, the per capita land footprint has drastically reduced that lacks places like yard and gardens where these nutrient-laden water could be used. Similarly, in periurban and rural setups, the soil qualities have been drastically deteriorated with overuse of chemical fertilizers and other harmful chemical inputs like pesticides and weedicides. Various routes of production of nitrogen (N)- and phosphorus (P)-based chemical fertilizers are depicted in Fig. 10.1. This has not only affected the food production but has also compromised with the environmental quality of the agri-sectors. Use of appropriate agricultural tools and approaches is the only way to revive and rejuvenate rapidly lost agri-landscapes and nutrient-rich soils (Mason 2003). The principles of sustainability and ancient agricultural philosophy hinge around eco-friendly methods and cost-effective cultivation strategies with the aid of native soil microflora. This further strengthens the views on

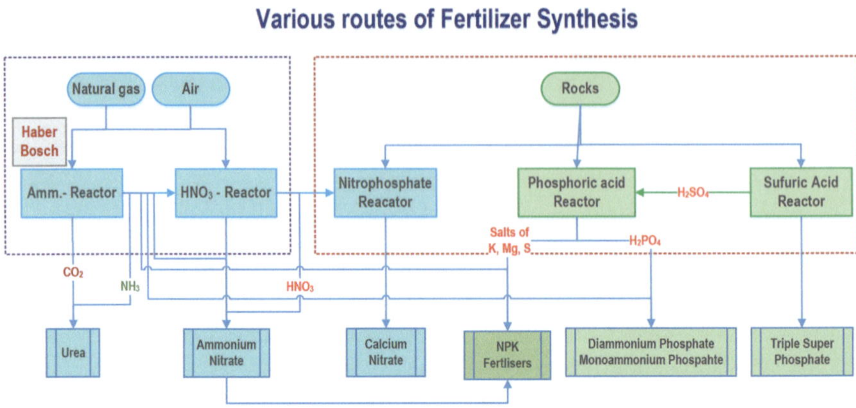

Fig. 10.1 Schematic representation of commercial synthesis of N- and P-based fertilizers

using natural resources optimally and usage of ecological bioprocesses that makes sure that there are no waste components left following a biorefinery approach. This essentially closes the nutrient loops within the system and at the same time increases the crop yield and agricultural productivity.

Soil microflora have been known to increase the soil fertility and help in higher biomass productivities (Koller et al. 2012). However, the algal microflora has been also realized as a true bio-based fertilizer for eco-friendly and pollution-free agricultural practices (Yamamguchi 1997; Benson et al. 2014; Mahapatra et al. 2015). Considering the fact that algal biomass are key to crucial ecosystem processes, they can play a pivotal role in improving crop yield and at the same time mitigation greenhouse gas (GHG) emissions (Singh 2011; Mahapatra et al. 2015). Studies conducted by Singh have reported essential roles of blue-green algae (BGA) in ecological restoration of degraded lands (Singh 2014). These algae although photosynthetic are adaptive to extremes of conditions and are known to survive at critical light intensities with minimal nutrients as C and N levels with a bare minimum water requirement (Castenholz 2001). Being efficient N-fixating agents, they are highly independent of perturbations in terrestrial N pools that produce key essential amino acids, growth promoters and hormones that are essential for a good growth in cropping systems (Pandey et al. 2004). Algal assemblages can derive most of their necessary nutrients from wastewaters, with capacities of breaking down highly complex, toxic and xenobiotic compounds (Cohen 2006).

Optimal plant growth and agricultural productivity are directly dependent on nutrient availability in balanced quantities (Chen 2006). In the developing world, the agricultural yield is dependent on nature and the type of soil, as most of the farmers belonging to the poorer section of the society cannot afford synthetic agricultural fertilizer unless it is subsidized. The lost fertility of the agricultural soils can be restored by usage of improved crop varieties and efficient nutrient management practices. This essentially suggests use of the abundantly present natural bioresources for increasing the nutrient status of the crops and allowing more potential biological nutrient transfer methods. In such cases, biofertilizers play a major role and are essential components of the controlled mineralization and fertilization process. Being cost-effective, eco-friendly and a renewable resource, this can altogether replace or augment the presently used cost- and energy-intensive chemical fertilizers. These are primarily live and/or dead cells with beneficial organisms applied to plant and soil systems. They rapidly colonize the rhizosphere and thus enhance plant growth and development, thereby converting unavailable mineral forms into essential nutrients via processes as N fixation, mineralization and solubilization of rock phosphates (Rokhzadi et al. 2008). These organisms eventually become a stable part of the root zone section that accelerates the agricultural yields and also safeguards the agri-systems from myriads of pests and pathogens (El-yazeid et al. 2007). There have been a number of studies and reviews on these beneficial aspects of the microorganism as biofertilizer (Wani et al. 1995; Yamaguchi 1997; Rai et al. 2000; Koler et al. 2012; Geisseler and Scow 2014).

10.2 Conventional Biofertilizers to Algae-Based Fertilizers: Definition and Scope

Biofertilizers can be defined as a composite of live or latent organisms with characteristic ability to fix atmospheric nitrogen, solubilize mineral phosphates or degrade organic matter that targets soil and composting regions to increase the abundance of a beneficial microflora that accelerates key nutrient mineralization and uptake processes to augment the quantum of nutrients in the medium. Biofertilizer terminology comprises of every organic resource for cultivation that is concentrated in available form, absorbed by the plants through microbes and plant association (Singh 2015). Traditionally composting and using the compostable materials (food materials, cow dung, agri-residues) mark the initiation of use of organic matter as plant nutrients that has been well known to the ancestral agricultural communities and has been passed on as heritage to the modern community. The realization of the true potential of these applications is the modern-day biofertilizers that accelerate the decomposition kinetics of the organic residues and agricultural byproducts (Singh 2014). In Southeast Asia that is the primary agrarian belt, industrial production of these beneficial microbes has been in progress. There are a number of reports on efficacy of algal community mostly cyanobacterial biofertilizers on crop productivity. One of the very widely used biofertilizers that has been often used for N fixation and assimilation into plants is the *Rhizobium* in legumes. On the other hand, the applications of mycorrhiza and *Azospirillum* are also studied to a great deal. It has been reported that the mycorrhiza inoculums have been rapidly adapted to the tropical Southeast Asian conditions as in India and Malaysia and are being actively used in agricultural industry. This encourages commercial production of biofertilizers that enhances soil fertility and ameliorates toxic effects of soil, checks growth of soil pests and restricts diseases with improved water and mineral absorption. Moreover, these microbial assemblages can be grown easily in substrates as mine sands and agricultural wastes, making them economically viable with a lower environmental footprint. However, biofertilizers are reported to be expensive than chemical fertilizers as it requires a great deal of skill for bioconversion of waste substrates and production of high-density inoculum. Furthermore, the tests on the land applications have shown that the activity of biofertilizers in terms of fertilization efficiency is slow compared to chemical fertilizers. During its production, great attention and care are taken for homogenizing, mixing and storage with the dried inoculum forms as powders to ensure viability of the inoculums for an extended use. The conventional biofertilizers as mentioned earlier deal with living microbes; therefore the growth activity and nutrient transformations also greatly depend on the surrounding environment and thus can have high inconsistencies in their activities. On the other hand, the presently reported algal biofertilizers are not based on the active algal biomass but are high-nutrient-enriched stable biomass mix that have active growth stimulants with nourishing vitamins, microelements and hormones that have been showing promising results. The algal biomass can itself act as both nutrient and carrier materials for biofertilizers, as it has a better shelf life (post drying) that is less thermo-sensitive and is robust for storage and transportation, making it a better choice as a biofertilizer.

Algal growth enhances the growth of rhizosphere bacteria through beneficial exudates and exopolysaccharides (Kapustka and DuBois 1987) that become the carbon (C) source for the soil bacteria. When the dried forms of algae are applied to the crop soils, after mineralization, the cell extracts from BGA and diatoms impart stimulatory effects on the growth of bacterium that produces siderophore (Amin et al. 2009). The mechanism involves the photolysis of Fe siderophore that triggers Fe assimilation in moist conditions eventually promoting the association of algae extracts and bacteria, where the organics released by algal members are used as food for bacterial members. This helps in bioavailability and assimilation of Fe through Fe chelators like siderophores under low Fe and N conditions that happens mostly in aquatic environments or under waterlogged conditions. Besides this BGA plays a major role in transformation of soil micronutrients (Fe and Mn) together with production of organic acids that have key chelating abilities that enhances the bioavailability of micronutrients for the growing plants. Algal species as *Anabaena flos-aquae*, *A. cylindrica*, *Synechococcus leopoliensis* and *Anacystis nidulans* have shown to excrete strong copper-complexing agents ($^{c}K > 10^{8}$), while eukaryotic algae have been only found to release relatively weak copper-complexing agents ($^{c}K < 10^{7.5}$) (McKnight and Morel 1980). Although these findings were mostly reported from studies in aquatic systems, the results provide a huge scope for further studies in an agricultural setup to understand micronutrient dynamics involving siderophore productions and connections between the soil algal and the plant species in the biofertilizer activity for soil nutrient translocation.

10.2.1 Production of Biofertilizers: The Algal Industry

The commercial production of biofertilizers considers several aspects as the microorganism's growth and nutrient profile, the assembly of the microbes, suitable conditions of maintaining an active biomass and formulation of the inoculum.

Box 10.1: Steps in Commercial Production of Biofertilizers

1. Exploration and identification of active microbial consortia
2. Isolation of select active organisms
3. Rapid screening and selection of beneficial target microbes
4. Efficacy of microbial target population
5. Selection of carrier material and the methods
6. Selection for the method of propagation
7. Investigations on nutrient studies and design of porotype and testing
8. Field trials and large-scale testing

For commercialization, the key emphasis is on inoculate formulation, nature of application and packaging and storage of products (Box 10.1).

The very first step in realizing the objective of the project is to identify what kind of organisms is required for the study. It is therefore crucial to check whether it is an active mineralizing bacteria or oxidative bacterial communities for conversion of organic acids or phosphorus-binding/phosphorus-solubilizing bacteria or N-fixing bacteria or a combination of them. According to the intended function of the microbial community, the microbial assemblage is explored and isolated, mostly from soil matrices, plant roots, marshes, water bodies, etc. This is followed by the routine laboratory-based culturing techniques where they are grown as plate cultures to flask-based cultures, and finally the best assemblages are selected. The bulk of the biofertilizer comprises of the carrier material and thus requires a rigorous selection of materials suitable for the purpose. The carrier material selection is again based on the final form of the biofertilizers, e.g. powder or liquid. In case of production of powdered biofertilizers, peat and tapioca flour have been used extensively. It is necessary to formulate the simplest and easiest ways to propagate the organism that are based on the experimental finding on environmental conditions and the growth suitability in various growth setups. The identification of the optimal propagation methods aids in finding out the best conditions for the organisms to grow. Prototype development and testing are carried out after testing the suitability of growth medium and the carrier material. During the final phase, the biofertilizers developed are tested on large-scale setup to understand the efficacy and limitations to various field conditions. After the development of the biofertilizers, it is necessary to ensure prolonged storage and avoid degradation and possible contamination by other types of bacteria for the reason which the carrier materials are sterilized with either irradiation with UV/gamma rays or autoclaving the materials. For these reasons, carrier material selection becomes the most crucial step for the design and development of the biofertilizers. Virtuous carrier material should be inexpensive and must be available in adequate quantities. At the same time, it should be nontoxic to the plants and must not have any side effects in a long run. As characteristics of appropriate carrier materials, the ideal material must have great moisture holding capacity and nice adhesion abilities to bind to seeds. Along with this, the carrier materials must have a good pH buffering ability, easy processing and sterilization with simple comminution.

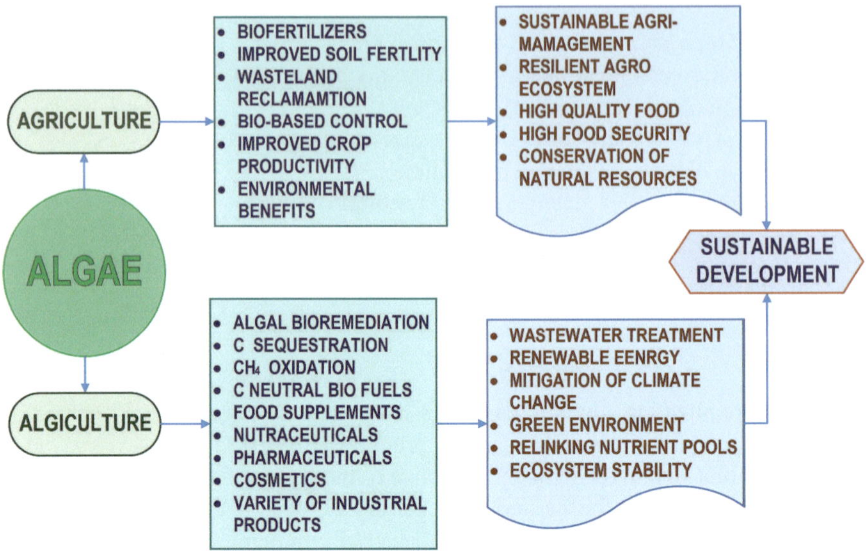

Fig. 10.2 Advantages and characteristic features of algae used as biofertilizers

Wastewater is being increasingly used to grow the algal assemblages rapidly that have immense potentials to immobilize nutrients and make the nutrients bioavailable later for a rapid growth of crops through controlled release of nutrients. The algal species can be grown together with the plants that helps in N fixation and balancing nutrient imbalances or can be used in the dried form to augment the fertility of the land with essential nutrients with an optimal ratio of NPK. A multi-tier algal cropping (algiculture) is envisioned that comprises of various algal communities as a function of light penetration and depth in the agricultural crops especially rice cultivation, which has standing waters during its growth and development (Mahapatra 2015). The various advantages of algal use in (a) agriculture and as (b) algiculture for sustainable and green future have been depicted in a schematic diagram in Fig. 10.2.

10.2.2 Microbial Assemblages as Biofertilizers

The most important class of biofertilizers includes (a) nitrogen fixers, (b) phosphorus solubilizers and (c) potassium solubilizers with a combination of various other organisms for symbiotic association or support as fungus and moulds. The bacterial assemblages or isolates used as biofertilizers have their association with plant roots. The most popular N-fixing bacteria *Rhizobium* has a symbiotic association with the root nodules of leguminous plants. *Rhizobacteria* on the other hand are found on the root surfaces and rhizosphere. As phosphorus is a limiting nutrient in any kind of

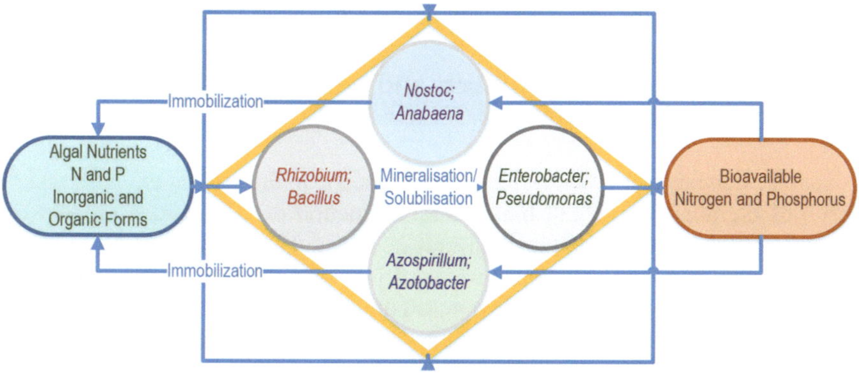

Fig. 10.3 Mineralization pathways of nutrients in algal-bacterial association

Table 10.1 Functions of major microbial class of biofertilizers with various species

Sl. no.	Microbial biofertilization	Microbial species
1	*Free-living N-fixing bacteria*	Obligate anaerobes: *Clostridium pasteurianum*
		Obligate aerobes: *Azotobacter*
		Facultative anaerobes
		Photosynthetic bacteria (*Rhodobacter*)
		Cyanobacteria
		Methanogens: *Methanococcus*
2	*P solubilizer*	*Bacillus megaterium*, *Bacillus circulans*, *Bacillus subtilis* and *Pseudomonas striata*
3	*K solubilizer*	*Bacillus mucilaginous*

ecosystem. The P-mineralizing bacteria and fungi mineralize the insoluble and unavailable phosphorus and make it bioavailable to the plants (Gupta 2004; Mahapatra 2015). This happens due to solubilizing the bound P in soil systems and organic matter by organic acid productions in bacteria that helps in lowering the soil pH and triggers release of the bound P forms. Many of the biofertilizers are targeted for specific type of the crop plants as *Rhizobium* for legumes, cyanobacteria for grains (rice and wheat) and *Azolla* for pulses and cereals. Many other categories of biofertilizers as vesicular arbuscular mycorrhiza (VAM), phosphorus-solubilizing bacteria (PSB), *Azospirillum* and *Azotobacter* can be considered for a wide range of crops and cultivars. The mineralization and solubilization routes across the algal-bacterial association are depicted in Fig. 10.3. These biofertilizers are associated with majority of the crops and thus enhance plant growth to suitable nutrient capture. In many cases, these microbes in association with the plant roots and soil systems ensure inhibition of pests and crop-damaging bacterial communities. Some of the extensively used biofertilizers are illustrated in Table 10.1.

10.3 Wastewater Algae as a Suitable Source of Biofertilizers

10.3.1 Algal Growth, Bioproducts and Nutrient Analysis from Urban Wastewaters: Case Study

Algal research has gained special impetus and scope in the present era. The exploitation of the algal whole cell biomass for extraction of value-added products in tandem with wastewater treatment ensures its techno-economic feasibility with no environmental externalities. A major area of interest in algal biotechnology is the economic benefits associated with the mass culture and biomass use of algal species thriving in tropical wastewaters. Most of the wastewater generated in the developing world remains unattended and enriches the surface and the ground waters with higher loads of nutrients rendering it unsuitable for any use (Mahapatra et al. 2011a, b, c; Mahapatra et al. 2017). The algal species can be used as robust treatment systems to recover nutrients from their wastewater and municipal landfill leachate (Naveen et al. 2016) by harvesting whole cell algal biomass (Mahapatra and Ramachandra 2013). The algal biomass have shown potential to remove nutrients as C, N and P as unialgal cultures or as microbial consortia with the production of valorizable algal biomass that shows significant quantity of lipids (Mahapatra and Ramachandra 2013; Mahapatra et al. 2013a, 2014; Ramachandra et al. 2015). Large-scale algae-based treatment systems have been also studied and have shown a higher techno-economic feasibility (Chanakya et al. 2012,2013) for algal single-cell proteins (SCP) (Mahapatra et al. 2013b, 2015) with a lower environmental impact (Ramachandra and Mahapatra 2015; Ramachandra et al. 2015). The algal modules designed at the Indian Institute of Science (IISc), Bangalore, have been successfully employed for nutrient (C, N and P) capture and recovery at the Jakkur Lake in Bangalore City and show promising augmentation to the conventional treatment, providing purified water that can be recycled and reused (Ramachandra et al. 2014; Mahapatra 2015).

In the present study, *Spirulina* species, predominant in urban wastewater-fed systems, have been investigated for their role in wastewater bioremediation and as nutrient sources for land applications as biofertilizers. *Spirulina* belong to blue-green algae (cyanobacteria) and are mostly spiral shaped in filaments comprising of multicellular trichrome arranged in an open helix. *Spirulina* sp. was isolated from the wastewater-fed lakes in Bangalore City, cultured in synthetic wastewater media (Mahapatra et al. 2013a) and then grown in concentrated and settled urban wastewaters (Mahapatra et al. 2013a). Various nutrient analyses were carried out by following standard protocols (APHA 2005). The macromolecular compound analysis for lipids, carbohydrates and proteins was performed by following standard protocols (Mahapatra 2015). Outdoor rooftop batch studies (10 litres) were performed in polypropylene open bioreactors (Mahapatra et al. 2014). The culture was maintained for 10 days under natural illumination at temperature ranging from 27 to 32 °C. Manual agitation was provided thrice a day. For algal and nutrient quantification, 200 ml of water was sampled every 2 days and was replaced with deionized water. Care was also taken to compensate the water loss due to evaporation. The algal biomass was

Table 10.2 Nutrient removal and treatment efficiency by algal cultivation

Sl. No	Treatment parameter	Initial concentration (mg/l)	Final concentration (mg/l)	Treatment efficiency (% removal)
1	*COD*	680 ± 12	28 ± 2.4	95.88
2	*TN*	96 ± 7.2	3.8 ± 0.35	96.04
3	*TP*	35 ± 4.3	2.2 ± 0.26	93.71
4	*Ortho-P*	22.5 ± 1.77	1.85 ± 0.28	91.78
5	*Amm.-N*	77 ± 4.78	1.04 ± 0.33	98.65
6	*Nitrate-N*	1.44 ± 0.26	1.3 ± 0.38	9.72
7	*Nitrite-N*	0.06	U.D.	–

tested for growth, nutrients and macromolecules. The cell morphologies and growth were checked with light and electron microscopy as per the protocols used in earlier studies (Mahapatra et al. 2014). The growth was quantified by both spectrophotometry and gravimetry.

10.3.2 Wastewater Treatment with Nutrient Recovery in Algal Biomass

The results for the cultivation studies showed thick growth of *Spirulina* spirals entangled with each other and were mostly concentrated on the upper layer of the reactors. Being large cells with very frequent entangling, the *Spirulina* biomass was quantified by spectrophotometry and was verified for incremental growth with gravimetry. Higher growth was observed at the end of the eighth day during the cultivation period (specific growth rate, 0.35/day) with a maximum biomass density of 1.04 g/l. The biomass productivity was 130 mg/l/d, and during the experiments, the pH varied from 7.66 to 8.16.

The growth experiments showed near complete C, N and P removal illustrated in Table 10.2. As high as 95.88% of COD was removed indicating mixotrophic intake of organic C in *Spirulina* species that proves its suitability for treatment of untreated wastewater. The COD dropped from 680 to <30 mg/l, highlighting positive prospects of *Spirulina* sp. for treatment of municipal wastewaters. Similarly, the TN dropped from 96 to <4 mg/l (~96% removal) and TP dropped from 35 to 2.2 mg/l (~94% removal) indicating its ability to recover essential inorganic nutrients. There were no significant changes in the nitrate-N concentration that were found, mostly consistent across the experiment, i.e. from 2.02 mg/l (initial concentration) to 1.72 mg/l (final concentration). The nitrites were mostly undetectable (UD). However, ammoniacal nitrogen (amm.-N) concentrations were greatly removed from 77 to 1.04 mg/l (~99%) that is suitable for aquaculture and thus shows enormous abilities to capture wastewater nitrogen. Moreover, the reactive phosphorus (ortho-P) was also significantly reduced from 22.5 to 1.85 mg/l (~92% removal). Being a non-heterocystous BGA, *Spirulina* species assimilates high N from the wastewaters that favours its growth and synthesis of valued proteins in the cells.

Table 10.3 Macromolecular and biochemical composition of algal biomass and fertilizers

Macromolecular composition	*Spirulina* sp. (cultivated)	*Spirulina* sp. (wastewater-fed lakes)	Organic fertilizer	Chemical fertilizer
Lipids (%)	11	9.8	–	–
Carbohydrate (%)	22	18.2	–	–
Protein (%)	48.125	45	–	–
Nitrogen (%)	7.7	7.2	1.09	12.4
Phosphorus (%)	0.88	0.71	0.7	6.6
Potassium (%)	1.76	2.44	1.27	12.5
Calcium (%)	0.67	0.52	0.48	0.1

The macromolecular composition of the algal biomass cultivated as rooftop cultures and collected from *Spirulina* sp. blooms in urban lakes is illustrated in Table 10.3. This has been compared with the organic fertilizers and chemical fertilizers available in the market. The results show higher protein quantities (~48%) due to high N content (7.7%) in present outdoor cultures compared to surface waters attributed to the difference in nutrient loads. Besides this, the cultivated *Spirulina* sp. showed relatively high lipids and carbohydrate contents as compared to the species collected directly from the lakes. This is possibly due to availability of high organic C in addition to inorganic nutrients in the wastewater feed. However, the N content of the cultivated algal species was almost 62% of that in commercially available chemical fertilizer. Similarly, lower levels of P (13.3%) and K (14%) were recorded in the cultivated algal species. However the Ca content in the *Spirulina* species cultivated was much higher (~6.7 folds) than chemical fertilizer. Moreover, the cultured algal species grown in wastewater nutrient content were far better than the organic fertilizers (Table 10.3). Albeit having a lower percentage of NPK compared to chemical fertilizers, the algal biofertilizers from wastewaters have shown to enhance plant growth to a similar extent as chemical fertilizers due to higher amounts of secondary and trace nutrients that are crucial and central in plant physiology and metabolism. This shows immense scope of facilitating *Spirulina* sp. cultivation in urban wastewaters for wastewater treatment and nutrient recovery as whole cell algal biomass that is suitable as biofertilizer. In this context, further potting experiments with wastewater-grown algal species are being carried out to check the efficacy of algal application as dried whole cell biomass as a biofertilizer for improvement in soil nutrients with plant growth and development.

10.3.3 Effectiveness of Algal Biofertilizers (*Spirulina* sp.) Grown in Wastewaters with a Biorefinery Approach: Scope for Further Research and Development

Table 10.3 shows the biochemical constitution of the wastewater-cultivated *Spirulina* species and its usefulness as a biofertilizer. There are many studies that have reported higher plant growth with soil enriched with algal input, i.e. *Spirulina* biomass, showing increased plant height (red Bayam, arugula and pak choy) in addition to greater fresh/dry weight when compared to controls without *Spirulina* amendment (Wuang et al. 2016). However, there were no apparent differences in root lengths, foliar number and overall chlorophyll content (Tripathi et al. 2008; Saadatnia and Riahi 2009; Wuang et al. 2016). The comparison of the nutrient value of the dried algal biomass after cultivation with the commercial fertilizer and organic fertilizer showed significant differences and indicates that the whole cell algal biomass can itself act as both a carrier and bio-stimulant (nutrients) for effective fertilization of crops. Studies conducted on growth of plants with algal biofertilizer (*Spirulina* sp.) and chemical fertilizer showed no significant difference in growth and development. However, the net biomass was high in case of chemical fertilizers. Therefore, such algal biomass can act as slow/passive fertilizers or as essential supplement to the chemical fertilizers to augment beneficial micronutrients that aid development of crop plants (Rai et al. 2000). Studies on wheat cultivation using wastewater-grown algal assemblage showed higher growth and yield with ~53% increase in leaf chlorophyll compared to controls (Renuka et al. 2017) and improved the nutritional status of plants (Prasanna et al. 2013). Moreover, algal biomass inputs as biofertilizers obtained from wastewaters aided in slow release of nutrients in a cropping study involving tomatoes that have also shown improved quality of the crop with high sugar and carotenoids. Improved dry weight through applications of green algae *Acutodesmus dimorphus* were observed together with enhanced germination rate, growth and floral production in tomato plant (Garcia-Gonzalez and Sommerfeld, 2015). Experiments on algal biofertilizer applications to soils also reveal production of growth-promoting compounds essentially comprising of plant growth hormones that profoundly help in growth and maintaining good plant health (Karthikeyan et al. 2007; Perez-Montano et al. 2014). More investigations on understanding the dynamics of nutrient translocation through wastewater-grown algae (various wastewater sources) and at various scales (plot trials to field scale) and its evaluation can aid in involving algal biomass as essential fertilization strategies for optimal nutrient management in tropical agricultural setups.

Rampant and imbalanced use of chemical fertilizers without proper scientific know-how of its dose response, effects and implications in land with reduced usage of agricultural crop residues has highly deteriorated the soil nutrient status globally. This has led to deficiency in the macro- and micronutrients, thus impairing the geomorphology, the soil chemistry and the biophysical conditions of the soil (Geisseler and Scow 2014; Murase et al. 2015; Li et al. 2017). Municipal wastewater (urban sewage) can be used as a suitable nutrient source for fertilization of nutrient-deficient soils for agriculture. However, wastewater comprises of a variety of pollutants as heavy metals and pathogenic microorganisms that restricts its direct use in agriculture. An ideal and scientific way for management of untreated nutrient reservoirs as urban wastewaters is by the use of algal communities proliferating in such wastewaters to (a) recover nutrients immobilized as algal whole cell biomass (Mahapatra 2015) and (b) at the same time aid in bioremediation of wastewater (Mahapatra et al. 2014) with (c) reducing GHG emissions (Ramachandra and Mahapatra 2015). Cultivating algal microflora in wastewaters and usage of algal biomass not only enhance the soil nutrient status but also help in providing growth-promoting triggers/compounds and restrict the growth of pathogenic undesirable bacterial communities due to an alkaline environment, through oxygenated conditions (Chaudhary et al. 2012; Prasanna et al. 2013).

Studies on biofortification of wheat by algal application (BGA) aid in improving soil fertility, improved crop yields and also fortify essential trace nutrients in plant tissues and whole grains (Rai et al. 2000; Renuka et al. 2016; Adak et al. 2016). It has been shown that application of algae grown in municipal wastewaters provided ~25% N savings during wheat productions (Renuka et al. 2016). Wastewater-grown algal species have increased the bioavailability to trace nutrients as Cu, Fe and Zn in soils. The N metabolism and fixation are directly influenced by the assimilation and availability of essential catalysts and cofactors as Zn and Fe. Counteractively, the expression of the transporters for Zn and Fe is also dependent on N status in the plant cells (Guo et al. 2014). Studies have shown 25–50% N replacement with cyanobacterial amendment (Prassana et al. 2013) and ~75% N replacement (Renuka et al. 2017) in wheat species. The applications of chemical fertilizers have shown lowest soil chlorophyll content that indicates harmful impacts on soil algal photosynthetic systems, and at the same time algal addition has improved chlorophyll content and N fixation in soil systems. Soil organic carbon (SOC) is an essential indicator of the soil status, health and suitability for agriculture or any cropping system (Lal 2004; Sa et al. 2017). Algal consortia amended to such systems with low organic carbon soils showed ~50% high SOC as compared to control.

Dehydrogenase enzyme activity is considered as one of the most crucial indicators in total soil microbial action effectiveness and plays a major role in organic matter bio-oxidation (Zhang et al. 2010; Leon et al. 2017). This enzyme activity is used as a significant index for soil fertility and health due to its impacts on the physicochemical and biochemical properties of the soil. Studies conducted on biofertilizer application on soil systems have shown a higher dehydrogenase activity with a strong correlation on soil organic carbon. But, during waterlogged condition and prevalence of anaerobic environment, there can be a very high dehydrogenase activity that impacts the transport and availability of Fe to the plants due to its utilization as Fe(III) as terminal electron acceptor (Das and Varma 2010).

Previous studies have shown dried BGA as soil inoculants that enhances the fertility of crops and rice cultivars (Watanabe and Konishi 1951; Venkataraman and Neelakantan 1967; Mishra and Pabbi 2004). These studies have shown increase in the rice grain yields (~20%) in agricultural field-based experiments. Studies on seed germination (seed viability) of BGA members as *Spirulina* sp. can be indicators of the performance of biofertilizers in the field. Studies conducted by rinsing and soaking the rice grains with BGA culture showed more effectiveness against sulphate-reducing mechanism and thus aid in rapid germination and faster growth of the rice seedlings (Nanda et al. 1991; Shariatmadari et al. 2011; Dineshkumar et al. 2017). Moreover, studies conducted on seed germination by Wuang et al. on three vegetable cultivars as Chinese cabbage, kai lan and white crown showed a threefold increase in the dry weight of the germinated seedling compared to the control with no algal inputs (Wuang et al. 2016). The key findings of the study showed *Spirulina* biofertilizers as seed germination enhancer, and their activities vary from species to

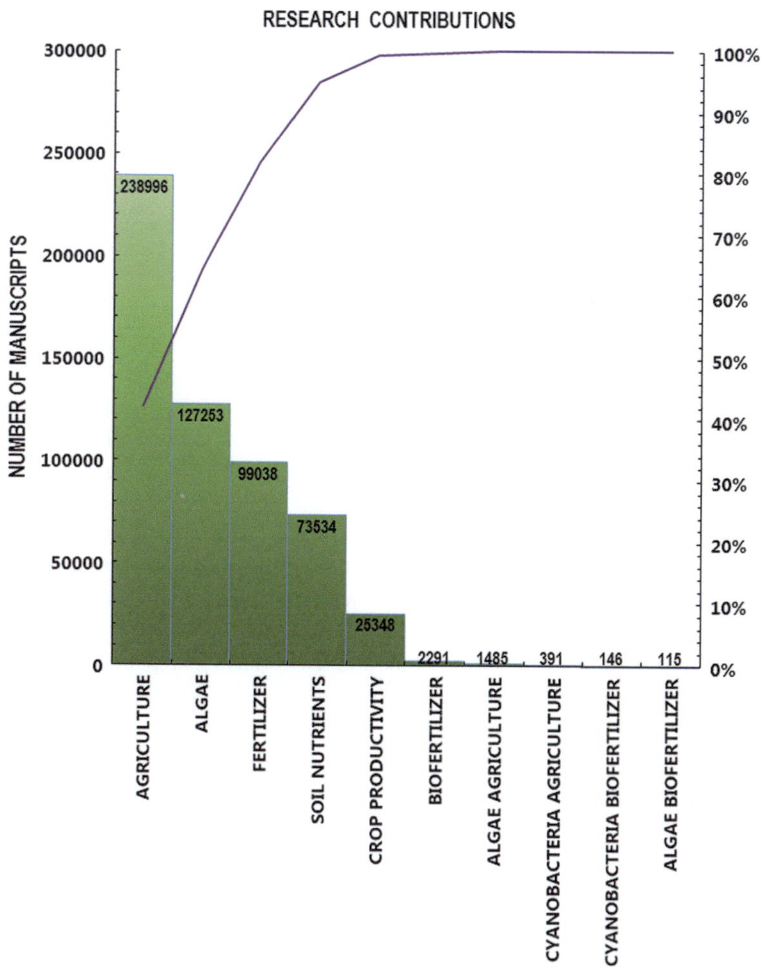

Fig. 10.4 Research contributions from agriculture to algal biofertilizer in the last 10 years

species. *Spirulina* amendment to soils growing pulses as green gram showed higher productivity (Aung 2011). Previous investigations on biofertilization have shown enhanced seed germination rates for sunflower (Jalaluddin and Hamid 2011); dill (Miri et al. 2013); and cumin, marigold and tomato (Garcia-Gonzalez and Sommerfeld 2015). Other studies have also reported higher growth enhancements in maize plants through biofertilization of red algae (Safinaz and Ragaa 2013). Besides BGA studies on green algae as *Chlorella vulgaris* showed higher plant development in lettuce (Faheed and Fattah 2008).

A review on the total number of peer-reviewed manuscripts published in the last 10 years based on Scopus and Thomson Reuters Indices on miscellaneous aspects of agriculture showed around ~2.4 lakh articles as illustrated in the Pareto chart (Fig. 10.4). However, the total articles on various aspects of algae, e.g. biotechnology, bioproducts, bioresources, engineering, nanosciences, chemical biology, biophysics, etc., were ~1.2 lakhs, i.e. almost half that of the manuscripts on various aspects of agriculture. The research on fertilizers mostly emphasizing crop studies, fertilizer mining, extraction and production, fertilizer amendments and economics and metallurgical engineering were around ~1 lakh. Articles on scientific findings in (a) soil sciences and nutrients, (b) crop productivity and (c) biofertilizers were ~73, 25 and 2 thousands, respectively. Nevertheless, studies on (a) algal use in agriculture, (b) BGA-based agriculture and c) BGA-based biofertilizer compared to the earlier figures were only 1485, 391 and 146, respectively. More importantly, very few studies, i.e. 115, have been conducted on the frontiers of algal biomass usage as nutrient-rich sources for biofertilizers mostly for land applications in agricultural setups.

Although these numbers might be indicative, there is definitely a huge scope for research in the field of algal biofertilizers in the context of increasing instances of climate change, decreasing soil organic matter and nutrients and unsustainable extraction and industrial fixing of inorganic nutrients as chemical fertilizers. In this context, the present study shows prospects of algal species as *Spirulina* as a byproduct of wastewater treatment that has potentially high nutrient values with optimal secondary and trace nutrient concentration suitable for land applications as biofertilizers. Wastewater treatment, nutrient recovery through sustainable algal bioprocesses and algal biomass application in agricultural fields demonstrate a biorefinery approach suitable for tropical agrarian economies to reach a zero-waste agricultural society that nurtures sustainable development.

10.4 Conclusion

The present communication shows the future prospects of biofertilizers in the agrarian economy and highlights the potential of wastewater-grown algae for enhancement of nutrient availability to plant systems. Cultivation of *Spirulina* sp. isolated from urban wastewater-fed lakes in outdoor rooftop batch cultures with concentrated wastewaters showed near-complete removal and recovery of essential macronutrients as C, N and P. *Spirulina* sp. was efficient in completely transforming the labile N and P forms as amm.-N and ortho-P into algal biomass and showed a mixotrophic growth evident from significant COD removal during the culturing experiments. The macromolecular composition revealed considerably high protein content useful for usage of whole cell algal biomass directly as N-rich biofertilizers. The macro-elemental composition showed significantly high NPK content better than available organic fertilizers and relatively low nutrient content when compared to chemical fertilizers. Despite this, a high micronutrient content as Ca in wastewater-grown algal biomass provides additional advantage for algal biomass to be used as biofertilizer. This enhances the soil macro- and micronutrient content and augments plant growth, for a sustainable and green agricultural future with a better food quality, realizing the biorefinery approach with a zero-waste concept.

Acknowledgement The authors sincerely acknowledge the laboratory facilities at Aquatic Ecology and Molecular Ecology labs in Centre for Ecological Sciences (CES), Inorganic and Physical Chemistry (IPC), Biochemistry (BC) and Molecular Biophysics Unit (MBU) at IISc, for their help during the culture experiments and biochemical composition analysis. The authors also deeply acknowledge the Science and Education Research Board (SERB), IUSSTF INDO-US Postdoctoral Fellowship, Government of India; Department of Biotechnology (DBT); Ministry of Science and Technology (DST); Ministry of Environment, Forest and Climate Change (MoEFCC), Government of India; and Indian Institute of Science for providing the financial and infrastructural support.

References

Adak A, Prasanna R, Babu S, Bidyarani N, Verma S, Pal M, Shivay YS, Nain L (2016) Micronutrient enrichment mediated by plant-microbe interactions and rice cultivation practices. J Plant Nutr 39:1216–1232

Amin SA, Green DH, Hart MC, Kupper FC, Sunda WG, Carrano CJ (2009) Photolysis of iron-siderophore chelates promotes bacterial-algal mutualism. Proc Natl Acad Sci U S A 106:17071–17076

APHA (2005) Standard method for examination of water and wastewater, 19th edn. American Public Health Association, Washington, DC

Aung KLN (2011) Effect of *Spirulina* biofertilizer suspension on growth yield of *Vigna radiata* (L.) Wilczek. Univ Res J 4:351–363

Benson D, Kerry K, Malin G (2014) Algal biofuels: impact significance and implications for EU multi-level governance. J Clean Prod 72:4–13. https://doi.org/10.1016/j.jclepro.2014.02.060

Castenholz RW (2001) Phylum BX. Cyanobacteria. In: Boone DR, Castenholz RW (eds) Bergey's manual of systematic bacteriology, 2nd edn. Springer, New York, pp 473–599

Chanakya HN, Mahapatra DM, Sarada R, Chauhan VS, Abitha R (2012) Sustainability of large-scale algal biofuel production in India. J Indian Inst Sci 92(1):63–98

Chanakya HN, Mahapatra DM, Sarada R, Abitha R (2013) Algal biofuel production and mitigation potential in India. Mitig Adapt Strat Glob Chang 18:113–136

Chaudhary V, Prasanna R, Nain L, Dubey SC, Gupta V, Singh R, Jaggi S, Bhatnagar AK (2012) Bioefficacy of novel cyanobacteria-amended formulations in suppressing damping off disease in tomato seedlings. World J Microbiol Biotechnol 28:3301–3310

Chen J (2006) The combined use of chemical and organic fertilizer and or biofertilizer for crop growth and soil fertility. In: International workshop on sustained management of the soil-rhizosphere system for efficient crop production and fertilizer use. October, Thailand, pp 16–20

Cohen RRH (2006) Use of microbes for cost reduction of metal removal from metals and mining industry waste streams. J Clean Prod 14:1146–1157. https://doi.org/10.1016/j.jclepro.2004.10.009

Das SK, Varma A (2010) Role of enzymes in maintaining soil health, soil Enzymology. Springer, Berlin/Heidelberg, pp 25–42

Dineshkumar R, Kumaravel R, Gopalsamy J, Mohammad N, Sikder A, Sampathkumar P (2017) Microalgae as bio-fertilizers for rice growth and seed yield productivity. Waste Biomass Valoriz:1–8. https://doi.org/10.1007/s12649-017-9873-5

El-Yazeid AA, Abou-Aly HA, Mady MA, Moussa SAM (2007) Enhancing growth, productivity and quality of squash plants using phosphate-dissolving microorganisms (bio phosphor) combined with boron foliar spray. Res J Agric Biol Sci 3(4):274–286

Faheed FA, Fattah AAE (2008) Effect of *Chlorella vulgaris* as bio-fertilizer on growth parameters and metabolic aspects of lettuce plant. J Agric Soc Sci 4:165–169

Garcia-Gonzalez J, Sommerfeld M (2015) Biofertilizer and biostimulant properties of the microalga *Acutodesmus dimorphus*, J Appl Phycol 28. 10.1007/s10811-015-0625-2

Geisseler D, Scow KM (2014) Long-term effects of mineral fertilizers on soil microorganisms – a review. Soil Biol Biochem 75:54–63

Guo X, Xiong H, Shen H, Qiu W, Ji C, Zhang Z, Zuo Y (2014) Dynamics in the rhizosphere and iron-uptake gene expression in peanut induced by intercropping with maize: role in improving iron nutrition in peanut. Plant Physiol Biochem 76:36–43

Gupta AK (2004) The complete technology book on biofertilizers and organic farming. National Institute of Industrial Research Press, New Delhi

Jalaluddin M, Hamid M (2011) Effect of adding inorganic, organic and microbial fertilizers on seed germination and seedling growth of sunflower. Pak J Bot 43:2807–2809

Kapustka LA, DuBois JD (1987) Dinitrogen fixation by cyanobacteria and associative rhizosphere bacteria in the arapaho prairie in the sand hills of Nebraska. Am J Bot 74:107–113

Karthikeyan N, Prasanna R, Nain L, Kaushik BD (2007) Evaluating the potential of plant growth promoting cyanobacteria as inoculants for wheat. Eur J Soil Biol 43:23–30

Koller M, Salernoa A, Tuffnerb P, Koiniggb M, Bochzeltb H, Schoberd S et al (2012) Characteristics and potential of micro algal cultivation strategies: a review. J Clean Prod 37:377–388. https://doi.org/10.1016/j.jclepro.2012.07.044

Lal R (2004) Soil carbon sequestration impacts on global climate change and food security. Science 304:1623–1627

Leon P, Espejo R, Gomez-Paccard C, Hontoria C, Mariscal I, Renella G, Benito M (2017) No tillage and sugar beet foam amendment enhanced microbial activity of degraded acidic soils in South West Spain. Appl Soil Ecol 109:69–74

Li R, Tao R, Ling N, Chu G (2017) Chemical, organic and bio-fertilizer management practices effect on soil physicochemical property and antagonistic bacteria abundance of a cotton field: implications for soil biological quality. Soil Tillage Res 167:30–38

Mahapatra DM (2015) Algal bioprocess development for sustainable wastewater treatment and biofuel production. Ph.D. thesis, Indian Institute of Science, Bangalore, India

Mahapatra DM, Ramachandra TV (2013) Algal biofuel: bountiful lipid from *Chlorococcum* sp. proliferating in municipal wastewater. Curr Sci 105:47–55

Mahapatra DM, Chanakya HN, Ramachandra TV (2011a) Assessment of treatment capabilities of Varthur Lake, Bangalore. Int J Environ Technol Manag 14(1–4):84–102

Mahapatra DM, Chanakya HN, Ramachandra TV (2011b) Role of macrophytes in urban sewage fed lakes. I Inte Omics Appl Biotech 2(7):1–9

Mahapatra DM, Chanakya HN, Ramachandra TV (2011c) C: N ratio of sediments in a sewage fed Urban Lake. Inte J Geol 5(3):86–92

Mahapatra DM, Chanakya HN, Ramachandra TV (2013a) *Euglena* sp. as a suitable source of lipids for potential use as biofuel and sustainable wastewater treatment. J Appl Phycol 25:855–865

Mahapatra DM, Chanakya HN, Ramachandra TV (2013b) Treatment efficacy of algae based sewage treatment plants. Environ Monit Assess 185:7145–7164

Mahapatra DM, Chanakya HN, Ramachandra TV (2014) Bioremediation and lipid synthesis of myxotrophic algal consortia in municipal wastewater. Bioresour Technol 168:142–150

Mahapatra DM, Chanakya HN, Ramachandra TV (2015) Book chapter: algae derived single-cell proteins: economic cost analysis and future prospects. In: Protein byproducts: transformation from environmental burden into value-added products, 1st edn. Elsevier, San Diego, pp 275–301

Mahapatra DM, Joshi NV, Ramachandra TV (2017) Insights to bioprocess and treatment competence of urban wetlands. J Environ Manag https://doi.org/10.1016/j.jenvman.2017.10.054

Mason J (2003) Sustainable agriculture, 2nd edn. Landlinks Press, Collingwood

McKnight DM, Morel FM (1980) Copper complexation by siderophores from filamentous blue-green algae. Limnol Oceanogr 25:62–71

Miri Y, Kochebagh SB, Mirshekari B (2013) Effect of inoculation with bio-fertilizers on germination and early growth, Dill (*Anethum graveolens*), Fennel (*Foeniculum vulgare*), Cumin (*Cuminum cyminum*) and Marigold (*Calendula officinalis*). Int J Agron Plant Prod 4:104–108

Mishra U, Pabbi S (2004) Cyanobacteria: a potential biofertilizer for rice. Resonance 9:6–10

Murase J, Hida A, Ogawa K, Nonoyama T, Yoshikawa N, Imai K (2015) Impact of long-term fertilizer treatment on the microeukaryotic community structure of a rice field soil. Soil Biol Biochem 80:237–243

Nanda S, Tripathy K, Padhi S (1991) Effect of algalization on seed germination of vegetable crops. World J Microbiol Biotechnol 7:622–623

Naveen BP, Mahapatra DM, Sitharam TG, Sivapullaiah PV, Ramachandra TV (2016) Physico-chemical and biological characterization of urban municipal landfill leachate. Environ Poll 220(Part A):2–12. https://doi.org/10.1016/j.envpol.2016.09.002

Pandey KD, Shukla SP, Skukla PN, Giri DD, Singh JS, Singh P et al (2004) Cyanobacteria in Antarctica: ecology, physiology and cold adaptation. Cell Mol Biol 50:574–584

Perez-Montano F, Alias-Villegas C, Bellogin RA, del Cerro P, Espuny MR, JimenezGuerrero I, Lopez-Baena FJ, Ollero FJ, Cubo T (2014) Plant growth promotion in cereal and leguminous agricultural important plants: from microorganism capacities to crop production. Microbiol Res 169:325–336

Prasanna R, Kumar A, Babu S et al (2013) Deciphering the biochemical spectrum of novel cyanobacterium-based biofilms for use as inoculants. Biol Agric Hortic 29:145–158

Rai AN, Soderback E, Bergman B (2000) Cyanobacterium-plant symbioses. Tansley review no. 116. New Phytol 147:449–481

Ramachandra TV, Mahapatra DM (2015) The science of carbon footprint assessment. In: The carbon footprint handbook. CRC Press, Taylor & Francis Group, Boca Raton, pp 1–44

Ramachandra TV, Mahapatra DM, Karthick B, Gordon R (2009) Milking diatoms for sustainable energy: biochemical engineering vs. gasoline secreting diatom solar panels. Ind Eng Chem 48(19):8769–8788

Ramachandra TV, Mahapatra DM, Samantray S, Joshi NV (2013) Biofuel from urban wastewater: scope and challenges. Renew Sust Energ Rev 21:767–777

Ramachanda TV, Mahapatra DM, Bhat SP, Asulabha KS, Varghese S, Aithal BH (2014) Integrated wetlands ecosystem: sustainable model to mitigate water crisis in Bangalore (2014), ENVIS Technical Report-76. Environmental Information System, CES, IISc, Bangalore

Ramachandra TV, Mahapatra DM, Bhat SP, Joshi NV (2015) Biofuel production along with remediation of sewage water through algae. In: Algae and environmental sustainability, developments in applied phycology, vol 7. Springer, New Delhi, pp 33–51

Renuka N, Prasanna R, Sood A, Ahluwalia AS, Bansal R, Babu S, Singh R, Shivay YS, Nain L (2016) Exploring the efficacy of wastewater-grown microalgal biomass as a biofertilizer for wheat. Environ Sci Pollut Res 23:6608–6620

Renuka N, Prasanna R, Sood A, Bansal R, Bidyarani N, Singh R, Shivay YS, Nain L, Ahluwalia AS (2017) Wastewater grown microalgal biomass as inoculants for improving micronutrient availability in wheat. Rhizosphere 3(Part 1):150–159. ISSN 2452-2198, https://doi.org/10.1016/j.rhisph.2017.04.005

Rokhzadi A, Asgharzadeh A, Darvish F, Nourmohammadi G, Majidi E (2008) Influence of plant growth-promoting rhizobacteria on dry matter accumulation and yield of chickpea (*Cicer arietinum* L.) under field condition. Am-Euras J Agric Environ Sci 3(2):253–257

Sa JCM, Lal R, Cerri CC, Lorenz K, Hungria M, de Faccio Carvalho PC (2017) Low-carbon agriculture in South America to mitigate global climate change and advance food security. Environ Int 98:102–112

Saadatnia H, Riahi H (2009) Cyanobacteria from paddy fields in Iran as a biofertilizer in rice plants. Plant Soil Environ 55:207–212

Safinaz AF, Ragaa AH (2013) Effect of some red marine algae as biofertilizers on growth of maize (*Zea mays* L.) plant. Int Food Res 20:1629–1632

Shariatmadari Z, Riahi H, Shokravi S (2011) Study of soil blue-green algae and their effect on seed germination and plant growth of vegetable crops. Bot J Iran 12:101–110

Singh JS (2011) Methanotrophs: the potential biological sink to mitigate the global methane load. Curr Sci 100:29–30

Singh JS (2014) Cyanobacteria: a vital bio-agent in eco-restoration of degraded lands and sustainable agriculture. Clim Change Environ Sustain 2:133–137

Singh JS (2015) Plant-microbe interactions: a viable tool for agricultural sustainability. Appl Soil Ecol 92:45–46. https://doi.org/10.1016/j.apsoil.2015.03.004

Tripathi RD, Dwivedi S, Shukla MK, Mishra S, Srivastava S, Singh R, Rai UN, Gupta DK (2008) Role of blue green algae biofertilizer in ameliorating the nitrogen demand and fly-ash stress to the growth and yield of rice (*Oryza sativa* L.) plants. Chemosphere 70:1919–1929

Venkataraman GS, Neelakantan S (1967) Effect of the cellular constituents of the nitrogen fixing blue green algae *Cylindrospermum muscicola* on the root growth of rice seedlings. J Gen Appl Microbiol 13:53–61

Wani SP, Rego TG, Rajeshwari S, Lee KK (1995) Effect of legume–based cropping systems on nitrogen mineralization potential of Vertisol. Plant Soil 175(2):265–274

Watanabe R, Konishi IC (1951) Effect of nitrogen-fixing blue-green algae on the growth of rice plants. Nature 168:748–749

Wuang SC, Khin MC, Chua PQD, Luo YD (2016) Use of *Spirulina* biomass produced from treatment of aquaculture wastewater as agricultural fertilizers. Algal Res 15:59–64

Yamaguchi K (1997) Recent advances in microalgal bioscience in Japan, with special reference to utilization of biomass and metabolites: a review. J Appl Phycol 8:487–502. https://doi.org/10.1007/BF02186327

Zhang N, He X-D, Gao Y-B, Li Y-H, Wang H-T, Ma D, Zhang R, Yang S (2010) Pedogenic carbonate and soil dehydrogenase activity in response to soil organic matter in *Artemisia ordosica* community. Pedosphere 20:229–235

Psychrotrophic Microbiomes: Molecular Diversity and Beneficial Role in Plant Growth Promotion and Soil Health

11

Ajar Nath Yadav, Priyanka Verma, Shashwati Ghosh Sachan, Rajeev Kaushik, and Anil Kumar Saxena

Abstract

Prospecting the cold habitats has led to the isolation of a great diversity of psychrotrophic microbes belonging to different groups. The cold-adapted microbes have potential biotechnological applications in agriculture, medicine, and industry as they can produce cold-adapted enzymes (amylase, cellulase, chitinase, laccase, lipase, pectinase, protease, xylanase, β-galactosidase, and β-glucosidase), antifreezing compounds, and antibiotics and possess diverse multifunctional plant growth-promoting attributes (production of ammonia, hydrogen cyanide, indole-3-acetic acid, and siderophores; solubilization of phosphorus, potassium, and zinc; 1-aminocyclopropane-1-carboxylate deaminase activity and biocontrol activity against plant pathogenic microbes). Cold-adapted microbes are ubiquitous in nature and have been reported from Antarctica, permanently ice-covered lakes, cloud droplets, ice cap cores from considerable depth, snow, glaciers, and those associated with plants growing in cold habitats. Cold-adapted microbial communities can be studied using culture-dependent and culture-independent

A. N. Yadav (✉)
Department of Biotechnology, Akal College of Agriculture, Eternal University, Baru Sahib, Sirmour, India
e-mail: ajarbiotech@gmail.com

P. Verma
Department of Microbiology, Akal College of Basic Science, Eternal University, Baru Sahib, Sirmour, India

S. G. Sachan
Department of Bio-Engineering, Birla Institute of Technology, Ranchi, India

R. Kaushik
Division of Microbiology, Indian Agricultural Research Institute, New Delhi, India

A. K. Saxena
ICAR-National Bureau of Agriculturally Important Microorganisms, Mau, India

D. G. Panpatte et al. (eds.), *Microorganisms for Green Revolution*, Microorganisms for Sustainability 7,
https://doi.org/10.1007/978-981-10-7146-1_11

techniques. Microbes recovered using both techniques revealed the occurrence of different and diverse major groups, *viz.*, Actinobacteria, Ascomycota, Bacteroidetes, Basidiomycota, Chlamydiae, Chloroflexi, Cyanobacteria, Euryarchaeota, Firmicutes, Gemmatimonadetes, Mucoromycota, Nitrospirae, Planctomycetes, Proteobacteria, Spirochaetes, Thaumarchaeota, and Verrucomicrobia. On the review of isolated cold-adapted microbes, it was found that Proteobacteria was the most dominant phylum followed by Firmicutes and Actinobacteria. This book chapter deals with the isolation, characterization, and biodiversity of cold-tolerant microbes from Antarctica; Himalayan cold desert; glaciers; ice-capped rivers; plant-associated, subalpine region of Uttarakhand; and different sub-glacial lakes. The biotechnological applications of cold-adapted microbes have been discussed. The benificial and potential cold-adapted microbes may have applications in diverse processes in agriculture, industry, and allied sectors.

Keywords
Antarctica · Biofertilizers · Biodiversity · Cold desert · Indian Himalayas · Psychrotrophic microbes · Sub-glacial lakes

11.1 Introduction

The microbiomes of cold environments are of particular importance in global ecology since the majority of terrestrial and aquatic ecosystems of our planet are permanently or seasonally submitted to cold temperatures. Earth is primarily a cold, marine planet with 90% of the ocean's waters being at 5 °C or lower. Permafrost soils, glaciers, polar sea ice, and snow cover make up 20% of the Earth's surface environments. Microbial communities under cold habitats have undergone the physiological adaptations to low temperature and chemical stress. Recently, these communities have attained the focus of applied research not only in terms of biotechnological prospects but also to understand the use of primitive analogues of biomolecules existed during early Earth environments (Yadav 2015; Saxena et al. 2016). The microbiomes of cold environments have been extensively investigated in the past few years with a focus on culture-dependent and culture-independent techniques. Cold-adapted microorganisms have been reported from Antarctic sub-glacial and permanently ice-covered lakes, cloud droplets, ice cap cores from considerable depth, snow, and glaciers (Srivastava et al. 2014; Yadav et al. 2015a, b, c). Many novel microbes have been sorted out from cold environments including *Halobacterium lacusprofundi* (Franzmann et al. 1988), *Sphingobacterium antarcticus* (Shivaji et al. 1992), *Octadecabacter arcticus* (Gosink et al. 1997), *Hymenobacter roseosalivarius* (Hirsch et al. 1998), *Cellulophaga algicola* (Bowman 2000), *Flavobacterium frigidarium* (Humphry et al. 2001), *Oleispira antarctica* (Yakimov et al. 2003), *Flavobacterium psychrolimnae* (Van Trappen et al. 2005), *Psychromonas ingrahamii* (Auman et al. 2006), *Exiguobacterium soli* (Chaturvedi et al. 2008), *Pseudomonas extremaustralis* (López et al. 2009), *Cryobacterium roopkundense* (Reddy et al. 2010), *Sphingomonas glacialis* (Zhang et al. 2011), *Pedobacter arcticus* (Zhou et al.

2012), *Sphingobacterium psychroaquaticum* (Albert et al. 2013), *Lacinutrix jangbogonensis* (Lee et al. 2014), *Massilia eurypsychrophila* (Shen et al. 2015), *Glaciimonas frigoris* (Margesin et al. 2016), and *Psychrobacter pocilloporae* (Zachariah et al. 2017). There are several reports on whole genome sequences of novel and potential psychrotrophic microbes (Kim et al. 2012; Singh et al. 2016).

The novel species of psychrotrophic microbes have been isolated worldwide and reported from different phylum of domain Archaea (Euryarchaeota and Thaumarchaeota), bacteria (Actinobacteria, Bacteroidetes, Chlamydiae, Chloroflexi, Cyanobacteria, Firmicutes, Gemmatimonadetes, Mucoromycota, Nitrospirae, Planctomycetes, Proteobacteria, Spirochaetes, and Verrucomicrobia), and fungi (Ascomycota, Basidiomycota, and Mucoromycota) (Franzmann et al. 1988; Shivaji et al. 1992; Gosink et al. 1997; Hirsch et al. 1998; Bowman 2000; Humphry et al. 2001; Yakimov et al. 2003; Van Trappen et al. 2005; Auman et al. 2006; Chaturvedi et al. 2008; López et al. 2009; Reddy et al. 2010; Zhang et al. 2011; Zhou et al. 2012; Albert et al. 2013; Lee et al. 2014; Shen et al. 2015; Margesin et al. 2016; Zachariah et al. 2017). Prospecting the cold habitats has led to the isolation of a great diversity of psychrotrophic microorganisms. The cold-adapted microbes have potential biotechnological applications in agriculture, medicine, and industry. The bacterial diversity from the cold environment could serve as a database for selection of bio-inoculants with PGP ability and could be used for improving the growth and yield of crops grown at high altitudes with prevailing low temperatures (Kumar et al. 2013; Yadav et al. 2014a, 2015d, 2017e; Kumar et al. 2016).

The psychrophilic/psychrotrophic/psychrotolerant microbes from Antarctica; Himalayan cold desert; glaciers; ice-capped rivers; plant-associated, subalpine region of Uttarakhand; and different sub-glacial lakes could be used in agriculture, medicine, and industry as they can produce cold-adapted enzymes (amylase, cellulase, chitinase, laccase, lipase, pectinase, protease, xylanase, β-galactosidase, and β-glucosidase), antifreezing compounds, and antibiotics and possess diverse multifunction plant growth-promoting attributes (production of ammonia, hydrogen cyanide, indole-3-acetic acid, and siderophore; solubilization of phosphorus, potassium, and zinc; 1-aminocyclopropane-1-carboxylate deaminase activity and biocontrol activity against plant pathogenic microbes). Psychrotrophic plant growth-promoting (PGP) microbes have been shown to promote plant growth either directly by biological N_2-fixation; solubilization of minerals such as phosphorus, potassium, and zinc; and production of siderophores and plant growth hormones or indirectly via production of antagonistic substances by inducing resistance against plant pathogens under the native conditions from which they have been isolated (Verma et al. 2015a, 2016; Yadav et al. 2016c, 2017c). The psychrotrophic PGP microbes can have an impact on plant growth providing the plant with compound(s) of microbial origin for facilitating the uptake of nutrients from the environment. Psychrotrophic PGP microbes were found in several genera, including *Arthrobacter*, *Bacillus*, *Brevundimonas*, *Burkholderia*, *Pseudomonas*, *Citricoccus*, *Exiguobacterium*, *Flavobacterium*, *Janthinobacterium*, *Kocuria*, *Lysinibacillus*, *Methylobacterium*, *Microbacterium*, *Paenibacillus*, *Providencia*, and *Serratia* (Yadav 2009; Saxena et al. 2015b; Verma et al. 2015d, 2016; Yadav et al. 2016b). Among these taxa, *Pseudomonas* and *Exiguobacterium* have been the best characterized for PGP at low temperatures (Yadav et al. 2013, 2015f, 2017h, i). There are several studies have demonstrated the benefits of PGP microbes on the growth and yield of different crops at different climates, soils, and temperatures. The use of PGP microbes improves plant growth by supplying plant nutrients, which can help sustain environmental health and soil productivity.

Psychrophilic/psychrotolerant microbes are important for many reasons, particularly because they produce cold-active enzymes. Enzymes from psychrophiles have become interesting for industrial applications, partly because of ongoing efforts to decrease energy consumption. These enzymes provide opportunities to study the adaptation of life to low temperature and the potential for biotechnological exploitation (Yadav et al. 2016a; Saxena et al. 2016). Most of the work that has been conducted on psychrophilic bacteria is focused on cold-adapted enzymes such as amylase, protease, lipase, pectinase, xylanase, cellulase, β-galactosidase, β-glucosidase, and chitinase (Yadav et al. 2016a). Cold-active enzymes are produced by psychrophilic microbes, namely, *Acinetobacter*, *Aquaspirillum*, *Arthrobacter*, *Bacillus*, *Carnobacterium*, *Clostridium*, *Cytophaga*, *Flavobacterium*, *Marinomonas*, *Moraxella*, *Moritella*, *Paenibacillus*, *Planococcus*, *Pseudoalteromonas*, *Pseudomonas*, *Psychrobacter*, *Shewanella*, *Vibrio*, and *Xanthomonas* (Gerday et al. 2000; Groudieva et al. 2004; Yadav et al. 2014b, 2016a, 2017h; Singh et al. 2016). Psychrophilic microbes can be applied for biodegradation of agro-wastes at low temperatures. Shukla et al. (2016) have developed psychrotrophic microbial consortium of *Eupenicillium crustaceum*, *Paecilomyces* sp., *Bacillus atrophaeus*, and *Bacillus* sp., for its potential applications toward degradation of agri-residues and conversion to a value-added product like compost for enhancing soil fertility and decreasing environmental pollution caused by burning of agro-wastes. Psychrotrophic microbes produced antifreezing compounds at low temperatures (Yadav 2015; Singh et al. 2016; Yadav et al. 2017e). Antifreezing compounds are useful in cryosurgery and also in the cryopreservation of whole organisms, isolated organs, cell lines, and tissues. In food industry, antifreezing proteins (AFPs) can be used to improve the quality of frozen food. Improved cold tolerance in fishes has been achieved in some cases by direct injection of AFPs and in another case by transgenic expression of an AFP.

Biotechnology has opened up new possibilities for potential applications of beneficial microbiomes to the soil for the PGP and biocontrol of soilborne pathogens. An understanding of microbial diversity and its potential applications in agriculture is important and useful to arrive at measures that can act as indicators of plant growth, yield, and soil health. Cold-adapted microbes have attracted the attention of the scientific community due to their ability to promote plant growth and produce cold-active enzymes, with potential biotechnological applications in a broad range of industrial, agricultural, and medical processes. Psychrotrophic microbes could be valuable in agriculture as bio-inoculants and biocontrol agents for low-temperature habitats. The use of psychrophiles as biofertilizers, biocontrol agent, and bioremediators would be of great use in agriculture under cold climatic conditions. The present book chapter describes the method of isolation of psychrotrophic microbes from extreme cold environments, its characterization, identification, biodiversity, and distributions and biotechnological applications in agriculture, medicine, and industry.

11.2 Isolation and Characterization of Cold-Adapted Microbes

Cold-adapted microbial communities have received much attention, because Earth is primarily a cold, marine planet with 90% of the ocean's waters being at 5 °C or lower, and microbes from cold habitat could be applied in agriculture as PGP microbes, in medicine as antifreezing compounds (AFCs) for cryosurgery, and in industry for natural drugs, cold-active enzymes, and valuable bioactive metabolites. Cold-adapted microbes can be characterized using culture-dependent and culture-independent techniques, and microbes from cold habitat and those associated with plants growing at hilly and low-temperature condition should be isolated using culturable and unculturable techniques to know their diversity and distribution. Cold-adapted microbes could be isolated using enrichment and serial dilution methods followed by spread or pour plate technique. The populations of *Bacillus* and *Bacillus*-derived genera (BBDG) in the different samples were enumerated through enrichment and heat treatment using the standard serial dilution plating technique (Yadav et al. 2015d). The different specific growth mediums were used to isolate the maximum possible culturable morphotypes of different genera of cold-adapted microbes such as Archaea (chemically defined medium, standard growth media, and halophilic medium), *Arthrobacter* (trypticase soy agar), BBDG (T3 agar), fungus (potato dextrose agar), *Methylobacterium* (ammonium mineral salt), N_2-fixing bacteria (Jensen N_2-free agar), *Pseudomonas* spp. (Kings' B agar), *Rhizobium* (Congo red yeast mannitol), and soil-specific bacteria (soil extract agar) (Table 11.1). Along with microbes in natural cold environments, also other abiotic stress conditions can be used for isolation of halophilic psychrotrophic (with 5–20% NaCl concentration), acidophilic psychrotrophic (pH 3–5), and alkaliphilic psychrotrophic (pH 8–11) microbes (Fig. 11.1). The isolated microbes should be screened for tolerances to different temperatures for grouping of these microbes in three categories as psychrophilic, psychrotrophic, and psychrotolerant (Yadav et al. 2016b).

To know the PGP ability and other industrial applications of cold-adapted microbes, purified microbes should be screened qualitatively for direct PGP attributes which included biological N_2-fixation (Boddey et al. 1995); production of phytohormones indole-3-acetic acid (Bric et al. 1991), gibberellic acid (Brown and Burlingham 1968), and 1-aminocyclopropane-1-carboxylate (ACC) deaminase (Jacobson et al. 1994); and solubilization of phosphorus (Pikovskaya 1948), potassium (Hu et al. 2006), and zinc (Fasim et al. 2002). The microbes should be also screened for qualitatively indirect PGP attributes which included production of ammonia (Cappucino and Sherman 1992), HCN (Bakker and Schippers 1987), siderophore (Schwyn and Neilands 1987), and lytic enzyme (Yadav et al. 2016a) and biocontrol against different fungal pathogens (Sijam and Dikin 2005) under different temperatures (5–40 °C). After qualitatively screening, the selected cold tolerant with PGP attributes should be quantitatively screened for N_2-fixing attributes by using the acetylene reduction assay (ARA) (Han and New 1998), P-solubilization (Mehta and Nautiyal 2001), K-solubilization (Verma et al. 2016), and IAA production (Patten and Glick 2002).

Table 11.1 The different media used for isolation of microbes from cold environments

Microbes	Media and composition per liter
Archaea	Chemically defined medium: 5 g casamino acids; 5 g yeast extract; 1 g sodium glutamate; 3 g trisodium citrate; 20 g $MgSO_4$; 2 g KCl; 100 g NaCl; 36 mg $FeCl_2$; 0.36 mg $MgCl_2$
	Standard growth media: 7.5 g casamino acids; 4 g $MgSO_4$; 2 g KCl; 150 g NaCl; 3 g trisodium citrate; 2.3 mg $FeCl_2$; 7 mg $CaCl_2$; 0.044 mg $MnSO_4$; 0.05 mg $CuSO_4$
	Halophilic medium: 100 g NaCl; 2 g KCl; 1 g $MgSO_4{\cdot}7H_2O$; 0.36 g $CaCl_2{\cdot}2H_2O$; 0.23 g NaBr; 0.06 g $NaHCO_3$; 5 g protease-peptone; 10 g yeast extract; 1 g glucose; trace FeCl3
Bacteria	Ammonium minerals salt: 0.70 g K_2HPO_4; 0.54 g KH_2PO_4; 1 g $MgSO_4$. $7H_2O$; 0.2 g $CaCl_2$. $2H_2O$; 4.0 mg $FeSO_4.7H_2O$; 0.5 g NH_4Cl; $ZnSO_4$. $7H_2O$; 30 μg $MnCl_2.4H_2O$; 300 μg H_3BO_3; 10 μg $CuCl_2$. $2H_2O$; 200 μg $CoCl_2$. $6H_2O$; 20 μg $NiCl_2.6H_2O$; 60 μg $Na_2MoO_4.2H_2O$
	Antarctic bacterial medium (ABM): 5 g peptone; 2 g yeast extract
	Jensen's agar: 20 g sucrose; 1 g K_2HPO_4; 0.5 g Mg_2SO_4; 0.5 g NaCl; 0.001 g Na_2MoO_4; 0.01 g $FeSO_4$; 2 g $CaCO_3$
	King's B agar: 20 g proteose peptone; 1.5 g K_2HPO_4; 1.5 $MgSO_4$.7H2O; 10 ml glycerol
	Nutrient agar: 5 g peptone; 5 g NaCl; 3 g beef extract
	R_2 agar: 0.5 g proteose peptone; 0.5 g casamino acids; 0.5 g yeast extract; 0.5 g dextrose; 0.5 g soluble starch; 0.3 g dipotassium phosphate; 0.05 g magnesium sulfate $7H_2O$; 0.3 g sodium pyruvate
	Sea water medium: 5 g tryptone; 10 g yeast extract; 5 g casamino acid; 3 g citrate solution; 2 g KCl; 20 g $MgSO_4 \cdot 7\ H_2O$
	Soil extract agar: 2 g glucose; 1 g yeast extract; 0.5 g K_2HPO_4; 100 ml soil extract (soil extract: 250 g soil from sampling site +1 L H_20, autoclave and filter)
	Starch yeast peptone agar: 2 g starch; 0.8 g yeast extract; 0.1 g peptone
	T_3 agar: 3 g tryptone; 2 g tryptose; 1.5 g yeast extract; 0.005 g $MnCl_2$; 0.05 sodium phosphate
	Tryptic soy agar: 17 g tryptone; 3 g soya meal; 2.5 g dextrose; 5 g NaCl; 2.5 g K_2HPO_4
	Yeast extract mannitol agar: 1 g yeast extract; 10 g mannitol; 0.5 g $K_2HPO_4.H_2O$; 0.002 g $MgSO_4.7H_2O$; 0.1 g NaCl
Fungi	Czapek Dox Agar: sucrose 30 g; sodium nitrate 2 g; dipotassium phosphate 1 g; magnesium sulfate 0.5 g; potassium chloride 0.5 g; ferrous sulfate 0.01
	Potato dextrose agar: dextrose 20 g; potato extract 4 g (200 g of potato infusion)
	Rose Bengal agar: papaic digest of soya bean meal 5 g; dextrose 10 g; monopotassium phosphate 1 g; magnesium sulfate 0.5 g; Rose Bengal 0.05 g
	Sabouraud dextrose agar: mycological peptone 10 g; dextrose 40 g

For identification and phylogenetic profiling of microbes from cold habitat, the genomic DNA should be isolated from purified pelleted microbial cells of 1.5 mL broth and should be washed two to three times in 1.0 mL of TE buffer (10 mM Tris-HCl and 1 mM EDTA pH 8.0). Microbial lysis should be performed using 0.5 mL SET buffer (75 mM NaCl, 25 mM EDTA, and 20 mM Tris) with 10 μL of lysozyme (10 mg mL^{-1}) for 30 min at 37 °C and 10% SDS with 20 mg mL^{-1} proteinase K for 1 h at 55 °C. DNA should be extracted using phenol/chloroform/isoamyl alcohol,

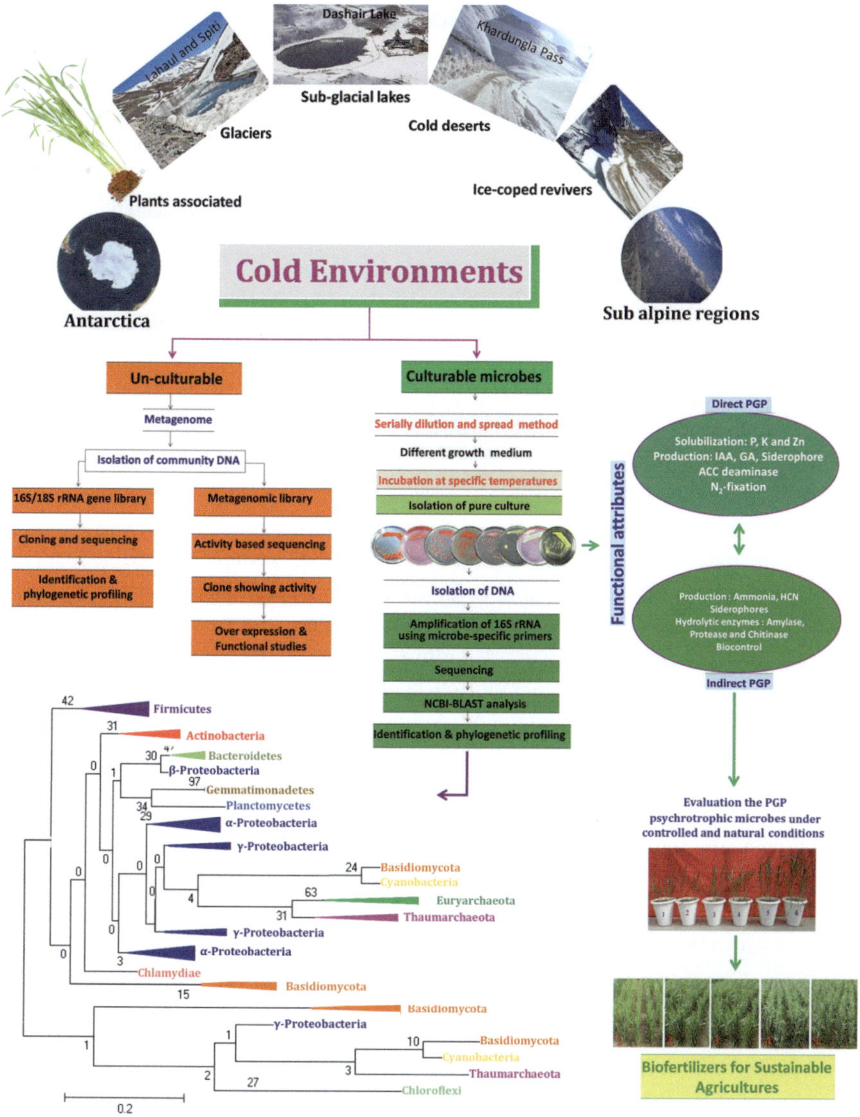

Fig. 11.1 A schematic representation of the isolation, identification, and potential applications of psychrotrophic microbes from different habitats of extreme cold environments

and aqueous phase can be transferred to a fresh tube. Finally, the washed DNA pellet should be incubated at 37 °C for 25–30 min to completely remove ethanol and then resuspended in 50 μL of TE buffer. The amount of DNA extracted should be electrophoresed on 0.8% agarose gel. The primers universal 16S rRNA/18S rRNA/ITS should be used for the amplification of conserved genes. The PCR-amplified 16S rRNA/18S rRNA/ITS should be purified with a Qiaquick purification kit (Qiagen).

The partial 16S rRNA/18S rRNA/ITS gene sequences should be compared with those available in the NCBI databases. Identification at the species level was determined using a 16S rRNA/18S rRNA/ITS gene sequence similarity of ≥97% with that of a prototype strain sequence in the GenBank. Sequence alignment and comparison can be performed, using the program CLUSTAL W. The phylogenetic tree can be constructed on the aligned datasets using the neighbor-joining method (Saitou and Nei 1987) implemented in the program MEGA 4.0.2 (Tamura et al. 2007).

11.3 Diversity and Distribution of Psychrotrophic Microbes

Earth is primarily a cold, marine planet with 90% of the ocean's waters being at 5 °C or lower. Permafrost soils, glaciers and ice sheets, polar sea ice, and snow cover make up 20% of the Earth's surface environments. Habitats for cold-adapted microorganisms represent a large proportion of the Earth's area. Much of the oceans, which cover some 70% of the Earth's surface, are at an average temperature of −1 to +5 °C. Alpine and Polar region constitute 25% of the world's land surface area. Man-made habitats (such as refrigeration and freezer systems) contribute only a small proportion of potential habitats for cold-adapted organisms. The diversity of microorganisms inhabiting cold environments has been extensively investigated in the past few years with a focus on culture-dependent techniques. The different groups of microbes have been reported such as archaea, eubacteria, and fungi, which included different phylum mainly Actinobacteria, Ascomycota, Bacteroidetes, Basidiomycota, Chlamydiae, Chloroflexi, Cyanobacteria, Euryarchaeota, Firmicutes, Gemmatimonadetes, Mucoromycota, Nitrospirae, Planctomycetes, Proteobacteria, Spirochaetes, Thaumarchaeota, and Verrucomicrobia (Fig. 11.2). Overall the distribution of psychrotrophic microbes varied in all bacterial phyla; Proteobacteria were the most dominant followed by Firmicutes and Actinobacteria. The least number of microbes was reported from phylum Mucoromycota followed by Gemmatimonadetes, Nitrospirae, and Thaumarchaeota (Fig. 11.3).

On the review of different extreme cold environments (Antarctica, permanently ice-covered lakes, ice-capped rivers, glaciers, and those associated with plants growing in cold habitat), it was found that 17 different phyla have been sorted out belonging to different domains of Archaea, Bacteria, and Eukarya. The overall percentages of different phyla included are Actinobacteria (12), Ascomycota (0.96), Bacteroidetes (5.18), Basidiomycota (4.22), Chlamydiae (1.15), Chloroflexi (0.96), Cyanobacteria (1.54), Euryarchaeota (1.34), Firmicutes (21.31), Gemmatimonadetes (0.38), Mucoromycota (0.19), Nitrospirae (0.38), Planctomycetes (0.77), Proteobacteria (48.08), Spirochaetes (0.58), Thaumarchaeota (0.38), and Verrucomicrobia (0.58) (Fig. 11.3). On the review of seven different extreme cold environments, Antarctica, plant associated [*Amaranthus*, *Brassica*, cabbage, garlic, maize, pea, and wheat], glacier [Gangotri Glacier, Kafni Glacier, Lahaul and Spiti, Pindari Glacier, Roopkund Glacier, Union Glacier, and Zangser Kangri Glacier], sub-glacial lakes [Chandratal Lake, Dal Lake, Dashair Lake, Gurudongmar Lake, and Pangong Lake], cold desert of Himalayas [Chumathang, Khardungla Pass, and

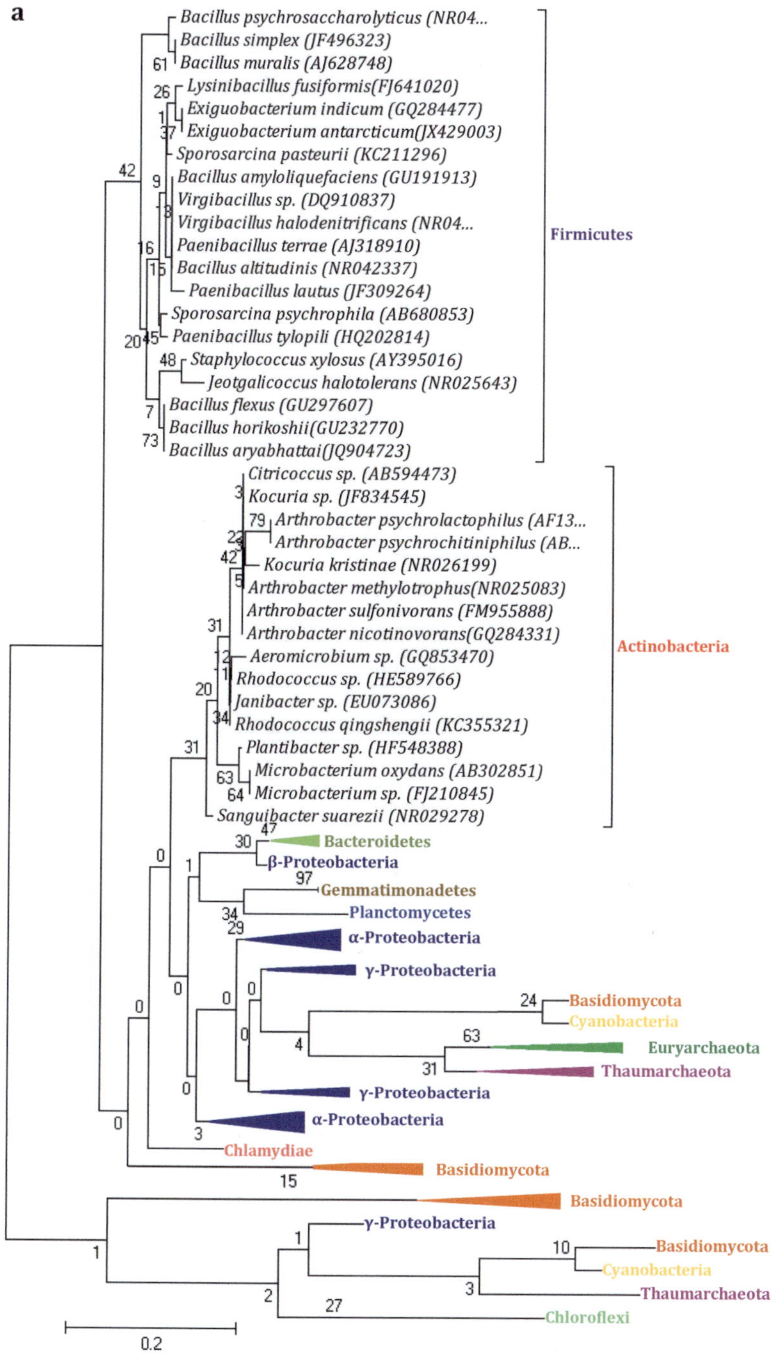

Fig. 11.2 (**a**, **b**, **c**) Phylogenetic tree showing the relationship among psychrotrophic microbes, isolated from diverse habitat of cold environments

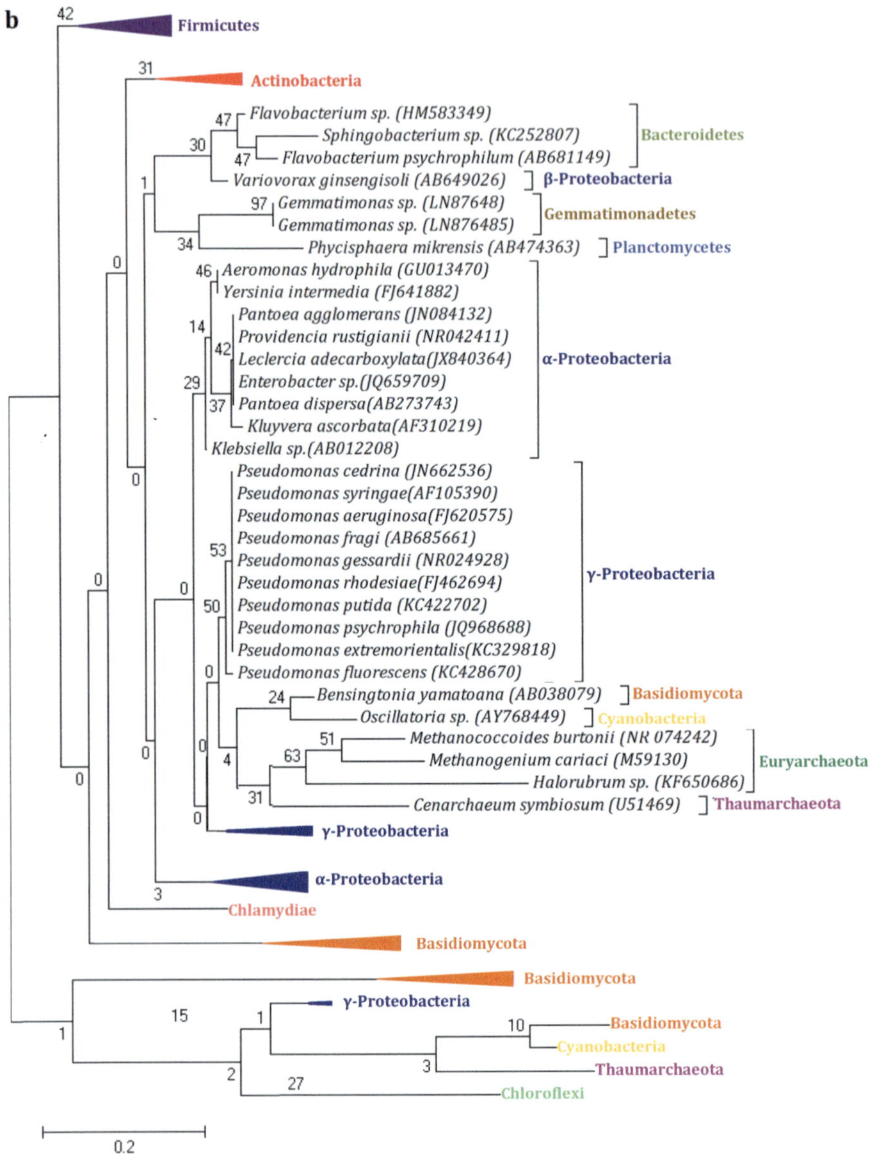

Fig.11.2 (continued)

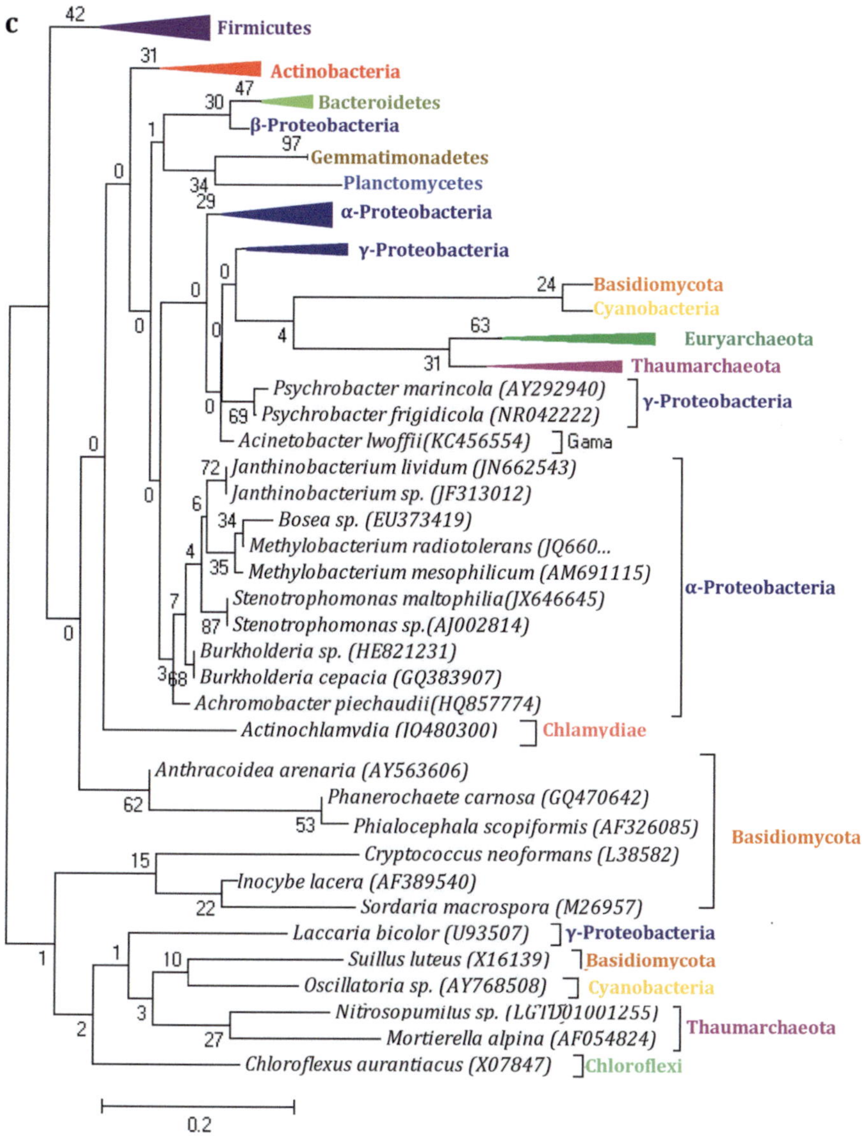

Fig.11.2 (continued)

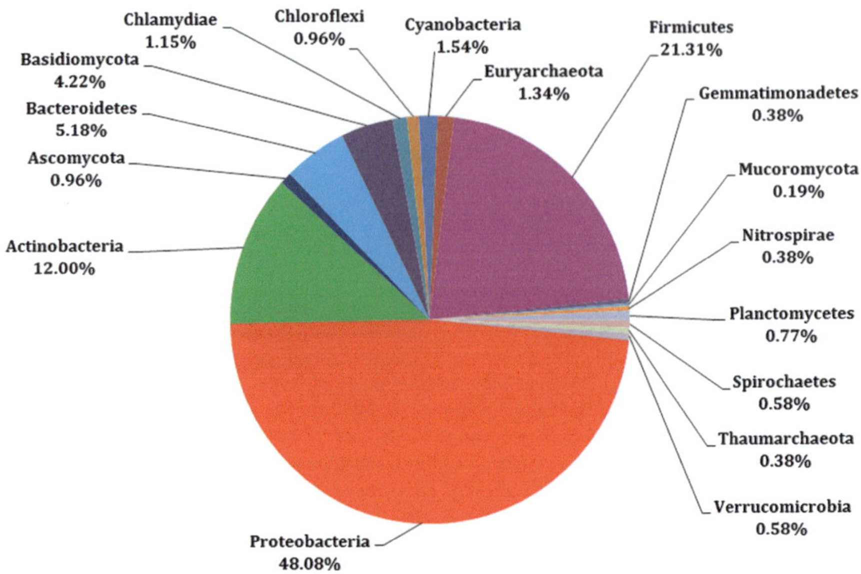

Fig. 11.3 Distribution of different phyla of psychrotrophic microbes, isolated from diverse habitat of cold environments

Rohtang Pass], ice-capped rivers [Indus River, Zanskar River, and Beas River], and subalpine Uttarakhand, it was found that, among 17 different phyla, four phyla, namely, Bacteroidetes, Cyanobacteria, Firmicutes, and Proteobacteria, are found at all site surveys (Fig. 11.4). Among all 17 phyla, the microbes of seven phyla, namely, Ascomycota, Euryarchaeota, Mucoromycota, Nitrospirae, Spirochaetes, Thaumarchaeota, and Verrucomicrobia, were reported of only one site to be niche-specific microbes (Fig. 11.4). On deciphering different habitats of extreme cold environments, it was found that more than 120 different genera of 17 different phyla have been sorted out and characterized. On the basis of different research, it may be concluded that among all reported cold-adapted genera, the five genera namely *Bacillus*, *Exiguobacterium*, *Pseudomonas*, *Psychrobacter*, and *Sphingobacterium* were ubiquitous in nature and have been reported from the all site survey (Fig. 11.5). The cold-adapted microbes have been characterized by both culture-dependent and culture-independent approach. It is possible to assess only a small fraction of the microbial diversity associated with plants using the isolation methods described above because few microbial species can be cultivated using traditional laboratory methods. The sizes of microbial communities as determined using culture-independent methods might be 100- to 1000-fold larger than communities uncovered via traditional isolation. Archaea were also reported to be associated with maize and rice using unculturable method only. There are very few reports for niche-/crop-specific microbes, but there were many reports on niche specificity of microbes from different extreme habitats (Kumar et al. 2014a, b; Pandey et al. 2013; Saxena et al. 2016; Suman et al. 2015; Yadav et al. 2015a, c, 2017h).

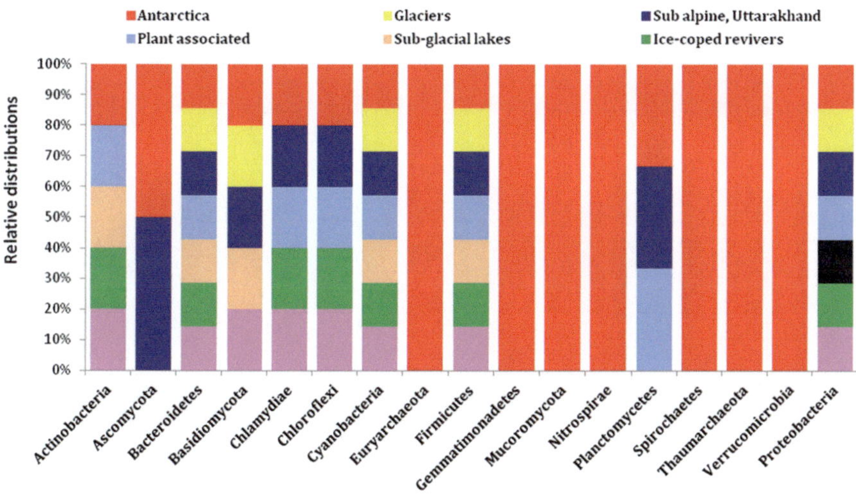

Fig. 11.4 Relative distribution of different phyla of psychrotrophic microbes, isolated from diverse habitat of cold environments

Many novel psychrotrophic microbes have been sorted out from cold environments including *Halobacterium lacusprofundi* (Franzmann et al. 1988), *Sphingobacterium antarcticus* (Shivaji et al. 1992), *Octadecabacter arcticus* (Gosink et al. 1997), *Hymenobacter roseosalivarius* (Hirsch et al. 1998), *Cellulophaga algicola* (Bowman 2000), *Flavobacterium frigidarium* (Humphry et al. 2001), *Oleispira antarctica* (Yakimov et al. 2003), *Flavobacterium psychrolimnae* (Van Trappen et al. 2005), *Psychromonas ingrahamii* (Auman et al. 2006), *Exiguobacterium soli* (Chaturvedi et al. 2008), *Pseudomonas extremaustralis* (López et al. 2009), *Cryobacterium roopkundense* (Reddy et al. 2010), *Sphingomonas glacialis* (Zhang et al. 2011), *Pedobacter arcticus* (Zhou et al. 2012), *Sphingobacterium psychroaquaticum* (Albert et al. 2013), *Lacinutrix jangbogonensis* (Lee et al. 2014), *Massilia eurypsychrophila* (Shen et al. 2015), *Glaciimonas frigoris* (Margesin et al. 2016), and *Psychrobacter pocilloporae* (Zachariah et al. 2017). There are several reports on whole genome sequences of novel and potential psychrotrophic microbes (Kim et al. 2012; Singh et al. 2016). The novel species of psychrotrophic microbes have been isolated worldwide and reported from different domain Archaea, Bacteria, and Eukarya which included members of phylum Actinobacteria, Proteobacteria, Bacteroidetes, Basidiomycota, Firmicutes, and Euryarchaeota. Along with novel species of psychrotrophic microbes, some microbial species including *Arthrobacter nicotianae*, *Brevundimonas terrae*, *Paenibacillus tylopili*, and *Pseudomonas cedrina* have been reported first time from cold deserts of NW Himalayas and exhibited multifunctional PGP attributes at low temperatures (Yadav et al. 2015a). In a study by Yadav et al. (2015b), the microbial species *Alishewanella* sp., *Aurantimonas altamirensis*, *Bacillus baekryungensis*, *Bacillus marisflavi*, *Desemzia incerta*, *Paenibacillus xylanexedens*, *Pontibacillus* sp., *Providencia* sp., *Pseudomonas frederiksbergensis*, *Sinobaca beijingensis*, and

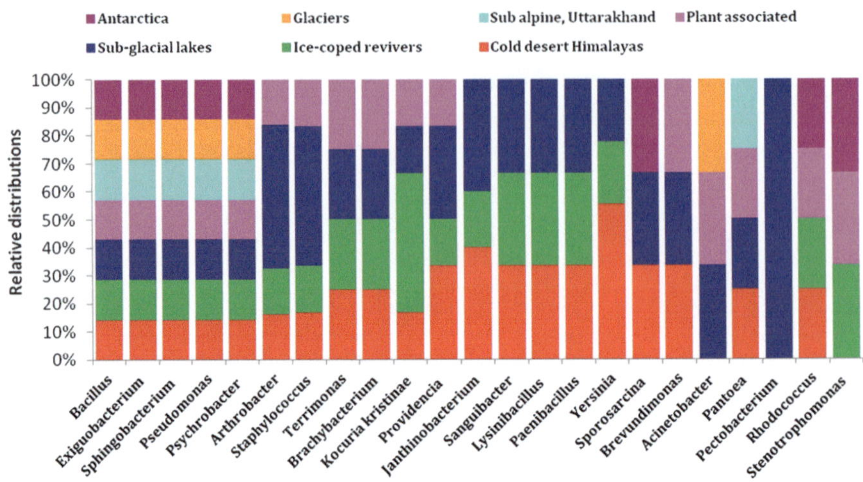

Fig. 11.5 Relative distribution of different genera of psychrotrophic microbes, isolated from diverse habitat of cold environments
Antarctica (Franzmann et al. 1992, 1997, 1988; Deming 2002; von Klein et al. 2002; Könneke et al. 2005; Marx et al. 2007; Chaturvedi et al. 2008; Kostadinova et al. 2009; Turchetti et al. 2011; Lee et al. 2016; Choi et al. 2016; He et al. 2017), *plant associated* [*Amaranthus*, *Brassica*, cabbage, garlic, maize, pea, and wheat] (Bisht et al. 2013; Verma et al. 2014b, 2015a, b, c, d, 2016, 2017a; Yadav et al. 2015a; Rana et al. 2016b, 2017; Kaur et al. 2017), *glacier* [Gangotri Glacier, Kafni Glacier, Lahaul and Spiti, Pindari Glacier, Roopkund Glacier, Union Glacier, and Zangser Kangri Glacier] (Baghel et al. 2005; Gulati et al. 2008; Malviya et al. 2009; Vyas et al. 2010; Pradhan et al. 2010; Shivaji et al. 2011; Srinivas et al. 2011; Barahona et al. 2016; Lee et al. 2017), *sub-glacial lakes* [Chandratal Lake, Dal Lake, Dashair Lake, Gurudongmar lake, and Pangong Lake] (Anupama et al. 2011; Yadav et al. 2012, 2013; Sahay et al. 2013; Pandey et al. 2013; Saxena et al. 2014; Srivastava et al. 2014; Yadav et al. 2014a, b, c, 2015a, b, d, e, f, 2016a, b, c, 2017h; Venkatachalam et al. 2015; Singh et al. 2016), *cold desert of Himalayas* [Chumathang, Khardungla Pass, and Rohtang Pass] (Pandey et al. 2013; Saxena et al. 2014, 2016; Srivastava et al. 2014; Yadav et al. 2015a, b, 2016a, b, c, 2017h; Shukla et al. 2016), *ice-capped rivers* [Indus River, Zanskar River, and Beas River] (Pandey et al. 2013; Saxena et al. 2014; Srivastava et al. 2014; Yadav et al. 2015a, b, 2016a, b, c, 2017h; Saxena et al. 2016), *subalpine, Uttarakhand* (Mishra et al. 2008; Selvakumar et al. 2008, 2009a, b, c, 2011; Mishra et al. 2009, 2011; Bisht et al. 2013; Nazir et al. 2017)

Vibrio metschnikovii have been reported first time from high-altitude and low-temperature environments of Indian Himalayas. Wheat-associated psychrotrophic bacteria *Arthrobacter methylotrophus* and *Pseudomonas rhodesiae* have been reported first time from wheat growing in north hills zone of India (Verma et al. 2015a). In a specific search of economically important *Bacillus* and *Bacillus*-derived genera (BBDG) at low temperature, various BBDG such as *Bacillus psychrosaccharolyticus*, *B. amyloliquefaciens*, *B. altitudinis*, *B. muralis*, *Paenibacillus tylopili*, *P. pabuli*, *P. terrae*, and *P. lautus* with efficient PGP attributes have been reported first time by Yadav et al. (2016b).

11.3.1 Archaea

Archaea are unique microorganisms that are present in ecological niches of high/low temperature and pH and high salinity. Archaea may be present freely or associated with plant rhizosphere. Archaea have been reported as ubiquitous and present in a wide range of environments, from cold environment (Dong and Chen 2012) and thermal springs (De León et al. 2013) and those associated with different plants (Saxena et al. 2015a; Yadav et al. 2015c, 2017b; Gaba et al. 2017). Archaea are a common component of different extreme environments. The largest proportion and greatest diversity of archaea exist in cold environments including Antarctic subglacial and permanently ice-covered lakes, cloud droplets, ice cap cores from considerable depth, snow deep sea, and glaciers. Most of the Earth's biosphere is cold, and archaea represent a significant fraction of the biomass. Although psychrophilic archaea have long been the neglected majority, the study of these microorganisms is beginning to come of age.

Psychrophilic archaea have diverse functional roles in a wide range of cold environments, and their environmental impact is well illustrated by their abundance in the ocean. There are no report of cold-adapted archaea in crop improvements, whereas as other groups of archaea including thermophilic, halophilic, and acidophilic have been reported for PGP under the respective abiotic stresses such as phosphorus solubilization by haloarchaea (Yadav et al. 2015c, 2017b), nitrogen fixation by methanogens (Leigh 2000), siderophore production (Dave et al. 2006), and IAA production (Saxena et al. 2015a; White 1987; Yadav et al. 2017b). It has become clear that cold-adapted archaea contribute to global energy cycles through the processing of organic and inorganic carbon and nitrogen compound.

A diverse group of methanogens and Crenarchaeota are among the most abundant members of the domain Archaea in the cold biosphere. The archaea that are most readily reported from naturally cold environments and that are most acquiescent to laboratory cultivation are methanogens, such as *Methanosarcina*, *Methanospirillum*, *Methanocorpusculum*, and *Methanomethylovorans* which can be isolated from Antarctic lake, freshwater lake, cold marine sediment, and the Baltic Sea (Franzmann et al. 1997; von Klein et al. 2002), and *Methanosarcina baltica*, isolated from the Gotland Deep of the Baltic Sea (von Klein et al. 2002). The eurypsychrophilic archaeon *Halorubrum lacusprofundi* was isolated from Deep Lake, Vestfold Hills, Antarctica (Franzmann et al. 1988).

Franzmann et al. (1992) reported a methylotrophic, methanogenic bacterium that was isolated from the anoxic hypolimnion of Ace Lake, Antarctica, which has been identified as *Methanococcoides methylutens*. The optimum and maximum temperatures for growth were 23.4 and 29.5 °C, respectively. The strain grew in artificial media at in situ lake temperature (1.7 °C), provided growth was first initiated in the media at higher temperatures. The strain had a theoretical minimum temperature for growth of −2.5 °C. The mol% G + C content of DNA from the strain was 39.6% (T_m).

Franzmann et al. (1997) reported *Methanogenium frigidum* from the perennially cold, anoxic hypolimnion of Ace Lake in the Vestfold Hills of Antarctica. The cells were psychrophilic, exhibiting the most rapid growth at 15 °C and no growth at temperatures above 18–20 °C. The cells were slightly halophilic; good growth occurred in medium supplemented with 350–600 mM Na^+, but no growth occurred with 100 or 850 mM Na^+. The pH range for growth was 6.5–7.9; no growth occurred at pH 6.0 or 8.5. Growth was slow (maximum specific growth rate, 0.24 day^{-1}; doubling time, 2.9 days). This is the first report of a psychrophilic methanogen growing by CO_2 reduction.

Wagner et al. (2013) reported that a methanogenic archaeon, strain SMA-21^T, was isolated from a permafrost-affected soil by serial dilution in a liquid medium. The cells were non-motile, were stained gram-negative, and grew as irregular cocci with a diameter of 1.3–2.5 μm. Optimal growth was observed at 28 °C, pH 7.8, and 0.02 M NaCl. The strain grew on H_2/CO_2, methanol, and acetate, but not on formate, ethanol, 2-butanol, 2-propanol, monomethylamine, dimethylamine, trimethylamine, or dimethyl sulfide. The 16S rRNA gene sequence was closely related to those of *Methanosarcina mazei* DSM 2053^T (similarity 99.9%) and *Methanosarcina horonobensis* HB-1^T (similarity 98.7%). On the basis of the level of DNA-DNA hybridization (22.1%) between strain SMA-21^T and *Methanosarcina mazei* DSM 2053^T as well as of phenotypic and genotypic differences, strain SMA-21^T was assigned to a novel species of the genus *Methanosarcina*, for which the name *Methanosarcinasoligelidi* sp. nov. is proposed. The type strain is SMA-21^T (=DSM 20065^T = JCM 18468).

11.3.2 Bacteria

Among different groups of microbes isolated and reported from extreme cold environments, the most dominant groups/domain was Bacteria which include different phylum mainly: Actinobacteria, Bacteroidetes, Chlamydiae, Chloroflexi, Cyanobacteria, Firmicutes, Gemmatimonadetes, Nitrospirae, Planctomycetes, Proteobacteria, Spirochaetes, and Verrucomicrobia (Figs. 11.3 and 11.4). The Actinobacteria is a phylum of gram-positive bacteria. The members of bacteria belonging to phylum Actinobacteria are classified into six classes, namely, Acidimicrobiia, Actinobacteria, Coriobacteriia, Nitriliruptoria, Rubrobacteria, and Thermoleophilia (Yadav et al. 2017f). Among six different classes, members of class Actinobacteria are the most dominant and contain one of the largest of bacterial genera, *Streptomyces*. The bacterial species of phylum Actinobacteria are ubiquitous in nature and have been isolated from different extreme environments (extreme temperatures, pH, salinities, pressure, and drought) and those associated with plants growing in different habitats (Yadav et al. 2017e, h, i). Based on a comprehensive literature analysis, psychrotrophic members of phylum Actinobacteria have been reported from different genera such as *Acidimicrobium*, *Actinomyces*, *Arthrobacter*, *Bifidobacterium*, *Cellulomonas*, *Clavibacter*, *Corynebacterium*, *Frankia*, *Microbacterium*, *Micrococcus*, *Mycobacterium*, *Nocardia*, *Propionibacterium*, *Pseudonocardia*, *Rhodococcus*, *Sanguibacter*, and *Streptomyces*. Among the members of class Actinobacteria, the species of *Arthrobacter* were the most dominant in cold environments which included *Arthrobacter antarcticus*, *A. aurescens*, *A. chlorophenolicus*, *A. defluvii*, *A. equi*, *A.*

globiformis, *A. ilicis*, *A. koreensis*, *A. methylotrophus*, *A. nicotianae*, *A. nicotinovorans*, *A. nitroguajacolicus*, *A. oryzae*, *A. oxydans*, *A. pascens*, *A. polychromogenes*, *A. psychrochitiniphilus*, *A. psychrolactophilus*, *A. ramosus*, *A. sulfonivorans*, *A. sulfurous*, and *A. ureafaciens* (Verma et al. 2015a; Yadav et al. 2015a, b, 2017c, g; Rana et al. 2016a, 2017; Singh et al. 2016; Suman et al. 2016b).

The bacterial species of *Bacillus* and *Bacillus*-derived genera (BBDG), which are associated with different plants, show different plant growth-promoting attributes. The PCR-denaturing gradient gel electrophoresis (DGGE) technique was developed to study the diversity of *Bacillus* (including the groups separated as *Paenibacillus*, *Alicyclobacillus*, *Aneurinibacillus*, *Virgibacillus*, *Salibacillus*, and *Gracilibacillus*). The genus *Bacillus* consists of a heterogenic group of gram-positive rods, able to form endospores that allow them to survive for extended periods under adverse environmental conditions, better than nonsporulating bacterial enteropathogens. Endospore formation is the dominant feature in the characterization of *Bacillus*. There is a boundary that separates this genus from other genera in which endospores are produced. The genus *Clostridium* is distinguished from *Bacillus* by its inability to grow on the surface of agar media in air, or, if growth does occur under these conditions, it is slight and does not lead to sporulation. There is also little or no catalase activity. *Sporosarcina* is sharply separated from *Bacillus* by the coccal form of its vegetative cells.

Many BBDG have been reported from cold environments such as *Bacillus aerius*, *B. aerophilus*, *B. stratosphericus*, *B. altitudinis* (Shivaji et al. 2006), *B. thuringiensis* (Das and Dangar 2008), *B. megaterium* (Trivedi and Pandey 2008), *Paenibacillus glacialis* (Kishore et al. 2010), and *Bacillus cereus*, *B. amyloliquefaciens*, *B. cibi*, *B. pumilus*, and *Lysinibacillus fusiformis* (Sahay et al. 2013, 2017). Among BBDG, several species of *Bacillus* have been sorted out from cold habitat, e.g., *Bacillus acidicola*, *B. altitudinis*, *B. amyloliquefaciens*, *B. anthracis*, *B. aryabhattai*, *B. asahii*, *B. baekryungensis*, *B. cereus*, *B. cibi*, *B. circulans*, *B. firmus*, *B. flexus*, *B. horikoshii*, *B. kochii*, *B. licheniformis*, *B. marisflavi*, *B. megaterium*, *B. mojavensis*, *B. muralis*, *B. pseudomycoides*, *B. psychrosaccharolyticus*, *B. pumilus*, *B. safensis*, *B. simplex*, *B. sonorensis*, *B. subtilis*, *B. thuringiensis*, and *B. weihenstephanensis*. Among the members of Firmicutes, *Bacillus* were reported as the most dominant genera from cold environments followed by *Exiguobacterium mexicanum*, *E. acetylicum*, *E. antarcticum*, *E. artemiae*, *E. aurantiacum*, *E. homiense*, *E. indicum*, *E. marinum*, *E. sibiricum*, and *E. soli*; *Paenibacillus amylolyticus*, *P. lautus*, *P. pabuli*, *P. terrae*, *P. tylopili*, and *P. xylanexedens*; and *Planococcus antarcticus*, *P. donghaensis*, and *P. kocurii* (Pandey et al. 2013; Verma et al. 2015a; Yadav et al. 2015a, b, c, d, e, f, 2016a, b, c, 2017e; Rana et al. 2016a, 2017; Saxena et al. 2015b; Singh et al. 2016; Kumar et al. 2017).

Verma et al. (2015d) investigated 41 endophytic bacteria that were isolated from surface-sterilized roots and culms of wheat *var.* HS507, growing in NW Indian Himalayas. These bacteria were screened *in vitro* for multifarious plant growth-promoting attributes such as solubilization of phosphorus, potassium, and zinc; production of indole-acetic acids, hydrogen cyanide, gibberellic acid, and siderophore; nitrogen fixation; ACC deaminase activity; and biocontrol against *Rhizoctonia solani* and *Macrophomina phaseolina* at low temperature (4 °C). One isolate, IARI-HHS2-30, showed that appreciable level of potassium solubilization was further characterized *in vivo* at control condition of low temperature. Based on 16S rRNA gene sequence analysis, this isolate was identified as *Bacillus amyloliquefaciens* assigned with accession number KF054757. Analysis of the phylogenetic characterization showed close homology with typical psychrotolerant bacteria *Bacillus amyloliquefaciens*, *Bacillus methylotrophicus*, *Bacillus polyfermenticus*, *Bacillus siamensis*, *Bacillus subtilis*, and *Bacillus vallismortis*. Endophytic nature and plant growth-promoting ability of IARI-HHS2-30 was tested qualitatively and followed by inoculation onto wheat seedlings in low-temperature conditions. At 30 days after inoculation, *Bacillus amyloliquefaciens* IARI-HHS2-30 to wheat plants resulted in significant increase in root/shoot length, fresh weight, and chlorophyll content. Plant growth-promoting features coupled with psychrophilic ability suggest that this endophytic bacterium may be exploited as bio-inoculants for various crops in low-temperature and high-altitude condition.

Yadav et al. (2016b) reported and characterized psychrotrophic bacilli from different sites in northwestern Indian Himalayas. A total of 247 morphotypes were obtained from different soil and water samples and were grouped into 43 clusters based on 16S rDNA-RFLP analysis. Sequencing of representative isolates from each cluster led to their identification, and 43 bacilli belonged to different species of 11 genera, *viz.*, *Desemzia*, *Exiguobacterium*, *Jeotgalicoccus*, *Lysinibacillus*, *Paenibacillus*, *Planococcus*, *Pontibacillus*, *Sinobaca*, *Sporosarcina*, *Staphylococcus*, and *Virgibacillus*. With an aim to develop microbial inoculants that can perform efficiently at low temperatures, all representative isolates were screened for different plant growth-promoting traits at low temperatures (5–15 °C). Among the strains, variations were observed for production of ammonia (22%), indole-3-acetic acid (20%), siderophores (8%), gibberellic acid (4%), and hydrogen cyanide (3%); solubilization of phosphate (12%), zinc (16%), and potassium (8%); 1-aminocycloprop ane-1-carboxylate deaminase activity (5%); and biocontrol activity (19%) against *Rhizoctonia solani* and *Macrophomina phaseolina*. Among all the strains, *Bacillus licheniformis*, *Bacillus muralis*, *Desemzia incerta*, *Paenibacillus tylopili*, and *Sporosarcina globispora* were found to be potent candidates to be developed as inoculants as they exhibited multiple PGP traits at low temperature.

The phylum Proteobacteria is a major group of gram-negative bacteria which included α/β/γ/δ-Proteobacteria. The α-Proteobacteria grows at very low levels of nutrients and includes agriculturally important bacteria capable of inducing *Azospirillum* and fixing nitrogen in symbiosis with plants. The β-Proteobacteria is highly metabolically diverse and contains chemolithoautotrophs, photoautotrophs, and generalist heterotrophs, while the γ-Proteobacteria is the largest class in terms of species, *Pseudomonas* and *Azotobacter*. Among the Proteobacteria, γ-Proteobacteria were the most dominant, in which the members of the genus *Pseudomonas* were predominant in extreme cold environments such as *Pseudomonas aeruginosa*, *P. antarctica*, *P. azotoformans*, *P. baetica*, *P. cedrina*, *P. corrugate*, *P. costantinii*, *P. deceptionensis*, *P. extremaustralis*, *P. extremorientalis*, *P. fluorescens*, *P. fragi*, *P. frederiksbergensis*, *P. geniculata*, *P. gessardii*, *P. graminis*, *P. jessani*, *P.*

kilonensis, *P. koreensis*, *P. lurida*, *P. mediterranea*, *P. moraviensis*, *P. orientalis*, *P. pavonaceae*, *P. peli*, *P. plecoglossicida*, *P. poae*, *P. psychrophila*, *P. putida*, *P. reactans*, *P. rhodesiae*, *P. simiae*, *P. stutzeri*, *P. syringae*, *P. teessidea*, *P. tolaasii*, *P. trivialis*, *P. vancouverensis*, and *P. xanthomarina* (Srivastava et al. 2014; Verma et al. 2015a; Saxena et al. 2015b; Singh et al. 2016; Rana et al. 2017; Yadav et al. 2017i; Kumar et al. 2017).

Verma et al. (2013) reported 135 wheat-associated plant growth-promoting bacteria from acidic soil; among all isolates, *Pseudomonas chlororaphis* IARI-THD-13, *Pseudomonas fluorescens* IARI-THD-21, *Pseudomonas rhodesiae* IARI-THD-11, and *Pseudomonas rhodesiae* IARI-THD-28 exhibited direct and indirect plant growth-promoting attributes. The plant growth-promoting (PGP) traits were done *in vitro* at different pH which included solubilization of phosphorus, potassium, and zinc; production of ammonia, hydrogen cyanide, indole-3-acetic acid, and siderophore; nitrogen fixation; 1-aminocyclopropane-1-carboxylate deaminase activity; and biocontrol against *Fusarium graminearum*, *Rhizoctonia solani*, and *Macrophomina phaseoli*. Acidotolerant isolates may have application as inoculant plant growth-promoting and biocontrol agents in crops growing under acidic condition.

Verma et al. (2014a, b) investigated thermotolerant wheat-associated plant growth-promoting bacteria which have been identified using 16S rRNA gene sequencing. A total of 348 isolates belonged to three phyla, namely, Actinobacteria, Firmicutes, and Proteobacteria, with 38 distinct species of 17 genera. *Bacillus* and *Pseudomonas* were dominant in rhizosphere while *Methylobacterium* were in phyllosphere. Different species of *Pseudomonas fuscovaginae* IARI-IIWP-29, *Pseudomonas lini* IARI-IIWP-33, *Pseudomonas monteilii* IARI-IIWP-27, *Pseudomonas stutzeri* IARI-IHD-4, and *Pseudomonas thivervalensis* IARI-IHD-3 have been sorted out from wheat as endophytic, rhizospheric, as well as epiphytic. *In vitro* plant growth-promoting activities of bacteria exposed more than three beneficial traits which may act independently or concurrently. P-solubilization and siderophores production are the predominant traits exhibited by these microbes. The many species of genera *Bacillus*, *Exiguobacterium*, *Micrococcus*, *Pseudomonas*, and *Psychrobacter* showed antagonistic properties against fungal pathogens *Fusarium graminearum*, *Rhizoctonia solani*, and *Macrophomina phaseoli*. These promising isolates showing a range of useful plant growth-promoting attributes insist to be explored for agricultural applications.

Verma et al. (2015a) reported psychrotolerant wheat-associated bacteria from northern hills zone of India. A total of 247 bacteria were isolated from 5 different sites. 16S rRNA gene-based phylogenetic analysis revealed that 65, 26, 8, and 1% bacteria belonged to four phyla, namely, Proteobacteria, Firmicutes, Actinobacteria, and Bacteroidetes, respectively. Overall 28% of the total morphotypes belonged to *Pseudomonas* followed by *Bacillus* (20%); *Stenotrophomonas* (9%); *Methylobacterium* (8%); *Arthrobacter* (7%) and *Pantoea* (4%); *Achromobacter*, *Acinetobacter*, *Exiguobacterium*, and *Staphylococcus* (3%); *Enterobacter*, *Providencia*, *Klebsiella*, and *Leclercia* (2%); and *Brevundimonas*, *Flavobacterium*, *Kocuria*, *Kluyvera*, and *Planococcus* (1%). Representative strains from each cluster were screened in vitro for plant growth-promoting traits, which included solubilization of phosphorus, potassium, and zinc; production of ammonia, hydrogen cyanide, indole-3-acetic acid, and siderophore; nitrogen fixation; 1-aminocyclopropane-1-carboxylate deaminase activity; and biocontrol against *Fusarium graminearum*, *Rhizoctonia solani*, and *Macrophomina phaseolina*. Cold-adapted isolates may be applied as inoculants for plant growth promotion and as biocontrol agents for crops growing under cold climatic condition.

11.3.3 Fungi

The psychrophilic fungi have an optimum growth temperature near 10 °C or below and that 10 °C was the minimum growth temperature required for most of the fungi. Many researchers agree with Morita's definition of psychrophiles that psychrophilic fungi grow well at 15 °C or lower, whereas psychrotrophic fungi require temperatures above 20 °C for their maximum growth. Microorganisms that can be recovered from the interior of deep-core samples of Arctic and Antarctic ice are expected to be millions of years old. Various ancient fungi, ranging from 10,000 to 140,000 years in age, have been isolated and documented from Arctic and Antarctic ice (Fisher et al. 1995; Babjeva and Reshetova 1998; Tosi et al. 2002; Malviya et al. 2009; Srivastava et al. 2014; He et al. 2017). Psychrophilic fungi exist in some of the coldest environments throughout the world because of their great efficiency of adaptation to cold environment of deep sea, glacier, the Arctic, and Alps regions. There is a wide range of natural habitats where low temperatures occur continuously or intermittently due to seasonal effects. These regions include oceans, the tundra, and subarctic regions. These fungi may be present either because they are true psychrophiles or because they are psychrotolerant with the ability to survive but not actively grow at temperatures <5 °C. *Penicillium* is a genus of ascomycetous fungi and it has an important role in various natural processes. A wide and ubiquitous presence of *Penicillium* species has been witnessed in several studies. Based on a comprehensive literature analysis Penicillium has been reported as one of the most common fungi occurring in various environments such as soils, air, extreme environments (temperature, salinity, water deficiency and pH) and also associated with plants and specific food products. It's huge diversity and existence in extreme environments provide great opportunity in exploring this fungus for various environmental, biotechnological and industrial applications (Yadav et al. 2017g). *Penicillium* and different groups of microbes have attracted significant attention because of the interest in cold-active enzymes and plant growth-promoting attributes at low temperatures (Yadav 2015; Yadav et al. 2015e, 2017g; Singh et al. 2016).

Cold-tolerant species of *Penicillium* have been primarily reported in connection with subarctic vegetation (Fisher et al. 1995; Babjeva and Reshetova 1998; Tosi et al. 2002), in snow and below snow-covered tundra (Vishniac 1993; Babjeva and Reshetova 1998; Pennisi 2003; Schadt et al. 2003), and in permafrost (Dmitriev et al. 1997; Babjeva and Reshetova 1998; Golubev 1998; Soina et al. 2000; Tosi et al. 2002) and offshore polar waters (Broady and Weinstein 1998). Very few studies describe their presence in Arctic glaciers. Viable fungi have, however, been isolated from Arctic and Antarctic ice, ranging in age from 10,000 and up to 140,000 years (Abyzov 1993; Ma et al. 1999, 2000; Christner et al. 2000, 2003). In these cases, the few isolated filamentous fungi were considered as randomly entrapped fungal Aeolian propagules originating from close and distant locations. Fungi have been only rarely isolated from glacial ice in extremely cold polar regions and were in these cases considered as random, long-term preserved Aeolian deposits. Fungal presence has so far not been investigated in polar sub-glacial ice, a recently discovered extreme habitat reported to be inhabited exclusively by heterotrophic bacteria.

Sonjak et al. (2006) reported on the very high occurrence (up to 9×10^3 CFU/L) and diversity of filamentous *Penicillium* spp. in the sediment-rich sub-glacial ice of three different polythermal Arctic glaciers. The dominant species was *P. crustosum*, representing the average half of all isolated strains from all three glaciers. The other most frequently isolated species were *P. bialowiezense*, *P. chrysogenum*, *P. thomii*, *P. solitum*, *P. palitans*, *P. echinulatum*, *P. polonicum*, *P. commune*, *P. discolor*, *P. expansum*, and new *Penicillium* species. This was the first report on the presence of large populations of *Penicillium* spp. in sub-glacial sediment-rich ice. *Penicillium crustosum* is an important and panglobal contaminant of lipid- and protein-rich foods and feeds. Although it is infrequent in extremely cold environments, we isolated a high number of *P. crustosum* strains from Arctic coastal, but particularly, sub-glacial environments in Svalbard, Norway, *P. crustosum* is extremely consistent in its phenotypic properties, including morphology, physiology, and secondary metabolite production. However, some Arctic isolates differed from other Arctic and non-Arctic strains in their weak growth on creatine and in the production of the secondary metabolite and rastin A. Penicillia strains that populate glacial ice must be physiologically adaptable and able to retain their viability throughout the dynamic processes of ice melting and freezing and extremes in pressure. Such enrichment would select for *Penicillium* populations best adapted to dark, cold, oligotrophic environments with shifting osmotic pressures (Yadav et al. 2017g).

Lyhne et al. (2006) reported that *Penicillium jamesonlandense* is a novel species from Greenland that grows exceptionally slow at 25 °C and has an optimum temperature for growth of 17–18 °C. The novel species is more psychrotolerant than any other *Penicillium* species described to date. Isolates of this novel species produce a range of secondary metabolites with a high chemical diversity, represented by kojic acid, penicillic acid, griseofulvin, pseurotin, chrysogine, tryptoquivalins, and cycloaspeptide. *Penicillium ribium*, another novel psychrotolerant species from the Rocky Mountains, Wyoming, USA, produces asperfuran, kojic acid, and cycloaspeptide.

Sonjak et al. (2007) reported that *Penicillium crustosum* is a panglobally distributed foodborne fungus, extremely consistent in its morphological and physiological properties. It is also known for its consistent production of several mycotoxins and other secondary metabolites such as penitrems, roquefortines, viridicatins, and terrestric acids. It is important in temperate regions as a contaminant of oil seeds, nuts, cheese, and meat and as producer of rot in apples. In spite of its ubiquitous nature and copious amounts of conidia produced, it has only rarely been reported from extremely cold environments. According to Sonjak et al. (2007), study of mycobiota in coastal Arctic environment resulted in isolation of a high number of *P. crustosum* strains from different niches, such as seawater and sea ice, but primarily from the sediment-rich subglacial ice of three different polythermal Arctic glaciers. Although phenotypic properties were very homogenous in all *P. crustosum* isolates, some Arctic isolates differed from all other Arctic and non-Arctic strains in their weak growth on creatine, used otherwise as a very reliable taxonomic marker, and furthermore in the production of andrastin A (Gunde-Cimerman et al. 2003; Sonjak et al. 2005; Yadav et al. 2017g).

Dhakar et al. (2014) isolated and identified 25 *Penicillium* species from the Indian Himalayan regions. Based on the phenotypic characters (colony morphology and microscopy), all the isolates were designated to the genus *Penicillium*. Exposure to low temperature resulted in enhanced sporulation in 23 isolates, while it ceased in the case of 2. The fungal isolates produced watery exudates in varying amounts that in many cases increased at low temperature. All the isolates could grow between 4 and 37 °C (optimum 24 °C), hence considered psychrotolerant. While all the isolates could tolerate pH from 2–14 (optimum 5–9), 7 isolates tolerated pH 1.5 as well. While all the fungal isolates tolerated salt concentration above 10%, ten isolates showed tolerance above 20%. Based on ITS region (ITS1-5.8S-ITS2) analysis, the fungal isolates belonged to 25 different species of *Penicillium*. Characters like tolerance for low temperature, wide range of pH, and high salt concentration and enhancement in sporulation and production of secondary metabolites such as watery exudates at low temperature can be attributed to the ecological resilience possessed by these fungi for survival under low-temperature environment of mountain ecosystem.

Barsainya et al. (2016) isolated the salt-tolerant and chromium-resistant fungal strains from Pangong Lake of Ladakh. They carried out on interaction of *Penicillium* sp. with chromium and NaCl and report the ability of *Penicillium* sp. to bind with chromium and NaCl in aqueous solution. It is demonstrated in the study that NaCl reduced the extent of Cr biosorption and promote the fungal strains of *Penicillium* for better growth. The present finding suggests that NaCl might reduce the toxicity of hexavalent chromium in *Penicillium* sp. and may be used as a potential organism for remediation of chromium in high-salinity environments.

11.4 Beneficial Role in Plant Growth Promotion, and Soil Health

The extreme environment of high salt, drought, and low temperature affects the productivity of several commercial crop plants. The low temperature plays a significant role in reducing agricultural production worldwide as 20% of the Earth's surface were covered with frozen soils (permafrost), glaciers and ice sheets, and snow. Cold-tolerant microorganisms are widely distributed in the agroecosystem and play a variety of roles extending from nitrogen fixation, plant growth promotion, and alleviation of cold stress in plants. Though most research work conducted so far has largely focused on psychrophilic/psychrotolerant bacteria, it is a welcome sign that many agriculturally important resourceful microbes are being described from various parts of the Earth.

Phosphate (P) is the major essential macronutrient for biological growth and development. In spite of that, these sources constitute the biggest reservoirs of P in soil because, under appropriate conditions, they can be solubilized and become available for plants. Microorganisms play a central role in the natural P cycle. There are considerable populations of P-solubilizing bacteria (PSB) in soil and in plant rhizospheres. PSB have the ability to solubilize inorganic phosphate compounds, such as tricalcium phosphate, hydroxyapatite, and rock phosphate. It has been reported that indolic compounds (indole-3-acetic acid, indolepyruvic acid, and indoleacetamide) have a positive effect on root growth and cause rapid establishment of roots beneficial for young seedlings as they increase their capacity to anchor themselves to the soil and to obtain water and nutrients from their environment. We have also tested the ability of isolates to produce IAA.

The capacity to synthesize IAA is widespread among soil- and plant-associated bacteria. It has been estimated that 80% of bacteria isolated from the rhizosphere can produce the plant growth regulator IAA. Iron, the fourth most abundant element in the Earth's crust, is largely required by all living organisms for direct microbial assimilation. In aqueous solution, iron can exist in either the ferrous (Fe^{2+}) or (Fe^{3+}) forms, the latter being the less soluble. However, in highly oxidized and aerated soils, the predominant form of iron is the ferric form, which is soluble in water (pH 7.4) at about 10^{-18} M. This is too low to support the growth of microorganisms, which generally need concentrations approaching 10^{-6} M for normal growth. Consequently, to survive in such environments, organisms secrete iron-binding ligands (siderophores) that can bind ferric iron and make it available to the host microorganisms. In the present study, 20 isolates were identified that could produce siderophore at low temperatures. Sustainable agriculture requires the use of strategies to increase or maintain the current rate of food production while reducing damage to the environment and human health. The use of microbial plant growth promoters is an alternative to conventional agricultural technologies. Plant growth-promoting microbe can affect plant growth directly or indirectly. The direct promotion of plant growth by PGP microbes, for the most part, entails providing the plant with a compound that is synthesized by the bacterium or facilitating the uptake of certain nutrients from the environment. The indirect promotion of plant growth occurs when PGP microbes decrease or prevent the deleterious effects of one or more phytopathogenic organisms (Verma et al. 2017b; Yadav et al. 2017a-i).

There are several ways in which different PGP microbes have been reported to directly facilitate the proliferation of their plant hosts. PGP microbes can fix atmospheric nitrogen and supply it to plants; they synthesize siderophores that can solubilize and sequester iron from the soil and provide it to the plant; they synthesize several different phytohormones that can act to enhance various stages of plant growth; they may have mechanisms for the solubilization of minerals such as phosphorus, potassium, and zinc that will become more available for plant growth; and they may synthesize some less well-characterized, low-molecular-mass compounds or enzymes that can modulate plant growth and development (Kloepper et al. 1989; Glick 1995; Quadt-Hallmann et al. 1997; Glick et al. 1999a, b; Yadav et al. 2017d). A particular PGP microbe may affect plant growth and development by using any one or more of these mechanisms (Yadav et al. 2017a) (Table 11.2).

Table 11.2 Psychrotrophic microbes with multifarious plant growth-promoting attributes

Microbes	Plant growth promoting attributes						References
	P	K	NF	IAA	Sidero.	ACC	
Actinobacteria							
Arthrobacter methylotrophus IARI-HHS1-25	55.9±1.4	–	+	21.4±1.3	4.9±0.1	+	Verma et al. (2015a)
Arthrobacter sulfonivorans IARI-L-16	25.6±1.2			27.6±0.7	4.7±0.9	–	Yadav et al. (2015b)
Cellulosimicrobium cellulans IARI-ABL-30	15.5±1.1			18.4±0.8	–	+	Yadav et al. (2015b)
Kocuria kristinae IARI-HHS2-64	64.0±1.0	–	–	20.4±1.1	2.1±0.1	–	Verma et al. (2015a)
Sanguibacter antarcticus IARI-L-33	20.1±0.1			9.3±0.9	10±0.8	+	Yadav et al. (2015b)
Sanguibacter suarezii IARI-R-7	18.1±0.5			76.8±0.3	5±0.15	+	Yadav et al. (2015a)
Firmicutes							
Bacillus altitudinis IARI-HHS2-2	43.9±0.7	–	+	6.6±1.0	–	–	Verma et al. (2015a)
Bacillus amyloliquefaciens IARI-HHS2-30	54.2±1.5	+	+	28.8±1.5	3.5±0.2	+	Verma et al. (2015d)
Bacillus amyloliquefaciens IARI-R-25	39.4±2.4			14.2±1.0	4.8±1.2	–	Yadav et al. (2015a)
Bacillus aryabhattai IARI-HHS1-30	45.6±1.0	–	+	15.6±0.7	–	–	Verma et al. (2015a)
Bacillus firmus IARI-L-21	35.2±3.3			35.2±1.0	6.0±0.8	+	Yadav et al. (2015b)
Bacillus licheniformis IARI-AL38	19.2±1.0	–		13.2±1.0	4.5±0.5		Yadav et al. (2016)
Bacillus muralis IARI-AR28	–	+		22.5±0.5	5.7±1.2	–	Yadav et al. (2016)
Bacillus pumilus IARI-L-54	36.1±0.8			32.3±1.2	4.3±0.9	–	Yadav et al. (2015b)
Bacillus subtilis IARI-L-69	19.8±0.5			27.7±0.9	5.3±0.5	+	Yadav et al. (2015b)
Desemzia incerta IARI-L-46	47.5±1.2			28.6±1.0	4.7±0.5	–	Yadav et al. (2015b)
Exiguobacterium antarcticum IARI-HHS2-49	–	+	–	22.8±0.8	3.1±0.1	–	Verma et al. (2015a)
Lysinibacillus sphaericus IARI-AR11		+		14.4±1.2	–	+	Yadav et al. (2016)
Paenibacillus tylopili IARI-AR36	48.4±2.4	+		39.4±2.4	3.8±1.2	–	Yadav et al. (2016)
Sporosarcina globispora IARI- AR111	–	+		37.6±0.3	–	+	Yadav et al. (2016)
Proteobacteria							
Acinetobacter rhizosphaerae BIHB 723	785 ± 1.2	–	–	15.6 ± 1.2	+	+	Gulati et al. (2009)
Aeromonas hydrophila IARI-R-6	31.5±1.8			21.4±1.0	2.4±1.5	–	Yadav et al. (2015a)

(continued)

Table 11.2 (continued)

Microbes	Plant growth promoting attributes						References
	P	K	NF	IAA	Sidero.	ACC	
Bordetella bronchiseptica IARI-HHS2-29	48.6±0.9	–	+	15.2±1.1	1.6±0.2	–	Verma et al. (2015a)
Pantoea agglomerans IARI-R-87	22.0±1.4			43.9±1.1	6.8±0.5	–	Yadav et al. (2015a)
Pantoea dispersa 1A	44.5±0.2	–	–	4.4±0.5	+	–	Selvakumar et al. (2008)
Providencia rustigianii IARI-R-91	131.7±1			51.0±2.0	5.2±0.7	+	Yadav et al. (2015a)
Pseudomonas cedrina IARI-R-53	182.6±1			9.99±1.0	5.2±1.2	+	Yadav et al. (2015a)
Pseudomonas fluorescens PPRs4	90.2 ± 1.7	–	–	9.4 ± 0.2	+	–	Mishra et al. (2011)
Pseudomonas fragi IARI-R-57	45.5±1			11.3±0.5	4.6±0.7	+	Yadav et al. (2015a)
Pseudomonas geniculata IARI-HHS1-19	45.0±1.2	–	+	66.7±0.5	2.5±0.2	–	Verma et al. (2015a)
Pseudomonas jessani PGRs1	7.9 ± 0.1	–	–	16.2 ± 0.3	+	–	Mishra et al. (2011)
Pseudomonas koreensis PBRs7	97.3 ± 1.9	–	–	15.8 ± 0.3	+	–	Mishra et al. (2011)
Pseudomonas lurida M2RH3	305.0±2.4	–	–	16.8 ± 0.2	+	–	Selvakumar et al. (2011)
Pseudomonas lurida NPRs3	69.7 ± 1.5	–	–	9.9 ± 0.2	+	–	Mishra et al. (2011)
Pseudomonas moraviensis IARI-R-132	44.2±2.1			154.6±1.	2.2±0.5	+	Yadav et al. (2015a)
Pseudomonas putida IARI-R-131	–			137.9±1	2.4±0.7	–	Yadav et al. (2015a)
Pseudomonas putida PGRs4	169.9 ± 3.9	–	–	10.1 ± 0.2	+	–	Mishra et al. (2011)
Pseudomonas reactans IARI-ABR-38	23.23±1			61.4±0.5	4.2±0.5	–	Yadav et al. (2015a)
Pseudomonas sp. NARs9	15.7±1.82	–	–	21.8±0.2	+	–	Mishra et al. (2009)
Pseudomonas sp. PGERs17	74.1±0.3	–	–	13.2±0.8	+	–	Mishra et al. (2008)
*Psychrobacter frigidicola*IARI-R-127	20.83±1			65.9±1.0	2.8±1.5	+	Yadav et al. (2015a)
Rahnella sp. BIHB 783	805.0±1.	–	–	24.5±1.5	88.0±3	+	Vyas et al. (2010)
Stenotrophomonas maltophilia IARI-HHS1-20	55.7±0.5	+	–	66.1±0.7	2.4±0.1	+	Verma et al. (2015a)
Bacteroidetes							
Flavobacterium psychrophilum IARI-HHS2-37	66.0±0.7	–	–	11.4±1.5	2.6±0.1	+	Verma et al. (2015a)

P Phosphorus, *K* Potassium, *NF* Nitrogen Fixation, *IAA* Indoleacetic acids, *Sidero.* Siderophore

11.4.1 Biological N_2-Fixation

Nitrogen is the major limiting factor for plant growth; the application of N_2-fixing bacteria as biofertilizers has emerged as one of the most efficient and environmentally sustainable methods for increasing the growth and yield of crop plants. Biological nitrogen fixation (BNF) is one of the possible biological alternatives to N fertilizers and could lead to more productive and sustainable agriculture without harming the environment. A variety of nitrogen-fixing bacteria like *Arthrobacter*, *Azoarcus*, *Azospirillum*, *Azotobacter*, *Bacillus*, *Enterobacter*, *Gluconacetobacter*, *Herbaspirillum*, *Klebsiella*, *Pseudomonas*, and *Serratia* have been reported to fix nitrogen under low-temperature condition (Verma et al. 2015a, b, c, d, 2016; Yadav 2015; Rana et al. 2016a, b, 2017).

Verma et al. (2015a) reported psychrotolerant wheat-associated bacteria from northern hills zone of India. A total of 247 bacteria were isolated from 5 different sites. Representative strains from each cluster were screened in vitro for plant growth-promoting traits, which included solubilization of phosphorus, potassium, and zinc; production of ammonia, hydrogen cyanide, indole-3-acetic acid, and siderophore; nitrogen fixation; 1-aminocyclopropane-1-carboxylate deaminase activity; and biocontrol against *Fusarium graminearum*, *Rhizoctonia solani*, and *Macrophomina phaseolina*. Among all, *Arthrobacter nicotinovorans* IARI-HHS1-1, *Arthrobacter methylotrophus* IARI-HHS1-25, *Bacillus flexus* IARI-HHS1-24, *Bacillus aryabhattai* IARI-HHS1-30, *Bacillus altitudinis* IARI-HHS2-2, *Bacillus amyloliquefaciens* IARI-HHS2-30, *Bordetella bronchiseptica* IARI-HHS2-29, *Providencia* sp. IARI-HHS1-3, *Pseudomonas fluorescens* IARI-HHS1-4, *Pseudomonas mediterranea* IARI-HHS1-5, *Pseudomonas peli* IARI-HHS1-8, *Pseudomonas geniculata* IARI-HHS1-19, *Pseudomonas azotoformans* IARI-HHS1-23, *Pseudomonas syringae* IARI-HHS1-29, and *Acinetobacter lwoffii* IARI-HHS2-26 show nitrogen fixation under low-temperature conditions. Cold-adapted isolates may be applied as inoculants for plant growth promotion and as biocontrol agents for crops growing under cold climatic condition.

Verma et al. (2016) reported the nitrogen-fixing bacilli from wheat growing in northern hills of India. A total of 395 bacilli were isolated by heat enrichment and employed in different growth media. Phylogenetic analysis based on 16S rRNA gene sequencing led to the identification of 55 distinct bacilli that could be grouped in 5 families, *Bacillaceae* (68%), *Paenibacillaceae* (15%), *Planococcaceae* (8%), *Staphylococcaceae* (7%), and *Bacillales incertae sedis* (2%), which included 8 genera, namely, *Bacillus*, *Exiguobacterium*, *Lysinibacillus*, *Paenibacillus*, *Planococcus*, *Planomicrobium*, *Sporosarcina*, and *Staphylococcus*. That study was the first to report for the presence of *Bacillus endophyticus*, *Paenibacillus xylanexedens*, *Planococcus citreus*, *Planomicrobium okeanokoites*, *Sporosarcina* sp., and *Staphylococcus succinus* in wheat rhizosphere which exhibit multifunctional PGP attributes. Among all screened bacteria, *Bacillus barbaricus* BSH5 (18.2 ± 1.6 nmol ethylene h^{-1} mg^{-1} protein), *Bacillus circulans* BSH11 (45.3 ± 1.5 nmol ethylene h^{-1} mg^{-1} protein), *Lysinibacillus fusiformis* BSH2 (43.5 ± 1.6 nmol ethylene h^{-1} mg^{-1} protein), and *Lysinibacillus sphaericus* BSH6 (19.5 ± 1.0 nmol ethylene

h^{-1} mg^{-1} protein) fixed the atmospheric nitrogen under the hilly area and low-temperature condition.

11.4.2 Solubilization of Phosphorus, Potassium, and Zinc

Phosphate (P) and potassium (K) are the major essential macronutrients for biological growth and development. However, the concentrations of soluble P and K in soil are usually very low, as the biggest proportions of P and K in soil are in soluble rocks, minerals, and other deposits (Goldstein 1994). Phosphorus is one of the major growth-limiting nutrients in plants and is often the limiting mineral nutrient for biomass production in natural ecosystems. It is important for the plant growth and promotes root development, tillering, and early flowering and performs other functions like metabolic activities, particularly in synthesis of protein. Thus, P is a key element in several metabolic pathways and biochemical reactions including glycolysis and numerous steps in the C3 and C4 pathways of the Calvin cycle. Phosphorus is an essential element for the establishment and development of plants because it improves the entire root system, consequently improving the shoot. In later stages of growth, the lack of phosphorus can lead to atrophy and death of the stems and leaves and can also delay fruit maturation. It is only taken up in monobasic or dibasic soluble forms. Phosphates applied to agricultural soils are rapidly immobilized and rendered inaccessible for plants. Due to this rapid immobilization, many agricultural soils have large reservoirs of phosphates in inaccessible forms. In this scenario, phosphorus-solubilizing activity is determined by the ability of microorganisms to release metabolites such as organic acids, which through their hydroxyl and carboxyl groups chelate the cation bound to phosphate, the latter being converted to soluble forms.

Gulati et al. (2008) reported fluorescent pseudomonads with phosphate-solubilizing ability from the cold desert region of Lahaul and trans-Himalayas, India. Among 216 phosphate-solubilizing isolates, 12 exhibiting high solubilization of tricalcium phosphate (TCP) in NBRIP liquid culture were identified as *Pseudomonas trivialis*, *P. poae*, *P. fluorescens*, and *Pseudomonas* spp. on the basis of phenotypic features, whole-cell fatty acid methyl ester (FAME) profiles, and 16S rDNA sequencing. These isolates also showed relatively high solubilization of North Carolina rock phosphate (NCRP) in comparison to the solubilization of Mussoorie rock phosphate (MRP) and Udaipur rock phosphate (URP). The solubilization of phosphate substrates by *P. trivialis* and *P. poae* is reported for the first time.

Vyas et al. (2010) deciphered phosphate-solubilizing bacteria isolated from *Hippophae rhamnoides* rhizosphere which was identified as *Rahnella* sp. based on its phenotypic features and 16S rRNA gene sequence. The bacterial strain showed the growth characteristics of a cold-adapted psychrotroph, with the multiple plant growth-promoting traits of inorganic and organic phosphate solubilization, 1-amino cyclopropane-1-carboxylate-deaminase activity, ammonia generation, and siderophore production. The strain also produced indole-3-acetic acid, indole-3-acetaldehyde, indole-3-acetamide, indole-3-acetonitrile, indole-3-lactic acid, and indole-3- pyruvic acid in tryptophan-supplemented nutrient broth. Gluconic, citric, and isocitric acids were the major organic acids detected during tricalcium phosphate solubilization. A rifampicin-resistant mutant of the strain exhibited high rhizosphere competence without disturbance to the resident microbial populations in pea rhizosphere. Seed bacterization with a charcoal-based inoculum significantly increased growth in barley, chickpea, pea, and maize under the controlled environment. The attributes of cold tolerance, high rhizosphere competence, and broad-spectrum plant growth-promoting activity exhibited the potential of *Rahnella* sp. BIHB 783 for increasing agriculture productivity.

Yadav et al. (2016b) investigated plant growth-promoting psychrotrophic bacilli from different sites in northwestern Indian Himalayas. A total of 247 morphotypes were obtained from different soil and water samples and were grouped into 43 clusters based on 16S rDNA-RFLP analysis with 3 restriction endonucleases. Sequencing of representative isolates from each cluster led to their identification, and 43 bacilli belonged to different species of 11 genera, viz., *Desemzia*, *Exiguobacterium*, *Jeotgalicoccus*, *Lysinibacillus*, *Paenibacillus*, *Planococcus*, *Pontibacillus*, *Sinobaca*, *Sporosarcina*, *Staphylococcus*, and *Virgibacillus*. With an aim to develop microbial inoculants that can perform efficiently at low temperatures, all representative isolates were screened for different plant growth-promoting traits at low temperatures (5–15 °C). Among the strains, variations were observed for production of ammonia (22%), indole-3-acetic acid (20%), siderophores (8%), gibberellic acid (4%), and hydrogen cyanide (3%); solubilization of phosphate (12%), zinc (16%), and potassium (8%); 1-aminocyclopropane-1-carboxylate deaminase activity (5%); and biocontrol activity (19%) against *Rhizoctonia solani* and *Macrophomina phaseolina*. Among all the strains, *Bacillus licheniformis*, *B. muralis*, *Desemzia incerta*, *Paenibacillus tylopili*, and *Sporosarcina globispora* were found to be potent candidates to be developed as inoculants as they exhibited multiple PGP traits at low temperature.

Verma et al. (2016) have reported 395 bacilli from wheat, and these bacteria have been screened for direct and indirect PGP traits, and result has been represented by 55 representative bacilli. Of 55 representatives, 39, 18, and 40 strains exhibited solubilization of phosphorus, potassium, and zinc, respectively. Among P, K, and Zn solubilizers, *Paenibacillus polymyxa* BNW6 solubilized the highest amount of phosphorus 95.6 ± 1.0 mg L^{-1} followed by *Sporosarcina* sp. BNW4 (75.6 ± 1.0 mg L^{-1}). *Planococcus salinarum* BSH13 (46.9 ± 1.2 mg L^{-1}) and *Bacillus pumilus* BCZ15 (7.5 ± 0.5 mg L^{-1}) solubilized the highest amount of potassium and zinc, respectively. Among plant growth-promoting activities, ammonia-producing bacilli were highest (79.0%), when compared to P solubilizer (73.9%), Zn solubilizers (67.1%), protease producers (56.7%), IAA producers (55.2%), siderophore producers (49.1%), biocontrol activity (47.8%), K solubilizers (39.2%), N2-fixers (31.4%), HCN producers (27.3%), and gibberellic acid producers (24.8%).

11.4.3 Phytohormones Production

Plant-associated microbes typically produce plant growth hormones such as cytokinins, auxins, and gibberellins. The gibberellin production is most typical for the root-associated microbes, cytokinins have been identified in some leaf isolates, and auxin production is common to all plant-associated microbes. Auxins are a group of indole derivatives that have various growth-promoting functions in plants, such as promotion of root formation, regulation of fruit ripening, and stimulation of cell division, extension, and differentiation. Indoleacetic acid (IAA) is the most well-known auxin. Auxins can promote the growth of roots and stems quickly (by increasing cell elongation) or slowly (through cell division and differentiation). The production of such growth regulators by endophytes provides numerous benefits to the host plant including the facilitation of root system expansion, which enhances the absorption of water and nutrients and improves plant survival. There are several types of bacterial auxins, and the well-studied of these is indoleacetic acid (IAA). IAA does not function as a hormone in microbial cells; therefore, the ability of bacteria to produce IAA may have evolved as the plant-microorganism relationship developed. The ability to synthesize these phytohormones is widely distributed among plant-associated microbes, and IAA may potentially be used to promote plant growth or suppress weed growth. Many studies have described the ability of endophytic bacteria to produce phytohormones and auxins, such as IAA (Hallmann et al. 1997), and the ability to produce IAA is considered to be responsible for plant growth promotion by beneficial bacteria, such as *Azospirillum*, *Alcaligenes faecalis*, *Klebsiella*, *Enterobacter*, *Acetobacter diazotrophicus*, and *Herbaspirillum seropedicae*. Cytokinins are a group of compounds with the backbone of adenine having a substitution at the N-6 atom of the purine ring. These compounds are important in many steps of plant development, as they stimulate plant cell division, induce germination of seeds, activate dormant buds, and play a role in apical dominance. Cytokinins also induce the biosynthesis of chlorophyll, nucleic acids, and chloroplast proteins at the early stages of leaf development. Both pathogenic and beneficial plant-associated bacterial species are capable of synthesizing cytokinins. Among plant-associated methylotrophs, species such as *Methylovorusmays* and *Methylobacterium mesophilicum* JCM2829 synthesize and excrete cytokinins (Ivanova et al. 2001, 2008).

Verma et al. (2015a) deciphered the biodiversity of wheat-associated bacteria from the northern hills zone of India. A total of 247 bacteria were isolated from 5 different sites. Analysis of these bacteria by amplified ribosomal DNA restriction analysis (ARDRA) using 3 restriction enzymes, *Alu*I, *Msp*I, and *Hae*III, led to the grouping of these isolates into 19–33 clusters for the different sites at 75% similarity index. Among all isolated bacteria, 14% showed IAA production in which strain IARI-HHS1-3 showed the highest IAA production (70.8 ± 1.5 μg mg^{-1} protein day^{-1}) followed by IARI-HHS1-8 (69.1 ± 0.5 μg mg^{-1} protein day^{-1}). On the basis of 16S rRNA gene sequencing, the strains IARI-HHS1-3 and IARI-HHS1-8 were identified as *Providencia* sp. and *Pseudomonas peli* respectively. The potential utility of such cold-active bacterial strains in the context of hill and mountain agro-ecosystems is immense, considering the unique crops growing in the climatic conditions of high altitude agricultural systems. Such systems require situation-specific microbial inoculants that withstand extremities of cold and retain their functional traits for PGP. The PGP potential of the bacterial strains dealt with in this study requires further evaluation and validation before their use as bio-inoculants.

The selection of native functional PGP microorganisms is a mandatory step for reducing the use of energy-intensive chemical fertilizers.

11.4.4 ACC Deaminase Activity

Ethylene is a stress-induced plant hormone that can inhibit plant growth. Some bacteria can lower the level of ethylene in the plant by cleaving the plant-produced ethylene precursor 1-aminocyclopropane-1-carboxylate (ACC). Inoculation of such bacteria can mitigate the effect of various stressors by sustaining plant growth in the face of ethylene. ACC deaminase-producing bacteria may play a role in regulating ethylene levels after such bursts, ensuring that ethylene levels stay below the point where growth is impaired (Glick 1995). Ethylene is a key regulator of the colonization of plant tissue by bacteria which in turn suggests that the ethylene-inhibiting effects of ACC deaminase may be a bacterial colonization strategy. Regardless of why plant-associated bacteria produce ACC deaminase, their application can clearly be a very useful strategy to mitigate the effects of various stressors on cultivated plants. Generally, ethylene is an essential metabolite for the normal growth and development of plants (Khalid et al. 2004, 2006). This plant growth hormone is produced endogenously by approximately all plants and is also produced by different biotic and abiotic processes in soils and is important in inducing multifarious physiological changes in plants. Apart from being a plant growth regulator, ethylene has also been established as a stress hormone. Under stress conditions like those generated by salinity, drought, water logging, heavy metals, and pathogenicity, the endogenous level of ethylene is significantly increased which negatively affects the overall plant growth. Plant growth-promoting bacteria which possess the enzyme ACC deaminase facilitate plant growth and development by decreasing ethylene levels, inducing salt tolerance and reducing drought stress in plants. Microbial strains exhibiting ACC deaminase activity have been identified in a wide range of genera such as *Acinetobacter*, *Achromobacter*, *Agrobacterium*, *Alcaligenes*, *Azospirillum*, *Bacillus*, *Burkholderia*, *Enterobacter*, *Pseudomonas*, *Ralstonia*, *Serratia*, *Rhizobium*, etc. (Khalid et al. 2006; Verma et al. 2014a, b, 2015a; Xu et al. 2014; Yadav and Saxena 2017).

In the study by Verma et al. (2015a) the wheat associated psychrotrophic bacteria from Indian Himalayas has been deciphered. The bacterial isolated exhibited multifunctional PGP attributes. Among 247 bacterial isolates, 15 strains showed ACC deaminase activity under the low temperature conditions. On the basis of 16S rRNA sequencing these bacteria were identified as *Arthrobacter methylotrophus* IARI-HHS1-25, *Flavobacterium psychrophilum* IARI-HHS2-37, *Bacillus horikoshii* IARI-HHS2-13, *Methylobacterium mesophilicum* IARI-HHS1-36, *Methylobacterium radiotolerans* IARI-HHS1-45, *Methylobacterium phyllosphaerae* IARI-HHS2-67, *Methylobacterium* sp. IARI-HHS2-69, *Providencia* sp. IARI-HHS1-3, *Pseudomonas fragi* IARI-HHS1-10, *Pseudomonas fluorescens* IARI-HHS1-4, *Stenotrophomonas maltophilia* IARI-HHS1-20, *Pseudomonas syringae* IARI-HHS1-29, *Pseudomonas aeruginosa* IARI-HHS2-12, *Leclercia adecarboxylata* IARI-HHS2-42 and *Enterobacter* sp. IARI-HHS2-44.

11.4.5 Antifungal Activity and Production of Siderophores, Antibiotics, and Lytic Enzymes

The indirect mechanism of plant growth occurs when microbes lessen or prevent the detrimental effects of pathogens on plants by production of inhibitory substances or by increasing the natural resistance of the host. The pathogenic microorganisms can be controlled by releasing siderophores; β-1,3-glucanase; chitinases; antibiotics; and fluorescent pigment or by hydogen cyanide production. World agriculture faces a great loss every year incurred from infection by pathogenic organisms. Application of microorganism for the control of diseases seems to be one of the most promising ways. Biocontrol systems are eco-friendly and cost-efficient and involved in improving the soil consistency and maintenance of natural soil flora. To act efficiently, the biocontrol agent should remain active under large range of conditions, *viz.*, varying pH, temperature, and concentrations of different ions. Biocontrol agents limit growth of pathogen as well as few nematodes and insects. Biocontrol bacteria can limit pathogens directly by producing antagonistic substances, competition for iron, detoxification, or degradation of virulence factors or indirectly by inducing systemic resistance (ISR) in plants against certain diseases, signal interference, competition for nutrients and niches, and interference with activity, survival, germination, and sporulation of the pathogen (Verma et al. 2017b). Recent studies have indicated that biological control of bacterial wilt disease could be achieved using antagonistic bacteria. Different bacterial species, namely, *Alcaligenes* sp., *Bacillus pumilus*, *B. subtilis*, *B. megaterium*, *Clavibacter michiganensis*, *Curtobacterium* sp., *Flavobacterium* sp., *Kluyvera* sp., *Microbacterium* sp., *Pseudomonas alcaligenes*, *P. putida*, and *P. fluorescens*, have been reported as endophytes and were inhibitory to plant pathogens (Inderiati and Franco 2008; Ramesh et al. 2009; Nagendran et al. 2013; Gholami et al. 2014; Purnawati 2014; Verma et al. 2015a, b, c, d, 2016; Suman et al. 2016b) (Table 11.2).

Selvakumar et al. (2009c) reported *Exiguobacterium acetylicum* strain 1P (MTCC 8707) from rhizospheric soil sample. The species level identification was achieved on the basis of 16S rRNA gene sequencing. The sequence showed 98% similarity with sequences of *E. acetylicum* available in the public domain. The strain was positive for siderophores and HCN production. In separate *in vitro* assays, it was found to inhibit the growth and development of *Rhizoctonia solani*, *Sclerotium rolfsii*, *Pythium*, and *Fusarium oxysporum*. The volatile compound produced by the bacterium was found to be the most potent in inhibiting the hyphal development of *R. solani*, *S. rolfsii*, *Pythium, and F. oxysporum* by 45.55, 41.38, 28.92, and 39.74%, respectively. Commonly observed deformities caused by the diffusible and volatile compounds produced by the bacterium included hyphal inhibition, constriction, and deformation. Under pot culture conditions, the bacterium improved the germination and early growth parameters of pea (*Pisum sativum*) in the presence of *R. solani* and *S. rolfsii*.

Growth enhancement through enzymatic activity is another mechanism used by plant growth-promoting bacteria. Plant growth-promoting bacterial strains can produce certain enzymes such as chitinases, dehydrogenase, β-glucanase, lipases, phosphatases, proteases, etc., and exhibit hyperparasitic activity, attacking pathogens by excreting cell wall hydrolases. Through the activity of these enzymes, plant growth-promoting bacteria play a very significant role in plant growth promotion particularly to protect them from biotic and abiotic stresses by suppression of pathogenic fungi including *Botrytis cinerea*, *Sclerotium rolfsii*, *Fusarium oxysporum*, *Phytophthora* sp., *Rhizoctonia solani*, and *Pythium ultimum* (Arora 2013; Yadav 2017).

The production of antibiotics that is considered to be one of the most powerful and studied biocontrol mechanisms of plant growth-promoting bacteria against phytopathogens has become increasingly better understood over the past two decades (Shilev 2013; Gupta et al. 2015). A variety of antibiotics have been identified, including compounds such as amphisin, 2,4-diacetylphloroglucinol (DAPG), oomycin A, phenazine, pyoluteorin, pyrrolnitrin, tensin, tropolone, and cyclic lipopeptides produced by pseudomonads and oligomycin A, kanosamine, zwittermicin A, and xanthobaccin produced by *Bacillus*, *Streptomyces*, and *Stenotrophomonas* sp. to prevent the proliferation of plant pathogens (generally fungi). *Bacillus amyloliquefaciens* is known for lipopeptide and polyketide production for biological control activity and plant growth promotion activity against soilborne pathogens (Ongena and Jacques 2008). Apart from the production of antibiotic, some bacteria are also capable of producing volatile compound known as hydrogen cyanide (HCN) for biocontrol of black root rot of tobacco, caused by *Thielaviopsis basicola* (Sacherer et al. 1994). Lanteigne et al. (2012) also reported the production of DAPG and HCN by *Pseudomonas* contributing to the biological control of bacterial canker of tomato.

Iron is a necessary cofactor for many enzymatic reactions and is an essential nutrient for virtually all organisms. In aerobic conditions, iron exists predominantly in its ferric state (Fe^{3+}) and reacts to form highly insoluble hydroxides and oxyhydroxides that are largely unavailable to plants and microorganisms. To acquire sufficient iron, siderophores produced by bacteria can bind to Fe^{3+} with a high affinity to solubilize this metal for its efficient uptake. Bacterial siderophores are low-molecular-weight compounds with high Fe^{3+} chelating affinities responsible for the solubilization and transport of this element into bacterial cells. Some bacteria produce hydroxamate-type siderophores, and others produce catecholate types. In a state of iron limitation, the siderophore-producing microorganisms are also able to bind and transport the iron-siderophore complex by the expression of specific proteins. The production of siderophores by microorganisms is beneficial to plants because it can inhibit the growth of plant pathogens. Siderophores have been implicated for both direct and indirect enhancement of plant growth by plant growth-promoting bacteria. The direct benefits of bacterial siderophores on the growth of plants that have been demonstrated by using radiolabeled ferric siderophores as a sole source of iron showed that plants are able to take up the labeled iron by a large number of plant growth-promoting bacteria including *Aeromonas*, *Azadirachta*, *Azotobacter*, *Bacillus*, *Burkholderia*, *Pseudomonas*, *Rhizobium*, *Serratia*, and *Streptomyces* sp. (Gulati et al. 2009; Mishra et al. 2011; Vyas et al. 2010; Yadav et al. 2015a, b; Verma et al. 2013, 2015a) (Table 11.2).

11.5 Biofertilizers: Microbes with Multifarious Plant Growth-Promoting Attributes

Biofertilizers are basically the microbes which bring about the enrichment of the nutrients of the soil by enhancement of the availability of the nutrients to the crops. Plant nutrients are one of the most essential components of the sustainable agriculture. The production of the healthy crops so as to meet the demands of the world's expanding population mainly relies on the type of the fertilizers which are basically used to supplement all the nutrients to the plants, but more reliability the chemical fertilizers are damaging the environmental ecology as well as affecting the human health with great severity. Thus, the use of the microbes as biofertilizers is considered as an alternative to chemical fertilizers so as to improve the fertility of the soil as well as increase the productivity of the crops in sustainable farming. These microbes are considered to be the biopotential and a novel tool for providing substantial benefits to the agriculture. These microbes basically colonize the roots and stimulate the growth, and these are referred to as the PGP microbes. PGP microbes stimulate the growth by various direct and indirect mechanisms such as the production of various plant growth regulators, biological nitrogen fixation, phosphorus solubilization, and production of the siderophores, HCN, and various lytic enzymes. Extensive work on the biofertilizers is available which reveals that these microbes have the capability of providing the required nutrients to the crops in amounts which are sufficient for the enhancement of yield of the crops (Suman et al. 2016a; Verma et al 2017c; Yadav et al. 2017a).

Selvakumar et al. (2008) reported and characterized a *Pantoea dispersa* strain 1A with multifarious PGP attributes. It is endowed with multiple plant growth-promoting attributes such as phosphate solubilization, IAA production, siderophore production, and HCN production, which are expressed differentially at suboptimal temperatures (15 and 4 °C). It was able to solubilize phosphate (17.6 μg of P_2O_5 ml^{-1} day^{-1}), and produce IAA (3.7 μg ml^{-1} day^{-1}), at 15 °C.

Mishra et al. (2008) deciphered a psychrotolerant, *Pseudomonas vancouverensis*, which possesses various PGP attributes. The isolate produced 8.33 and 1.38 μg/ml of IAA at 15 °C and 4 °C, respectively, on the third day after incubation. It solubilized 42.3, 66.3, and 74.1 μg/ml of tricalcium phosphate at 4, 15, and 28 °C, respectively, after 7 days of incubation. The strain also possessed HCN and siderophore production abilities at 4 °C. It exhibited inhibitory activity against several phytopathogenic fungi in three different bioassays. The maximum relative growth inhibition was recorded against *Sclerotium rolfsii* and *Rhizoctonia solani* (100%), followed by *Pythium* sp. (73.1%) and *Fusarium oxysporum* (19.7%), in volatile compound assays.

Gulati et al. (2009) isolated and characterized *Acinetobacter rhizosphaerae* strain BIHB 723 with multifunctional PGP activities of phosphate solubilization, auxin production, 1-aminocyclopropane-1-carboxylate deaminase activity, ammonia generation, and siderophore production. A significant increase in the growth of pea, chickpea, maize, and barley was recorded for inoculations under controlled conditions. Field testing with the pea also showed a significant increment in plant growth and yield.

Verma et al. (2015a) reported psychrotolerant wheat-associated bacteria from northern hills zone of India. A total of 247 bacteria were isolated from 5 different sites. Representative strains from each cluster were screened in vitro for plant growth-promoting traits, which included solubilization of phosphorus, potassium, and zinc; production of ammonia, hydrogen cyanide, indole-3-acetic acid, and siderophore; nitrogen fixation; 1-aminocyclopropane-1-carboxylate deaminase activity; and biocontrol against *Fusarium graminearum*, *Rhizoctonia solani*, and *Macrophomina phaseolina*. Cold-adapted isolates may be applied as inoculants for plant growth promotion and as biocontrol agents for crops growing under cold climatic condition.

11.6 Conclusion and Future Prospect

In conclusion, utility of such cold-active microbial strains in the context of hill and mountain agro-ecosystems is immense considering the unique crop-growing situations and the climatic conditions of the high-altitude agricultural systems. Such systems require situation-specific microbial inoculants that withstand extremities of cold and retain their functional traits for plant growth promotion. Microbial strains with multifarious plant growth promotion potential could be used as bio-inoculants (biofertilizers and biocontrol agents) in the hill and mountain agro-ecosystems, where temperature is a major determinant of plant and microbial activity. The selection of native functional plant growth-promoting microorganisms is a mandatory step for reducing the use of energy-intensive chemical fertilizers. The benificial and potential cold-adapted microbes may have applications in diverse processes in agriculture, industry, and allied sectors.

Acknowledgments The authors are grateful to the Division of Microbiology; ICAR-Indian Agricultural Research Institute (IARI), New Delhi; and National Agricultural Innovation Project on "Diversity analysis of *Bacillus* and other predominant genera in extreme environments and its utilization in Agriculture," Indian Council of Agricultural Research, for providing the facilities and financial support, to undertake the investigations. There are no conflicts of interest.

References

Abyzov S (1993) Microorganisms in the Antarctic ice. Ant Microbiol 1(1):265–296

Albert RA, Waas NE, Pavlons SC, Pearson JL, Ketelboeter L, Rosselló-Móra R, Busse H-J (2013) *Sphingobacterium psychroaquaticum* sp. nov., a psychrophilic bacterium isolated from Lake Michigan water. Int J Syst Evol Microbiol 63(3):952–958

Anupama P, Praveen K, Singh R, Kumar S, Srivastava A, Arora D (2011) A psychrophilic and halotolerant strain of *Thelebolus microsporus* from Pangong Lake, Himalaya. Mycosphere 2(5):601–609

Arora NK (2013) Plant microbe symbiosis: fundamentals and advances. Springer, New Delhi

Auman AJ, Breezee JL, Gosink JJ, Kämpfer P, Staley JT (2006) *Psychromonas ingrahamii* sp. nov., a novel gas vacuolate, psychrophilic bacterium isolated from Arctic polar sea ice. Int J Syst Evol Microbiol 56(5):1001–1007

Babjeva I, Reshetova I (1998) Yeast resources in natural habitats at polar circle latitude. Food Technol Biotechnol 36:1–6

Baghel V, Tripathi R, Ramteke P, Gopal K, Dwivedi S, Jain R, Rai U, Singh S (2005) Psychrotrophic proteolytic bacteria from cold environment of Gangotri glacier, Western Himalaya, India. Enzym Microb Technol 36(5):654–659

Bakker AW, Schippers B (1987) Microbial cyanide production in the rhizosphere in relation to potato yield reduction and *Pseudomonas spp* mediated plant growth-stimulation. Soil Biol Biochem 19(4):451–457

Barahona S, Yuivar Y, Socias G, Alcaíno J, Cifuentes V, Baeza M (2016) Identification and characterization of yeasts isolated from sedimentary rocks of Union Glacier at the Antarctica. Extremophiles 20(4):479–491

Barsainya M, Chandra P, Singh DP (2016) Investigation of Cr (VI) Uptake in saline condition using psychrophilic and mesophilic *Penicillium* sp. Int J Curr Microbiol App Sci 5(1):274–288

Bisht SC, Mishra PK, Joshi GK (2013) Genetic and functional diversity among root-associated psychrotrophic Pseudomonad's isolated from the Himalayan plants. Arch Microbiol 195(9):605–615

Boddey R, De Oliveira O, Urquiaga S, Reis V, De Olivares F, Baldani V, Döbereiner J (1995) Biological nitrogen fixation associated with sugar cane and rice: contributions and prospects for improvement. Plant Soil 174(1–2):195–209

Bowman JP (2000) Description of *Cellulophaga algicola* sp. nov., isolated from the surfaces of Antarctic algae. Int J Syst Evol Microbiol 50(5):1861–1868

Bric JM, Bostock RM, Silverstone SE (1991) Rapid *in situ* assay for indoleacetic acid production by bacteria immobilized on a nitrocellulose membrane. Appl Environ Microbiol 57(2):535–538

Broady PA, Weinstein RN (1998) Algae, lichens and fungi in la Gorce Mountains, Antarctica. Antarct Sci 10(04):376–385

Brown ME, Burlingham SK (1968) Production of plant growth substances by *Azotobacter chroococcum*. J Gen Microbiol 53(1):135–144

Cappucino JC, Sherman N (1992) Nitrogen cycle. In: Microbiology: a laboratory manual, 4th edn. Benjamin/Cumming Publishing, New York, pp 311–312

Chaturvedi P, Prabahar V, Manorama R, Pindi PK, Bhadra B, Begum Z, Shivaji S (2008) *Exiguobacterium soli* sp. nov., a psychrophilic bacterium from the McMurdo Dry Valleys, Antarctica. Int J Syst Evol Microbiol 58(10):2447–2453

Choi S-B, Kim J-G, Jung M-Y, Kim S-J, Min U-G, Si O-J, Park S-J, Hwang CY, Park J, Lee S (2016) Cultivation and biochemical characterization of heterotrophic bacteria associated with phytoplankton bloom in the Amundsen sea polynya, Antarctica. Deep-Sea Res II Top Stud Oceanogr 123:126–134

Christner BC, Mosley-Thompson E, Thompson LG, Zagorodnov V, Sandman K, Reeve JN (2000) Recovery and identification of viable bacteria immured in glacial ice. Icarus 144(2):479–485

Christner BC, Mosley-Thompson E, Thompson LG, Reeve JN (2003) Bacterial recovery from ancient glacial ice. Environ Microbiol 5(5):433–436

Das J, Dangar T (2008) Microbial population dynamics, especially stress tolerant Bacillus thuringiensis, in partially anaerobic rice field soils during post-harvest period of the Himalayan, island, brackish water and coastal habitats of India. World J Microbiol Biotechnol 24(8):1403–1410

Dave B, Anshuman K, Hajela P (2006) Siderophores of halophilic Archaea and their chemical characterization. Indian J Exp Biol 44(4):340

De León KB, Gerlach R, Peyton BM, Fields MW (2013) Archaeal and bacterial communities in three alkaline hot springs in Heart Lake Geyser Basin, Yellowstone National Park. Front Microbiol 4:330

Deming JW (2002) Psychrophiles and polar regions. Curr Opin Microbiol 5(3):301–309

Dhakar K, Sharma A, Pandey A (2014) Cold, pH and salt tolerant *Penicillium* spp. inhabit the high altitude soils in Himalaya, India. World J Microbiol Biotechnol 30(4):1315–1324

Dmitriev V, Gilichinsky D, Faizutdinova R, Shershunov I, Golubev W, Duda V (1997) Occurrence of viable yeasts in 3-million-year-old permafrost in Siberia. Mikrobiologiya 66:655–660

Dong X, Chen Z (2012) Psychrotolerant methanogenic archaea: diversity and cold adaptation mechanisms. Sci China Life Sci 55(5):415–421

Fasim F, Ahmed N, Parsons R, Gadd GM (2002) Solubilization of zinc salts by a bacterium isolated from the air environment of a tannery. FEMS Microbiol Lett 213(1):1–6

Fisher P, Graf F, Petrini L, Sutton B, Wookey P (1995) Fungal endophytes of Dryas octopetala from a high arctic polar semidesert and from the Swiss Alps. Mycologia 87:319–323

Franzmann P, Stackebrandt E, Sanderson K, Volkman J, Cameron D, Stevenson P, McMeekin T, Burton H (1988) *Halobacterium lacusprofundi* sp. nov., a halophilic bacterium isolated from Deep Lake, Antarctica. Syst Appl Microbiol 11(1):20–27

Franzmann P, Springer N, Ludwig W, De Macario EC, Rohde M (1992) A methanogenic archaeon from Ace Lake, Antarctica: *Methanococcoides burtonii* sp. nov. Syst Appl Microbiol 15(4):573–581

Franzmann PD, Liu Y, Balkwill DL, Aldrich HC, De Macario EC, Boone DR (1997) *Methanogenium frigidum* sp. nov., a psychrophilic, H2-using methanogen from Ace Lake, Antarctica. Int J Syst Evol Microbiol 47(4):1068–1072

Gaba S, Singh RN, Abrol S, Yadav AN, Saxena AK, Kaushik R (2017) Draft genome sequence of *Halolamina pelagica* CDK2 isolated from natural salterns from Rann of Kutch, Gujarat, India. Genome Announc 5(6):1–2

Gerday C, Aittaleb M, Bentahir M, Chessa J-P, Claverie P, Collins T, D'Amico S, Dumont J, Garsoux G, Georlette D (2000) Cold-adapted enzymes: from fundamentals to biotechnology. Trends Biotechnol 18(3):103–107

Gholami M, Khakvar R, Niknam G (2014) Introduction of some new endophytic bacteria from *Bacillus* and *Streptomyces* genera as successful biocontrol agents against *Sclerotium rolfsii*. Arch Phytopathol Plant Protect 47(1):122–130

Glick BR (1995) The enhancement of plant growth by free-living bacteria. Can J Microbiol 41(2):109–117

Glick BR, Patten CL, Holguin G, Penrose DM (1999a) Overview of plant growth-promoting bacteria. In: Glick BR, Patten CL, Holguin G, Penrose DM (eds) Biochemical and genetic mechanisms used by plant growth promoting bacteria. World Scientific, London, pp 1–13

Glick BR, Patten CL, Holguin G, Penrose DM (1999b) Biochemical and genetic mechanisms used by plant growth promoting bacteria. World Scientific, London

Goldstein AH (1994) Involvement of the quinoprotein glucose dehydrogenase in the solubilization of exogenous phosphates by gram-negative bacteria. In: Torriani-Gorini A, Yagil E, Silver S (eds) Phosphate in microorganisms: cellular and molecular biology. ASM Press, Washington, DC, pp 197–203

Golubev W (1998) New species of basidiomycetous yeasts, *Rhodotorula creatinovora* and *R. yakutica*, isolated from permafrost soils of Eastern-Siberian Arctic. Mykol Phytopathol 32:8–13

Gosink J, Herwig R, Staley J (1997) *Octadecabacter arcticus* gen. nov., sp. nov., and *O. antarcticus*, sp. nov., nonpigmented, psychrophilic gas vacuolate bacteria from polar sea ice and water. Syst Appl Microbiol 20(3):356–365

Groudieva T, Kambourova M, Yusef H, Royter M, Grote R, Trinks H, Antranikian G (2004) Diversity and cold-active hydrolytic enzymes of culturable bacteria associated with Arctic sea ice, Spitzbergen. Extremophiles 8(6):475–488

Gulati A, Rahi P, Vyas P (2008) Characterization of phosphate-solubilizing fluorescent pseudomonads from the rhizosphere of seabuckthorn growing in the cold deserts of Himalayas. Curr Microbiol 56(1):73–79

Gulati A, Vyas P, Rahi P, Kasana RC (2009) Plant growth-promoting and rhizosphere-competent *Acinetobacter rhizosphaerae* strain BIHB 723 from the cold deserts of the Himalayas. Curr Microbiol 58(4):371–377

Gunde-Cimerman N, Sonjak S, Zalar P, Frisvad JC, Diderichsen B, Plemenitaš A (2003) Extremophilic fungi in arctic ice: a relationship between adaptation to low temperature and water activity. Phys Chem Earth ABC 28(28):1273–1278

Gupta G, Parihar S, Ahirwar N, Snehi S, Singh V (2015) Plant Growth Promoting Rhizobacteria (PGPR): current and future prospects for development of sustainable agriculture. J Microb Biochem Technol 7:96–102

Hallmann J, Quadt-Hallmann A, Mahaffee W, Kloepper J (1997) Bacterial endophytes in agricultural crops. Can J Microbiol 43(10):895–914

Han SO, New P (1998) Variation in nitrogen fixing ability among natural isolates of *Azospirillum*. Microb Ecol 36(2):193–201

He L, Mao Y, Zhang L, Wang H, Alias SA, Gao B, Wei D (2017) Functional expression of a novel α-amylase from Antarctic psychrotolerant fungus for baking industry and its magnetic immobilization. BMC Biotechnol 17(1):22

Hirsch P, Ludwig W, Hethke C, Sittig M, Hoffmann B, Gallikowski C (1998) *Hymenobacter roseosalivarius* gen. nov., sp. nov. from continental Antarctic soils and sandstone: bacteria of the Cytophaga/Flavobacterium/Bacteroides line of phylogenetic descent. Syst Appl Microbiol 21(3):374–383

Hu X, Chen J, Guo J (2006) Two phosphate- and potassium-solubilizing bacteria isolated from Tianmu Mountain, Zhejiang, China. World J Microbiol Biotechnol 22(9):983–990. https://doi.org/10.1007/s11274-006-9144-2

Humphry DR, George A, Black GW, Cummings SP (2001) *Flavobacterium frigidarium* sp. nov., an aerobic, psychrophilic, xylanolytic and laminarinolytic bacterium from Antarctica. Int J Syst Evol Microbiol 51(4):1235–1243

Inderiati S, Franco CM (2008) Isolation and identification of endophytic actinomycetes and their antifungal activity. J Biotechnol Res 1:1–6

Ivanova E, Doronina N, Trotsenko YA (2001) Aerobic methylobacteria are capable of synthesizing auxins. Microbiology 70(4):392–397

Ivanova E, Pirttilä A, Fedorov D, Doronina N, Trotsenko Y (2008) Association of methylotrophic bacteria with plants: metabolic aspects. In: Sorvari S, Pirttilä AM (eds) Prospects and applications for plant associated microbes a laboratory manual, part A: bacteria. Biobien Innovations, Turku, pp 225–231

Jacobson CB, Pasternak J, Glick BR (1994) Partial purification and characterization of 1-aminocyclopropane-1-carboxylate deaminase from the plant growth promoting rhizobacterium *Pseudomonas putida* GR12-2. Can J Microbiol 40(12):1019–1025

Kaur R, Saxena A, Sangwan P, Yadav AN, Kumar V, Dhaliwal HS (2017) Production and characterization of a neutral phytase of *Penicillium oxalicum* EUFR-3 isolated from Himalayan region. Nusantara Biosci 9(1):68–76

Khalid A, Arshad M, Zahir Z (2004) Screening plant growth-promoting rhizobacteria for improving growth and yield of wheat. J Appl Microbiol 96(3):473–480

Khalid A, Akhtar M, Mahmood M, Arshad M (2006) Effect of substrate-dependent microbial ethylene production on plant growth. Microbiology 75(2):231–236

Kim SJ, Shin SC, Hong SG, Lee YM, Choi I-G, Park H (2012) Genome sequence of a novel member of the genus *Psychrobacter* isolated from Antarctic soil. J Bacteriol 194(9):2403–2403

Kishore KH, Begum Z, Pathan AAK, Shivaji S (2010) *Paenibacillus glacialis* sp. nov., isolated from the Kafni glacier of the Himalayas, India. Int J Syst Evol Microbiol 60(8):1909–1913

Kloepper JW, Lifshitz R, Zablotowicz RM (1989) Free-living bacterial inocula for enhancing crop productivity. Trends Biotechnol 7(2):39–44

Könneke M, Bernhard AE, José R, Walker CB, Waterbury JB, Stahl DA (2005) Isolation of an autotrophic ammonia-oxidizing marine archaeon. Nature 437(7058):543–546

Kostadinova N, Krumova E, Tosi S, Pashova AM (2009) Isolation and identification of filamentous fungi from island Livingston, Antarctica. Biotechnol Biotechnol Equip 23:267–270

Kumar V, Singh P, Jorquera MA, Sangwan P, Kumar P, Verma A, Agrawal S (2013) Isolation of phytase-producing bacteria from Himalayan soils and their effect on growth and phosphorus uptake of Indian mustard (*Brassica juncea*). World J Microbiol Biotechnol 29(8):1361–1369

Kumar M, Yadav A, Tiwari R, Prasanna R, Saxena A (2014a) Evaluating the diversity of culturable thermotolerant bacteria from four hot springs of India. J Biodivers Biopros Dev 1:127

Kumar M, Yadav AN, Tiwari R, Prasanna R, Saxena AK (2014b) Deciphering the diversity of culturable thermotolerant bacteria from Manikaran hot springs. Ann Microbiol 64(2):741–751
Kumar V, Yadav AN, Saxena A, Sangwan P, Dhaliwal HS (2016) Unravelling rhizospheric diversity and potential of phytase producing microbes. J Biol 2(1):1009
Kumar V, Yadav AN, Verema P, Sangwan P, Abhishake S, Singh B (2017) β-Propeller phytases: diversity, catalytic attributes, current developments and potential biotechnological applications. Int J Biol Macromol 98:595–609
Lanteigne C, Gadkar VJ, Wallon T, Novinscak A, Filion M (2012) Production of DAPG and HCN by *Pseudomonas* sp. LBUM300 contributes to the biological control of bacterial canker of tomato. Phytopathology 102(10):967–973
Lee YM, Hwang CY, Lee I, Jung Y-J, Cho Y, Baek K, Hong SG, Kim J-H, Chun J, Lee HK (2014) *Lacinutrix jangbogonensis* sp. nov., a psychrophilic bacterium isolated from Antarctic marine sediment and emended description of the genus *Lacinutrix*. Antonie Van Leeuwenhoek 106(3):527–533
Lee J, Jung Y-J, Lee HK, Hong SG, Kim O-S (2016) Complete genome sequence of *Pedobacter cryoconitis* PAMC 27485, a CRISPR-Cas system-containing psychrophile isolated from Antarctica. J Biotechnol 226:74–75
Lee CW, Park S-H, Lee SG, Shin SC, Han SJ, Kim H-W, Park HH, Kim S, Kim HJ, Park H (2017) Crystal structure of the inactive state of the receiver domain of Spo0A from *Paenisporosarcina* sp. TG-14, a psychrophilic bacterium isolated from an Antarctic glacier. J Microbiol 1:1–11
Leigh JA (2000) Nitrogen fixation in methanogens: the archaeal perspective. Curr Issues Mol Biol 2:125–131
López NI, Pettinari MJ, Stackebrandt E, Tribelli PM, Põtter M, Steinbüchel A, Méndez BS (2009) *Pseudomonas extremaustralis* sp. nov., a poly (3-hydroxybutyrate) producer isolated from an Antarctic environment. Curr Microbiol 59(5):514–519
Lyhne E, Jarvis B, Fettinger J, Overy D (2006) Four psychrotolerant species with high chemical diversity consistently producing cycloaspeptide A, *Penicillium jamesonlandense* sp. nov., *Penicillium ribium* sp. nov., *Penicillium soppii* and *Penicillium lanosum*. Int J Syst Evol Microbiol 56:14271437
Ma L, Catranis CM, Starmer WT, Rogers SO (1999) Revival and characterization of fungi from ancient polar ice. Mycologist 13(2):70–73
Ma L-J, Rogers SO, Catranis CM, Starmer WT (2000) Detection and characterization of ancient fungi entrapped in glacial ice. Mycologia 92:286–295
Malviya MK, Pandey A, Trivedi P, Gupta G, Kumar B (2009) Chitinolytic activity of cold tolerant antagonistic species of *Streptomyces* isolated from glacial sites of Indian Himalaya. Curr Microbiol 59(5):502–508
Margesin R, Zhang D-C, Frasson D, Brouchkov A (2016) *Glaciimonas frigoris* sp. nov., a psychrophilic bacterium isolated from ancient Siberian permafrost sediment, and emended description of the genus *Glaciimonas*. Int J Syst Evol Microbiol 66(2):744–748
Marx J, Collins T, D'Amico S, Feller G, Gerday C (2007) Cold-adapted enzymes from marine Antarctic microorganisms. Mar Biotechnol 9(3):293–304
Mehta S, Nautiyal CS (2001) An efficient method for qualitative screening of phosphate-solubilizing bacteria. Curr Microbiol 43(1):51–56
Mishra PK, Mishra S, Selvakumar G, Bisht SC, Bisht JK, Kundu S, Gupta HS (2008) Characterisation of a psychrotolerant plant growth promoting *Pseudomonas* sp. strain PGERs17 (MTCC 9000) isolated from North Western Indian Himalayas. Ann Microbiol 58(4):561–568
Mishra PK, Mishra S, Bisht SC, Selvakumar G, Kundu S, Bisht J, Gupta HS (2009) Isolation, molecular characterization and growth-promotion activities of a cold tolerant bacterium *Pseudomonas* sp. NARs9 (MTCC9002) from the Indian Himalayas. Biol Res 42(3):305–313
Mishra PK, Bisht SC, Ruwari P, Selvakumar G, Joshi GK, Bisht JK, Bhatt JC, Gupta HS (2011) Alleviation of cold stress in inoculated wheat (*Triticum aestivum* L.) seedlings with psychrotolerant Pseudomonads from NW Himalayas. Arch Microbiol 193(7):497–513

Nagendran K, Karthikeyan G, Peeran MF, Raveendran M, Prabakar K, Raguchander T (2013) Management of bacterial leaf blight disease in rice with endophytic bacteria. World Appl Sci J 28(12):2229–2241

Nazir R, Hayat W, Rehman P, Iqbal A, Irshad U (2017) Novel P-Solubilizers from calcium bound phosphate rich pine forest of Lower Himalaya. Geomicrobiol J 34(2):119–129

Ongena M, Jacques P (2008) *Bacillus lipopeptides*: versatile weapons for plant disease biocontrol. Trends Microbiol 16(3):115–125

Pandey S, Singh S, Yadav AN, Nain L, Saxena AK (2013) Phylogenetic diversity and characterization of novel and efficient cellulase producing bacterial isolates from various extreme environments. Biosci Biotechnol Biochem 77(7):1474–1480

Patten CL, Glick BR (2002) Role of *Pseudomonas putida* indoleacetic acid in development of the host plant root system. Appl Environ Microbiol 68(8):3795–3801

Pennisi E (2003) Neither cold nor snow stops tundra fungi. Science 301(5638):1307–1307

Pikovskaya R (1948) Mobilization of phosphorus in soil in connection with vital activity of some microbial species. Mikrobiologiya 17:362–370

Pradhan S, Srinivas T, Pindi PK, Kishore KH, Begum Z, Singh PK, Singh AK, Pratibha M, Yasala AK, Reddy G (2010) Bacterial biodiversity from Roopkund glacier, Himalayan mountain ranges, India. Extremophiles 14(4):377–395

Purnawati A (2014) Endophytic bacteria as biocontrol agents of tomato bacterial wilt disease. J Trop Life Sci 4(1):33–36

Quadt-Hallmann A, Kloepper J, Benhamou N (1997) Bacterial endophytes in cotton: mechanisms of entering the plant. Can J Microbiol 43(6):577–582

Ramesh R, Joshi A, Ghanekar M (2009) Pseudomonads: major antagonistic endophytic bacteria to suppress bacterial wilt pathogen, Ralstonia solanacearum in the eggplant (*Solanum melongena* L.) World J Microbiol Biotechnol 25(1):47–55

Rana KL, Kour D, Yadav AN, Kumar V, Dhaliwal HS (2016a) Biotechnological applications of endophytic microbes associated with barley (*Hordeum vulgare* L.) growing in Indian Himalayan regions. In: Proceeding of 86 annual session of NASI and symposium on "Science, Technology and Entrepreneurship for Human Welfare in The Himalayan Region", pp 80

Rana KL, Kour D, Yadav AN, Kumar V, Dhaliwal HS (2016b) Endophytic microbes from wheat: diversity and biotechnological applications for sustainable agriculture. In: Proceeding of 57 association of microbiologist of India and international symposium on "Microbes and biosphere: what's new what's next", pp 453

Rana KL, Kour D, Verma P, Yadav AN, Kumar V, Singh DH (2017) Diversity and biotechnological applications of endophytic microbes associated with maize (*Zea mays* L.) growing in Indian Himalayan regions. In: Proceeding of national conference on advances in food science and technology, pp 120

Reddy G, Pradhan S, Manorama R, Shivaji S (2010) *Cryobacterium roopkundense* sp. nov., a psychrophilic bacterium isolated from glacial soil. Int J Syst Evol Microbiol 60(4):866–870

Sacherer P, Défago G, Haas D (1994) Extracellular protease and phospholipase C are controlled by the global regulatory gene gacA in the biocontrol strain *Pseudomonas fluorescens* CHA0. FEMS Microbiol Lett 116(2):155–160

Sahay H, Babu BK, Singh S, Kaushik R, Saxena AK, Arora DK (2013) Cold-active hydrolases producing bacteria from two different sub-glacial Himalayan lakes. J Basic Microbiol 53(8):703–714

Sahay H, Yadav AN, Singh AK, Singh S, Kaushik R, Saxena AK (2017) Hot springs of Indian Himalayas: potential sources of microbial diversity and thermostable hydrolytic enzymes. 3 Biotech 7:118

Saitou N, Nei M (1987) The neighbor-joining method: a new method for reconstructing phylogenetic trees. Mol Biol Evol 4(4):406–425

Saxena AK, Yadav AN, Kaushik R, Tyagi S, Kumar M, Prasanna R, Shukla L (2014) Use of microbes from extreme environments for the benefits of agriculture. In: Afro-Asian Congress on microbes for human and environmental health, pp 102. https://doi.org/10.13140/RG.2.1.3479.1841

Saxena AK, Yadav AN, Kaushik R, Tyagi SP, Shukla L (2015a) Biotechnological applications of microbes isolated from cold environments in agriculture and allied sectors. In: International conference on "Low Temperature Science and Biotechnological Advances", p 104. https://doi.org/10.13140/RG.2.1.2853.5202

Saxena AK, Kaushik R, Yadav AN, Gulati S, Sharma D (2015b) Role of Archaea in sustenance of plants in extreme saline environments. In: Proceeding of 56th annual conference of association of microbiologists of India and international symposium on "Emerging discoveries in microbiology". IA06. https://doi.org/10.13140/RG.2.1.2073.9925

Saxena AK, Yadav AN, Rajawat MVS, Kaushik R, Kumar R, Kumar M, Prasanna R, Shukla L (2016) Microbial diversity of extreme regions: an unseen heritage and wealth. Indian J Plant Genet Resour 29(3):246–248

Schadt CW, Martin AP, Lipson DA, Schmidt SK (2003) Seasonal dynamics of previously unknown fungal lineages in tundra soils. Science 301(5638):1359–1361

Schwyn B, Neilands J (1987) Universal chemical assay for the detection and determination of siderophores. Anal Biochem 160(1):47–56

Selvakumar G, Kundu S, Joshi P, Nazim S, Gupta A, Mishra P, Gupta H (2008) Characterization of a cold-tolerant plant growth-promoting bacterium *Pantoea dispersa* 1A isolated from a sub-alpine soil in the North Western Indian Himalayas. World J Microbiol Biotechnol 24(7):955–960

Selvakumar G, Joshi P, Mishra PK, Bisht JK, Gupta HS (2009a) Mountain aspect influences the genetic clustering of psychrotolerant phosphate solubilizing Pseudomonads in the Uttarakhand Himalayas. Curr Microbiol 59(4):432–438

Selvakumar G, Joshi P, Nazim S, Mishra P, Bisht J, Gupta H (2009b) Phosphate solubilization and growth promotion by *Pseudomonas fragi* CS11RH1 (MTCC 8984), a psychrotolerant bacterium isolated from a high altitude Himalayan rhizosphere. Biologia 64(2):239–245

Selvakumar G, Joshi P, Nazim S, Mishra PK, Kundu S, Gupta HS (2009c) *Exiguobacterium acetylicum* strain 1P (MTCC 8707) a novel bacterial antagonist from the North Western Indian Himalayas. World J Microbiol Biotechnol 25(1):131–137

Selvakumar G, Joshi P, Suyal P, Mishra PK, Joshi GK, Bisht JK, Bhatt JC, Gupta HS (2011) *Pseudomonas lurida* M2RH3 (MTCC 9245), a psychrotolerant bacterium from the Uttarakhand Himalayas, solubilizes phosphate and promotes wheat seedling growth. World J Microbiol Biotechnol 27(5):1129–1135

Shen L, Liu Y, Gu Z, Xu B, Wang N, Jiao N, Liu H, Zhou Y (2015) *Massilia eurypsychrophila* sp. nov. a facultatively psychrophilic bacteria isolated from ice core. Int J Syst Evol Microbiol 65(7):2124–2129

Shilev S (2013) Soil rhizobacteria regulating the uptake of nutrients and undesirable elements by plants. In: Plant microbe symbiosis: fundamentals and advances. Springer, New Delhi, pp 147–167

Shivaji S, Ray M, Rao NS, Saisree L, Jagannadham M, Kumar GS, Reddy G, Bhargava PM (1992) *Sphingobacterium antarcticus* sp. nov., a psychrotrophic bacterium from the soils of Schirmacher Oasis, Antarctica. Int J Syst Evol Microbiol 42(1):102–106

Shivaji S, Chaturvedi P, Suresh K, Reddy G, Dutt C, Wainwright M, Narlikar J, Bhargava P (2006) *Bacillus aerius* sp. nov., *Bacillus aerophilus* sp. nov., *Bacillus stratosphericus* sp. nov. and *Bacillus altitudinis* sp. nov., isolated from cryogenic tubes used for collecting air samples from high altitudes. Int J Syst Evol Microbiol 56(7):1465–1473

Shivaji S, Pratibha M, Sailaja B, Kishore KH, Singh AK, Begum Z, Anarasi U, Prabagaran S, Reddy G, Srinivas T (2011) Bacterial diversity of soil in the vicinity of Pindari glacier, Himalayan mountain ranges, India, using culturable bacteria and soil 16S rRNA gene clones. Extremophiles 15(1):1–22

Shukla L, Suman A, Yadav AN, Verma P, Saxena AK (2016) Syntrophic microbial system for ex-situ degradation of paddy straw at low temperature under controlled and natural environment. J App Biol Biotech 4(2):30–37

Sijam K, Dikin A (2005) Biochemical and physiological characterization of *Burkholderia cepacia* as biological control agent. Int J Agric Biol 7(3):385–388

Singh RN, Gaba S, Yadav AN, Gaur P, Gulati S, Kaushik R, Saxena AK (2016) First high quality draft genome sequence of a plant growth promoting and cold active enzyme producing psychrotrophic *Arthrobacter agilis* strain L77. Stand Genomic Sci 11:54

Soina V, Vorobyova E, Suzina N, Dmitriev V, Faizutdinova R, Duda V (2000) Viable ancient permafrost yeast. Delft University Press, Delft

Sonjak S, Frisvad JC, Gunde-Cimerman N (2005) Comparison of secondary metabolite production by *Penicillium crustosum* strains, isolated from Arctic and other various ecological niches. FEMS Microbiol Ecol 53(1):51–60

Sonjak S, Frisvad JC, Gunde-Cimerman N (2006) *Penicillium mycobiota* in Arctic subglacial ice. Microb Ecol 52(2):207–216

Sonjak S, Frisvad JC, Gunde-Cimerman N (2007) Genetic variation among *Penicillium crustosum* isolates from arctic and other ecological niches. Microb Ecol 54(2):298–305

Srinivas TNR, Singh SM, Pradhan S, Pratibha MS, Kishore KH, Singh AK, Begum Z, Prabagaran SR, Reddy GSN, Shivaji S (2011) Comparison of bacterial diversity in proglacial soil from Kafni Glacier, Himalayan mountain ranges, India, with the bacterial diversity of other glaciers in the world. Extremophiles 15(6):673

Srivastava A, Kumar S, Kaushik R, Saxena A, Padaria J, Gupta A, Pal K, Gujar G, Sharma A, Singh P (2014) Diversity analysis of Bacillus and other predominant genera in extreme environments and its utilization in Agriculture. Technical report 414402/C30026. https://doi.org/10.13140/2.1.1357.3927

Suman A, Verma P, Yadav AN, Saxena AK (2015) Bioprospecting for extracellular hydrolytic enzymes from culturable thermotolerant bacteria isolated from Manikaran thermal springs. Res J Biotechnol 10:33–42

Suman A, Verma P, Yadav AN, Srinivasamurthy R, Singh A, Prasanna R (2016a) Development of hydrogel based bio-inoculant formulations and their impact on plant biometric parameters of wheat (*Triticum aestivum* L.) Int J Curr Microbiol App Sci 5:890–901

Suman A, Yadav AN, Verma P (2016b) Endophytic microbes in crops: diversity and beneficial impact for sustainable agriculture. In: Singh D, Abhilash P, Prabha R (eds) Microbial inoculants in sustainable agricultural productivity, research perspectives. Springer-Verlag, India, pp 117–143. https://doi.org/10.1007/978-81-322-2647-5_7

Tamura K, Dudley J, Nei M, Kumar S (2007) MEGA4: Molecular Evolutionary Genetics Analysis (MEGA) software version 4.0. Mol Biol Evol 24(8):1596–1599

Tosi S, Casado B, Gerdol R, Caretta G (2002) Fungi isolated from Antarctic mosses. Polar Biol 25(4):262–268

Trivedi P, Pandey A (2008) Plant growth promotion abilities and formulation of *Bacillus megaterium* strain B 388 (MTCC6521) isolated from a temperate Himalayan location. Indian J Microbiol 48(3):342–347

Turchetti B, Hall SRT, Connell LB, Branda E, Buzzini P, Theelen B, Müller WH, Boekhout T (2011) Psychrophilic yeasts from Antarctica and European glaciers: description of *Glaciozyma* gen. nov., *Glaciozyma martinii* sp. nov. and *Glaciozyma watsonii* sp. nov. Extremophiles 15(5):573

Van Trappen S, Vandecandelaere I, Mergaert J, Swings J (2005) *Flavobacterium fryxellicola* sp. nov. and *Flavobacterium psychrolimnae* sp. nov., novel psychrophilic bacteria isolated from microbial mats in Antarctic lakes. Int J Syst Evol Microbiol 55(2):769–772

Venkatachalam S, Gowdaman V, Prabagaran SR (2015) Culturable and culture-independent bacterial diversity and the prevalence of cold-adapted enzymes from the Himalayan Mountain Ranges of India and Nepal. Microb Ecol 69(3):472–491

Verma P, Yadav AN, Kazy SK, Saxena AK, Suman A (2013) Elucidating the diversity and plant growth promoting attributes of wheat (*Triticum aestivum*) associated acidotolerant bacteria from southern hills zone of India. Natl. J Life Sci 10(2):219–227

Verma P, Yadav AN, Kazy SK, Saxena AK, Suman A (2014a) Evaluating the diversity and phylogeny of plant growth promoting bacteria associated with wheat (*Triticum aestivum*) growing in central zone of India. Int J Curr Microbiol App Sci 3(5):432–447

Verma P, Yadav AN, Khannam KS, Panjiar N, Mishra S, Kumar S, Saxena AK, Suman A (2014b) Assessment of genetic bacterial diversity and plant growth promoting attributes of drought tolerant K-solubilizing bacteria allied with wheat (*Triticum aestivum*). In: 84th annual session of NASI and symposium on "Desert Science-Opportunities and Challenges" held at Jai Narain Vyas University, Jodhpur, Rajasthan

Verma P, Yadav A, Khannam K, Panjiar N, Kumar S, Saxena A, Suman A (2015a) Assessment of genetic diversity and plant growth promoting attributes of psychrotolerant bacteria allied with wheat (*Triticum aestivum*) from the northern hills zone of India. Ann Microbiol 65(4):1885–1899

Verma P, Yadav AN, Khannam KS, Kumar S, Saxena AK, Suman A (2015b) Growth promotion and yield enhancement of wheat (*Triticum aestivum* L.) by application of potassium solubilizing psychrotolerant bacteria. In: Proceeding of 56th annual conference of association of microbiologists of India and international symposium on "Emerging Discoveries in Microbiology". https://doi.org/10.13140/RG.2.1.1648.0080

Verma P, Yadav AN, Khannam KS, Kumar S, Saxena AK, Suman A (2015c) Plant growth promotion and mitigation of cold stress in inoculated wheat (*Triticum aestivum* L.) by K-solubilizing psychrotolerant *Methylobacterium phyllosphaerae* strain IARI-HHS2-67. In: Proceeding of international conference on "Low temperature science and biotechnological advances", pp 175. https://doi.org/10.13140/RG.2.1.4885.1362

Verma P, Yadav AN, Shukla L, Saxena AK, Suman A (2015d) Alleviation of cold stress in wheat seedlings by *Bacillus amyloliquefaciens* IARI-HHS2-30, an endophytic psychrotolerant K-solubilizing bacterium from NW Indian Himalayas. Natl J Life Sci 12(2):105–110

Verma P, Yadav AN, Khannam KS, Kumar S, Saxena AK, Suman A (2016) Molecular diversity and multifarious plant growth promoting attributes of Bacilli associated with wheat (*Triticum aestivum* L.) rhizosphere from six diverse agro-ecological zones of India. J Basic Microbiol 56(1):44–58

Verma P, Yadav AN, Kumar V, Kumar K, Dhaliwal HS (2017a) Microbes mediated biofortification of wheat (*Triticum aestivum* L.) for micronutrients by Fe-chelating and Zn-solubilizing bacteria. In: Proceeding of national conference on advances in food science and technology. pp 199–200

Verma P, Yadav AN, Kumar V, Khan MA, Saxena AK (2017b) Microbes in termite management: potential role and strategies. In: Khan MA, Ahmad W (eds) Sustainable termite management. Springer, Cham, pp 197–217. https://doi.org/10.1007/978-3-319-68726-1_9

Verma P, Yadav AN, Kumar V, Singh DP, Saxena AK (2017c) Beneficial plant-microbes interactions: biodiversity of microbes from diverse extreme environments and its impact for crops improvement. In: Singh DP, Singh HB, Prabha R (eds) Plant-microbe interactions in agro-ecological perspectives. Springer, Singapore. https://doi.org/10.1007/978-981-10-6593-4_22

Vishniac H (1993) The microbiology of Antarctic soils. Ant Microbiol 1:297–341

von Klein D, Arab H, Völker H, Thomm M (2002) *Methanosarcina baltica*, sp. nov., a novel methanogen isolated from the Gotland Deep of the Baltic Sea. Extremophiles 6(2):103–110

Vyas P, Joshi R, Sharma K, Rahi P, Gulati A, Gulati A (2010) Cold-adapted and rhizosphere-competent strain of *Rahnella* sp. with broadspectrum plant growth-promotion potential. J Microbiol Biotechnol 20(12):1724–1734

Wagner D, Schirmack J, Ganzert L, Morozova D, Mangelsdorf K (2013) *Methanosarcina soligelidi* sp. nov., a desiccation-and freeze-thawresistant methanogenic archaeon from a Siberian permafrost-affected soil. Int J Syst Evol Microbiol 63(8):2986–2991

White RH (1987) Indole-3-acetic acid and 2-(indol-3-ylmethyl) indol-3-yl acetic acid in the thermophilic archaebacterium *Sulfolobus acidocaldarius*. J Bacteriol 169(12):5859–5860

Xu M, Sheng J, Chen L, Men Y, Gan L, Guo S, Shen L (2014) Bacterial community compositions of tomato (*Lycopersicum esculentum* Mill.) seeds and plant growth promoting activity of ACC deaminase producing *Bacillus subtilis* (HYT-12-1) on tomato seedlings. World J Microbiol Biotechnol 30(3):835–845

Yadav AN (2009) Studies of methylotrophic community from the phyllosphere and rhizosphere of tropical crop plants. M.Sc. Thesis, Bundelkhand University, pp 66. https://doi.org/10.13140/2.1.5099.0888

Yadav AN (2015) Bacterial diversity of cold deserts and mining of genes for low temperature tolerance. PhD Thesis, Indian Agricultural Research Institute, New Delhi and Birla Institute of Technology, Mesra, Ranchi, pp 234. https://doi.org/10.13140/RG.2.1.2948.1283/2

Yadav AN (2017) Agriculturally important microbiomes: biodiversity and multifarious PGP attributes for Amelioration of diverse abiotic stresses in crops for sustainable agriculture. Biomed J Sci Tech Res 1:1–4

Yadav AN, Saxena AK (2017) Biodiversity and potential biotechnological applications of halophilic microbes for sustainable agriculture. J Appl Biol Biotechnol 5(6):58–65

Yadav AN, Kumar R, Kumar S, Kumar V, Sugitha T, Singh B, Chauhan VS, Dhaliwal HS, Saxena AK (2017a) Beneficial microbiomes: biodiversity and potential biotechnological applications for sustainable agriculture and human health. J Appl Biol Biotechnol 5(6):45–57

Yadav AN, Verma P, Kaushik R, Dhaliwal HS, Saxena AK (2017b) Archaea endowed with plant growth promoting attributes. EC Microbiol 8:294–298

Yadav AN, Verma P, Kour D, Rana KL, Kumar V, Singh B, Chauahan VS, Sugitha T, Saxena AK, Dhaliwal HS (2017c) Plant microbiomes and its beneficial multifunctional plant growth promoting attributes. Int J Environ Sci Nat Resour 3:1–8. 10.19080/IJESNR.2017.03.555601

Yadav AN, Verma P, Kumar R, Kumar V, Kumar K (2017d) Current applications and future prospects of eco-friendly microbes. EU Voice 3:21–22

Yadav AN, Verma P, Kumar V, Sachan SG, Saxena AK (2017e) Extreme cold environments: a suitable niche for selection of novel psychrotrophic microbes for biotechnological applications. Adv Biotechnol Microbiol 2(2):1–4

Yadav AN, Verma P, Kumar S, Kumar V, Kumar M, Singh BP, Saxena AK, Dhaliwal HS (2017f) Actinobacteria from rhizosphere: molecular diversity, distributions and potential biotechnological applications. In: Singh BP, Gupta VK, Passari AK (eds) Actinobacteria: diversity and biotechnological applications. Elsevier, USA, pp 1–28. https://doi.org/10.1016/B978-0-444-63994-3.00002-3

Yadav AN, Verma P, Kumar V, Sangwan P, Mishra S, Panjiar N, Gupta VK, Saxena AK (2017g) Biodiversity of the Genus *Penicillium* in Different Habitats. In: Gupta VK, Rodriguez-Couto S (eds) New and future developments in microbial biotechnology and bioengineering, *Penicillium* system properties and applications. Elsevier, Amsterdam, pp 3–18. https://doi.org/10.1016/B978-0-444-63501-3.00001-6

Yadav AN, Verma P, Sachan SG, Saxena AK (2017h) Biodiversity and biotechnological applications of psychrotrophic microbes isolated from Indian Himalayan regions. EC Microbiol ECO 01:48–54

Yadav AN, Verma P, Singh B, Chauhan VS, Suman A, Saxena AK (2017i) Plant growth Promoting bacteria: biodiversity and multifunctional attributes for sustainable agriculture. Adv Biotechnol Microbiol 5:1–16

Yadav AN, Verma P, Sachan SG, Kaushik R, Saxena A (2012) Diversity of culturable psychrotrophic bacteria from Leh Ladakh and bioprospecting for cold-active extracellular enzymes. In: Proceeding of national seminar on "Biotechnological interventions for the benefit of mankind", pp 32

Yadav AN, Verma P, Sachan SG, Kaushik R, Saxena A (2013) Diversity and phylogeny of low temperature plant growth promoting fluorescent *Pseudomonas* isolated from cold desert of north western Himalayas. In: Proceeding of 54 AMI, international symposium on "Frontier discoveries and innovations in microbiology and its interdisciplinary relevance", pp 396–397

Yadav AN, Sachan SG, Kaushik R, Saxena AK (2014a) Cold tolerant microbes from Indian Himalayas: diversity and potential applications in biotechnology and agriculture. In: 84th annual session of NASI and symposium on "Desert Science-Opportunities and challenges". Jai Narain Vyas University, Jodhpur, Rajasthan, pp 17

Yadav AN, Sachan SG, Verma P, Kaushik R, Saxena AK (2014b). Diversity and bioprospecting of cold active enzymes produced by psychrotrophic bacilli from three sub-glacial lakes of north

western Indian Himalayas. In: 55th AMI national conference on "Empowering mankind with microbial technologies" AM-78
Yadav AN, Verma P, Sachan SG, Singh S, Saxena AK (2014c) Indian cold deserts: exploration of bacterial diversity and novel potassium and phosphorus solubilizing psychrotolerant bacteria. In: Proceeding of 55th association of microbiologist and national conference on "Empowering mankind with microbial technologies" pp 75
Yadav AN, Sachan SG, Verma P, Tyagi SP, Kaushik R, Saxena AK (2015a) Culturable diversity and functional annotation of psychrotrophic bacteria from cold desert of Leh Ladakh (India). World J Microbiol Biotechnol 31(1):95–108
Yadav AN, Sachan SG, Verma P, Saxena AK (2015b) Prospecting cold deserts of north western Himalayas for microbial diversity and plant growth promoting attributes. J Biosci Bioeng 119(6):683–693
Yadav AN, Sharma D, Gulati S, Singh S, Kaushik R, Dey R, Pal KK, Saxena AK (2015c) Haloarchaea endowed with phosphorus solubilization attribute implicated in phosphorus cycle. Sci Rep 5:12293
Yadav AN, Verma P, Kumar M, Pal KK, Dey R, Gupta A, Padaria JC, Gujar GT, Kumar S, Suman A (2015d) Diversity and phylogenetic profiling of niche-specific Bacilli from extreme environments of India. Ann Microbiol 65(2):611–629
Yadav AN, Verma P, Sachan SG, Kaushik R, Saxena AK (2015e) Microbes mediated alleviation of cold stress for growth and yield of wheat (*Triticum aestivum* L.). In: Proceeding of international conference on "Low temperature science and biotechnological advances", pp 179. https://doi.org/10.13140/RG.2.1.2374.2883
Yadav AN, Verma P, Sachan SG, Kaushik R, Saxena AK (2015f) Mitigation of cold stress for growth and yield of wheat (*Triticum aestivum* L.) by psychrotrophic pseudomonads from cold deserts of Indian Himalayas. In: Proceeding of 56 AMI 20105 and international symposium on "Emerging discoveries in microbiology". https://doi.org/10.13140/RG.2.1.4367.7523
Yadav AN, Sachan SG, Verma P, Kaushik R, Saxena AK (2016a) Cold active hydrolytic enzymes production by psychrotrophic Bacilli isolated from three sub-glacial lakes of NW Indian Himalayas. J Basic Microbiol 56:294–307
Yadav AN, Sachan SG, Verma P, Saxena AK (2016b) Bioprospecting of plant growth promoting psychrotrophic Bacilli from cold desert of north western Indian Himalayas. Indian J Exp Biol 52(2):142–150
Yadav AN, Verma P, Sachan SG, Kaushik R, Saxena AK (2016c) Microbiome of Indian Himalayan regions: molecular diversity, phylogenetic profiling and biotechnological applications. In: Proceeding of 86th annual session of NASI and symposium on "Science, technology and entrepreneurship for human welfare in the Himalayan region", pp 58
Yakimov MM, Giuliano L, Gentile G, Crisafi E, Chernikova TN, Abraham W-R, Lünsdorf H, Timmis KN, Golyshin PN (2003) *Oleispira antarctica* gen. nov., sp. nov., a novel hydrocarbonoclastic marine bacterium isolated from Antarctic coastal sea water. Int J Syst Evol Microbiol 53(3):779–785
Zachariah S, Kumari P, Das SK (2017) *Psychrobacter pocilloporae* sp. nov., isolated from a coral, Pocillopora eydouxi. Int J Syst Evol Microbiol 66(12):5091–5098
Zhang D-C, Busse H-J, Liu H-C, Zhou Y-G, Schinner F, Margesin R (2011) *Sphingomonas glacialis* sp. nov., a psychrophilic bacterium isolated from alpine glacier cryoconite. Int J Syst Evol Microbiol 61(3):587–591
Zhou Z, Jiang F, Wang S, Peng F, Dai J, Li W, Fang C (2012) *Pedobacter arcticus* sp. nov., a facultative psychrophile isolated from Arctic soil. Int J Syst Evol Microbiol 62(8):1963–1969

Role of Microbes in Organic Farming for Sustainable Agro-Ecosystem

12

Sarita K. Yadav, Ruchi Soni, and Ajay Singh Rajput

Abstract

Sustainability is the key for continued existence of any civilization. Agriculture may be considered sustainable when the system is resource conserving, socially supportive, commercially competitive, and environmentally sound. In farming systems, microbes play a pivotal role as the main dynamic forces. The health of the planet relies on the development of efficient and sustainable agricultural systems. Soil is the base of many of the biological processes, viz., biological nitrogen fixation, residue decomposition, mineralization/immobilization turnover, nutrient cycling, and denitrification, which are regulated by the microbes. In today's scenario, wherein the addition of chemicals is done to an extreme level, the soil has lost rich flora of beneficial microbes. It is high time to reintroduce environmentally friendly microflora into the field, so as to accomplish the goal of sustainable agriculture. In this chapter we will attempt to explain the application of soil beneficial microbes through traditional and advanced approaches using organic manures in organic farming system so that human race and nature may be benefited by maintaining sustainable agroecosystem.

Keywords

Organic farming · Sustainability · Biofertilizers · Microbes

S. K. Yadav (✉) · R. Soni · A. S. Rajput
Regional Centre of Organic Farming, Department of Agriculture and Cooperation, Ministry of Agriculture and Farmers Welfare, Government of India, Nagpur, Maharashtra, India
e-mail: yadavsarita24@rediffmail.com

D. G. Panpatte et al. (eds.), *Microorganisms for Green Revolution*, Microorganisms for Sustainability 7,
https://doi.org/10.1007/978-981-10-7146-1_12

12.1 Introduction

India witnessed an era of green revolution during which enormous enhancement in the food grain production took place, but it did lead to the insufficient concern for agricultural sustainability. The chemical fertilizers are based on fossil fuel; hence their availability and affordability at farm level in developing countries are ensured through imports and subsidies which are largely dependent on GDP of the country. Indiscriminate use of synthetic fertilizers has polluted the biotic and abiotic systems of soil and water basins and destroyed microbes and other friendly worms and insects. Critical conditions arise when phenomenon like eutrophication takes place, which is capable of demolishing the whole aquatic ecosystem (Pindi 2012).

After these experiences one has to go back to sustainable growth of agriculture. Sustainable agricultural growth faces threat due to climatic changes, which is enhanced by the consistent accumulation of greenhouse gas emissions in the atmosphere, produced anthropogenically through tremendous industrial growth and stiff shift in the lifestyle. The whole world is undergoing environmental changes and requires support from the entire human race so as to ensure ecological sustainability in its socioeconomic development. Agriculture is the basis for ensuring food and livelihood security of any country; hence this sector becoming resilient to increasing climatic variabilities is most important. A lot of changes are going to be taking place; if there is as little as 1 °C rise in temperature, for instance, irrigation requirements in arid and semiarid regions are estimated to increase by 10% for every 1 °C rise in temperature. The whole world shall be vulnerable in the event of climate change, but India in particular, being highly dependent on agriculture, shall fall prey to it to a greater extent. Rise in sea level would also likely to have adverse effects on the livelihoods of fishermen and coastal communities (NMSA 2010).

An ecosystem with an integrated region of agricultural production is agroecosystem. These are complex units where several biotic and abiotic factors like soil, water, air, wildlife, insects, pathogens, plants, and humans interact. The agroecosystem concept provides a background to analyze food production systems, including their complex sets of inputs and outputs and interconnections of their component parts. If a sustainable agroecosystem is to be created, then the challenges faced are same as that lies in creating a natural ecosystem-like characteristics while maintaining harvest output. To achieve sustainability it is important to work toward the system of ecosystem concept. Less dependency on nonrenewable sources can be utilized for energy flow and a better balance achieved to maintain the internal processes of the system and that which is available for export as harvestable goods. Maintaining nutrient cycles which are as closed as possible may lead to lower nutrient losses from the system. Population regulation mechanisms can depend more on system-level resistance to pests, through an assay of mechanisms that range from increasing habitat diversity to ensuring the presence of natural enemies and antagonists. The qualities of resilience, stability, productivity, and balance, if incorporated in an agroecosystem, will better ensure the maintenance of the dynamic equilibrium necessary to establish an ecological basis for sustainability. Sustainability refers to productive performance of a system over time. It implies use of natural resource to meet the present needs without jeopardizing the future potential (Bagyaraj 2014).

As the use of external human inputs for control of agroecosystem processes is reduced, we can expect a shift from systems dependent on synthetic inputs to systems designed to make use of natural ecosystems processes and interactions and materials derived from within the system (Bagyaraj 2014).

Organic farming is a tool for attaining sustainable agroecosystem. In the present scenario, organic farming has attained a lot of attention due to its necessity and demands. One may look into the development and maintenance of "closed" nutrient cycles, so as to lower nutrient losses from the system and to search for sustainable ways to return exported nutrients to the farm (Bagyaraj 2014).

12.2 Organic Farming: Tool for Sustainable Agriculture

Organic farming can be considered as the most effective tool for attaining sustainable agroecosystem. In the present scenario, organic farming has attained a lot of attention due to its necessity and demands. Organic farming assists to maintain sustainable ecosystem, boost in biological cycles within farming system, usage of renewable resources, harmonious balance between animal husbandry and crop production, increased fertility of soil, and the genetic diversity of the production system and its surroundings, including the protection of plant and wildlife habitats; these are the necessities to maintain sustainable ecosystem. At the same time, there is a demand from the society for providing chemical-free, nutritious food, so as to help the generations to come and provide a better and healthy future (Roychowdhury et al. 2013).

12.2.1 Organic Farming Has Gained So Much of Emphasis Due to the Following Reasons (Roychowdhury et al. 2013)

1. Organic farming predates all its approaches to "environmentally friendly" agriculture.
2. It is a rapidly developing agricultural sector in India and in many other countries.

12.2.2 Organic Farming Aims

1. To produce high-quality food in sufficient quantity
2. To create a constructive way of enhancing life with natural systems and cycles
3. To encourage the involvement of microorganisms, soil, flora and fauna, plants, and animals, within biological cycles
4. To improve, maintain, and enhance the fertility of soils
5. To maintain and prevent loss of genetic diversity of plant and wildlife habitats
6. To establish proper usage of water and water resources

7. To establish a relation between the crop production and animal husbandry
8. To minimize all forms of pollution
9. To maximize the use of renewable resources and minimize the usage of renewable resources

Since time immemorial our ancestors have been using various organic technologies to make agriculture sustainable while conserving soil, water, energy, and biological resources. Among the benefits of organic technologies are higher soil organic matter and nitrogen, lower fossil energy inputs, yields similar to those of conventional systems, and conservation of soil moisture and water resources (especially advantageous under drought conditions). Conventional agriculture can be made more sustainable and ecologically sound by adopting some traditional organic farming technologies (Roychowdhury et al. 2013).

12.3 Importance and Creditability of Microbes in Sustainable Agriculture

If agriculture is to be done in terms of agroecology, a complete approach toward farming is to be followed. It is copiously considered that microbes play an important role in various forms, viz., processes of soil formation, plant nutrition, and the suppression of plant pathogens, pests, and weeds. In recent times it has been found that certain well-established conventional farming techniques like intensive tillage, spray of pesticide and fertilizer, and monocropping are directly or indirectly quite harmful to soil microbes. The heavy usage of manufactured inputs rather than natural resources has led to threaten the soil microbial life.

The soil microbial flora may vary depending on biotic and abiotic conditions, soil texture and structure, pH, air/moisture content, and soil temperature (Campbell et al. 1999). The flora may include both eukaryotic (fungi, yeasts, protozoa, and algae) and prokaryotic organisms (Eubacteria, actinomycetes, and archaea) (Shannon et al. 2002).

Soil enrichment may be through the degradation or hydrolysis of organic nitrogen, through the urease enzymes present in some of the microorganisms (Hasan 2000).

The type and amount of organic material that enters the soil ecosystem determine the soil microbial status (Shannon et al. 2002). Hence, it is determined that the optimum input of organic inputs modifies microbial populations, the soil food web, and biological processes involved in nutrient transformation (Nannipieri et al. 1990; Vanhala and Ahtiainen 1994; Stockdale et al. 2002).

The use of vermiculture also helps in restoring the quality of soil (Soni and Sharma 2016).

There are various microbes which in various ways help to achieve the objective of sustainable agriculture.

12.4 Microbes as Biofertilizers

Biofertilizers or microbial inoculants are carrier-based ready-to-use live bacterial or fungal formulations, which, on application to plants, soil, or composting pits, help in mobilization of various nutrients by their biological activity. Biofertilizers which are carrier-based have shown incredible effect on the agriculture production globally in comparison with the chemical fertilizers in the last 20 years (Pindi 2012). For easy application, biofertilizers are packed in suitable carrier such as lignite, peat, or charcoal. Carrier also plays an important role in maintaining sufficient shelf life (Singh et al. 1999a, b). In recent times, as an alternative for carrier-based biofertilizer, liquid biofertilizer has been introduced which has several advantages over carrier-based biofertilizer (Pindi 2012).

Biofertilizers are referred to as biologically active products or microbial inoculants of bacteria, algae, and fungi (separately or in combination), which may help biological nitrogen fixation for the benefit of plants (Higa and Parr 1994; Vessey 2003; Pindi 2012; Bhattacharjee and Dey 2014). Their mode of action differs and can be used alone or in combination.

The main sources of biofertilizers can be classified into the groups of bacteria, fungi, and cyanobacteria. *Rhizobium* acts in symbiotic association, present in the nodules on the roots of leguminous plants, which acts as biofertilizers. These bacteria fix atmospheric nitrogen into organic forms, which is used by the plant as nutrient. *Azospirillum* and *Azotobacter* can fix atmospheric nitrogen while free-living in the soil, hence enhancing the nitrogen content of the soil (Bhattacharjee and Dey 2014).

Mycorrhiza is known to associate symbiotically with roots of higher plants, for instance, *Glomus*. The fungal symbiont in these associations absorbs phosphorus from soil and passes it to the plant. Plants having such associations are benefitted by resistance to root-borne pathogens, tolerance to salinity and drought, and an overall increase in plant growth and development (Bhattacharjee and Dey 2014).

Cyanobacteria are autotrophic microbes widely distributed in aquatic and terrestrial environments, many of which can fix atmospheric nitrogen, e.g., *Anabaena*, *Nostoc*, *Oscillatoria*, etc. In paddy fields, cyanobacteria serve as important biofertilizers. Blue-green algae also add organic matter to the soil and increase its fertility (Bhattacharjee and Dey 2014).

Biofertilizers can also be classified based on the activity carried out by them, as represented in Table 12.1 (Bhattacharjee and Dey 2014):

Advantages of Biofertilizers (Abd El-Lattief 2016):

1. Reduction in the usage of chemical fertilizers.
2. Reduction in environmental pollution.
3. Increase the strength of nutrients and easily absorbed.
4. Excretion of doping substances for growth.
5. Improvisation in the physical, chemical, and biological characteristics of the soil.
6. Excretion of some antibiotics that are resistant to some plant diseases.
7. Microorganisms convert complex organic material into simple compounds and hence increase the uptake by plants.

Table 12.1 Types of biofertilizers

S. no.	Groups	Examples
N_2-fixing biofertilizers		
1.	Free-living	*Azotobacter, Beijerinckia, Clostridium, Klebsiella, Anabaena, Nostoc*
2.	Symbiotic	*Rhizobium, Frankia, Anabaena azollae*
3.	Associative symbiotic	*Azospirillum*
P-solubilizing biofertilizers		
1.	Bacteria	*Bacillus megaterium* var. *phosphaticum, Bacillus subtilisBacillus circulans, Pseudomonas striata*
2.	Fungi	*Penicillium* sp., *Aspergillus awamori*
P-mobilizing biofertilizers		
1.	Arbuscular mycorrhiza	*Glomus* sp., *Gigaspora* sp., *Acaulospora* sp., *Scutellospora* sp., and *Sclerocystis* sp.
2.	Ectomycorrhiza	*Laccaria* sp., *Pisolithus* sp., *Boletus* sp., *Amanita* sp.
3.	Ericoid mycorrhizae	*Pezizella ericae*
4.	Orchid mycorrhiza	*Rhizoctonia solani*
Biofertilizers for micronutrients		
1.	Silicate and zinc solubilizers	*Bacillus* sp.
Plant growth-promoting rhizobacteria		
1.	Pseudomonas	*Pseudomonas Fluorescens*

8. Enhances root proliferation due to release of growth-promoting hormones.
9. Environmentally friendly and cleanse the plant from precipitated chemical fertilizers.

Disadvantages:
1. Biofertilizers are sensitive to temperature and humidity changes, making them difficult to be stored.
2. Their results are slower than chemical fertilizer.
3. Difficult to find a selling retailer.
4. Biofertilizers complement other fertilizers, but they cannot totally replace them.
5. Shortages of particular strains of microorganisms or of the best growing medium reduce the availability of some biofertilizers.

12.4.1 Rhizobium

Rhizobium is the most widely studied and important genus as far as nitrogen fixation is concerned (Odame 1997). This bacterium is associated with legume plants and, hence, plays an important role in pulse production. Legumes have been an important component of agriculture since ancient times. It is widely believed that legumes improve soil fertility because of their N_2-fixing ability (Laranjo et al. 2014) (Table 12.2).

Table 12.2 Crop specificity of *Rhizobium*

Host group	*Rhizobium* species	Crops	N fix kg/ha
Pea group	*Rhizobium leguminosarum*	Green pea, lentil	62–132
Soybean group	*R. Japonicum*	Soybean	57–105
Lupini group	*R. Lupine orinthopus*	Lupinus	70–90
Alfalfa group	*R. meliloti, Medicago, Trigonella*	Melilotus	100–150
Beans group	*R. Phaseoli*	Phaseoli	80–110
Clover group	*R. Trifoli*	Trifolium	130
Cowpea group	*R.* Species	Moong, redgram, cowpea, groundnut	57–105
Cicer group	*R.* Species	Bengal gram	75–117

Desirable Characters of *Rhizobium* Strains to Be Used in Commercial Inoculants (Halliday 1984):

- To form nodules and fix N_2 in target legume
- To compete in nodule formation; ability to fix N_2 across a range of environmental conditions
- To form nodules and fix N_2 in the presence of soil nitrate; ability to grow well in artificial media in inoculants carrier and in soil
- To persist in soil particularly for annually regenerating legumes
- To migrate from initial site of inoculation
- To colonize in the absence of a legume host
- To tolerate environmental stresses; ability to fix N_2 with a wide range of host genotypes
- To maintain and control genetic stability
- To control low mortality on inoculated seed
- To colonize the rhizosphere of host plant

12.4.2 Azospirillum

After anonymity of 50 long years, *Azospirillum* was rediscovered in the mid-1970s by J. Döbereiner and her colleagues in Brazil. *Azospirillum* are free-living bacteria and fix atmospheric bacteria in cereal crops without any symbiosis. They fix 20–40 kg ha^{-1} nitrogen per year (Bashan 1993).

Azospirillum seek their representation in agronomic utilization as biofertilizers in the crops of corn, rice, wheat, and sorghum, which are economically important crops (Döbereiner 1997; Reinhold and Hurek 1988; Sundaram et al. 1988).

Inoculation of *Azospirillum* exerts beneficial effect on yield with varying physiological activities, including synthesis of plant growth-promoting substances (Motsara et al. 1995) (Table 12.3).

Table 12.3 N_2-fixing capacity of *Azospirillum* in the roots of several plants and the amount of N_2 fixed by them

Plant	N_2 fixed/g of substrate
Oryza sativa (paddy)	28
Sorghum bicolor (Sorghum)	20
Zea mays (maize)	20
Panicum sp.	24
Cynodon Dactylon	36
Setaria sp.	12
Amaranthus Spinosus	16

12.4.3 Azotobacter

Along with N_2 fixation, *Azotobacter* also carries out the functions of synthesizing and secretion of considerable amounts of biologically active substances like B vitamins, nicotinic acid, pantothenic acid, biotin, heteroxins, gibberellins, etc. which enhance root growth of plants. These species show sensitivity toward pH, high salts, and temperature above 35 °C (Rao 1986). Among the important species of *Azotobacter* are *A. chroococcum*, *A. agilis*, *A. paspali*, and *A. vinelandii*. *Azotobacter* shows significant response in crop growth, viz., all agricultural crops about 10–12% (Jaga and Singh 2010). Growth and grain yield in wheat crop can be enhanced with Azotobacter (Kader et al. 2002; Abd El-Lattief 2016). *Azotobacter* when inoculated with yeast shows much superior results (Ahmed et al. 2011). Along with rice and other cereals, *Azotobacter* may apply by seed sipping and seedling root dipping methods (Kannaiyan, et al. 1980; Kannaiyan 1999; Singh et al. 1999a, b; Rüttimann, et al. 2003).

12.4.4 Phosphorus-Solubilizing and Phosphorus-Mobilizing Microbes

Phosphorus (P) stands to be the second major plant nutrient limiting factor for crop productivity.

Phosphorus-solubilizing and phosphorus-mobilizing microbes in the form of biofertilizers can make available the accumulated phosphates for plant growth (Goldstein 1986). Soil nutrient status and structure are vitally altered with the use of PSB as bio-inoculants (Blake 1993). Microbes belonging to the groups of bacteria, fungi, and actinomycetes fall to be in this specific group of organisms and enumerated from different sources such as soil (Roychowdhury et al. 2013). Various bacteria belonging to this group are *Bacillus* sp., *B. megaterium*, *B. circulans*, *Pseudomonas* sp., and *P. striata*, and fungi are *Penicillium* sp., *P. digitatum*, *P. oxysporum*, *Aspergillus* sp., *A. niger*, *A. flavus*, and A. *awamori*.

Mycorrhizae are phosphorus-mobilizing microbes, which enact to make available the phosphorus to the plants and hence support the development of plants (Roychowdhury et al. 2013) (Table 12.4).

Table 12.4 Recommended liquid biofertilizers and their application method and quantity to be used for different crops

Crop	Recommended biofertilizer	Application method	Quantity to be used
Field crops			
Pulses			
Chickpea, pea, groundnut, soybean, beans, lentil, lucerne, berseem, green gram, black gram, cowpea, and pigeon pea	*Rhizobium*	Seed treatment	200 ml/ acre
Cereals			
Wheat, oat, barley	*Azotobacter*/*Azospirillum*	Seed treatment	200 ml/ acre
Rice	*Azospirillum*	Seed treatment	200 ml/ acre
Oil seeds			
Mustard, sesame, linseeds, sunflower, castor	*Azotobacter*	Seed treatment	200 ml/ acre
Millets			
Pearl millets, finger millets, Kodo millet	*Azotobacter*	Seed treatment	200 ml/ acre
Maize and sorghum	*Azospirillum*	Seed treatment	200 ml/ acre
Forage crops and grassesBermuda grass, Sudan grass, Napier grass, Para grass, star grass etc.	*Azotobacter*	Seed treatment	200 ml/ acre
Other misc. plantation cropstobacco	*Azotobacter*	Seedling treatment	500 ml/ acre
Tea, coffee	*Azotobacter*	Soil treatment	400 ml/ acre
Rubber, coconuts	*Azotobacter*	Soil treatment	2–3 ml/ plant
Agro-forestry/fruit plantsall fruit/ agro-forestry (herb, shrubs, annuals, and perennial) plants for fuel wood, fodder, fruit, gum, spice, leaf, flower, nut, and seed purposes	*Azotobacter*	Soil treatment	2–3 ml/ plant at nursery
Leguminous plants/trees	*Rhizobium*	Soil treatment	1–2 ml/ plant

12.5 Bacterial Endophytes

The microbes which live inside the host microenvironment and receive protection from environmental stresses, hence, face quite low competition from microbes. At the same time, they have superior access to nutrients (Dutta 2014; Yadav and Yadav 2017). Endophytes are also found to associate in supplying biologically fixed nitrogen in nonlegumes, and these associations can increase the nitrogen economy of a crop, reducing the requirement for N fertilizers. Also prevention from various diseases is carried out through endophyte-mediated de novo synthesis of structural compounds and fungi toxic metabolites (Sturz et al. 2000). Also endophytes improve the seedling performance and survival by providing resistance against insects and nematodes and drought, improving nitrogen assimilation and thus yielding higher seed set (Fescue 1990). Small molecular compounds called siderophores are produced by some endophytes, which are chelating compounds that can benefit iron to plants and divest pathogen of iron (Compant 2005).

12.6 Microbes for Bioremediation of Agricultural Soils

Toxic metals accumulated in agricultural soils can be easily transferred to plants and food and thereby to their food chain. Chemical extractants may be used to recover the metal-polluted soils, but it turns out to be difficult and incomplete. Other than the usage of chemicals, with the help of microbes, having the ability to biosorb heavy metals, the transfer of heavy metals from soil to plant can be reduced (Bagot et al. 2005).

12.7 Microbes in Decomposition of Agricultural Waste

Plant waste contains high content of cellulose, which takes abundant time to decompose by naturally occurring microorganisms. Due to this, the agricultural wastes are burned which leads to environmental pollution and various respiratory diseases. Hence a method has to be incorporated to fasten the activity of composting. This can be done by the usage of microbes belonging to bacteria, actinomycetes, fungi, algae, and protozoa. Due to high organic matter and biological activity of compost, it is effective in various activities like soil erosion, biofiltration, bioremediation, improvisation of soil structure, and moisture retention. Major pollution problems, viz., air and water, can be minimized by utilizing the composting technology (Anyanwu et al. 2015).

References

Abd El-Lattief EA (2016) Use of *Azospirillum* and *Azotobacter* bacteria as biofertilizers in cereal crops: A review. Int J Res Eng Appl Sci (IJREAS) 6(7):36–44

Ahmed MA, Amal GA, Magda MH, Tawafik MM (2011) Integrated effect of organic and biofertilizers on wheat productivity in new reclaimed sandy soil. Res J Agric Biol Sci 7(1):105–114

Anyanwu1 CF, Ngohayon SL, Ildefonso RL and Ngohayon JL (2015) Application of Indigenous Microorganisms (IMO) for Bio-Conversion of Agricultural Waste (Online). Int J Sci Res 4(5):778–784

Bagot D, Lebeau T, Jezequel K, Fabre B (2005) Selection of microorganisms for bioremediation of agricultural soils contaminated by cadmium. In: Lichtfouse E, Schwarzbauer J, Robert D (eds) Environmental chemistry. Springer, Berlin/Heidelberg. https://doi.org/10.1007/3-540-26531-7_20

Bagyaraj DJ (2014) Microorganisms in Sustainable Agriculture. Proc Indian Nat Sci Aca 80(2):357
Bashan Y (1993) Potential Use of *Azospirillum* as Biofertilizer. Turrialba 43(4):286–291
Bhattacharjee R, Dey U (2014) Biofertilizers: a way towards organic agriculture. Afr J Microbiol Res 8(24):2332–2342. https://doi.org/10.5897/AJMR2013.6374.
Blake F (1993) In: Cartwright (ed) Organic food production. In-World Agriculture. Strerling Publication Ltd., Hong Kong, pp 22–24
Campbell CA, Lafond GP, Biederbeck VO, Wen NG, Schoenau J, Hahn D (1999) Seasonal trends in soil biochemical attributes: effects of crop management on a black chernozem. Can J Soil Sci 79:85–97
Compant S (2005) Use of plant growth-promoting bacteria for biocontrol of plant diseases: principles, mechanisms of action, and future prospects. Appl Environ Microbiol 71(9):4951–4959
Döbereiner J (1997) A importância da fixação biológica de nitrogênio para a agricultura sustentável. Biotecnologia Ciência and Desenvolvimento - Encarte especial 1:2–3
Dutta D (2014) Endophytes: exploitation as a tool in plant protection. Braz Arch Biol Technol 57:621–629
Fescue T (1990) A review of the agronomic characteristics of endophyte-free and endophyte-infected. Appl Agric Res 5(3):188–194
Goldstein AH (1986) Bacterial solubilization of phosphates: historical perspective and future prospects. Amer J AlterAgric 1:51–57
Halliday J (1984) Principles of *Rhizobium* strain selection. In: Alexander M (ed) Biological nitrogen fixation. Springer, Boston. https://doi.org/10.1007/978-1-4613-2747-9-7.
Hasan HAH (2000) Ureolytic microorganisms and soil fertility. Commun Soil Sci Plant Anal 31(15–16):2565–2589
Higa T, Parr JF (1994) Beneficial and effective microorganisms for a sustainable agriculture and environment. International Nature Farming Research Center, Atami
Jaga PK, Singh V (2010) Effect of biofertilizer, nitrogen and sulphur on sorghum-mustard cropping system. Proceedings of National Seminar on Soil Security for Sustainable Agriculture held at College of Agriculture, Nagpur
Kader MA, Miar MH, Hoque MS (2002) Effects of *Azotobacter* inoculants on the yield and nitrogen uptake by wheat. J Biol Sci 2(4):259–261
Kannaiyan S (1999) Biological fertilizers for rice production. In: Kannaiyan S (ed) Bioresources technology for sustainable agriculture. Associate Publishing Company, New Delhi, pp 1–29
Kannaiyan S, Govindarajan K, Lewin HD (1980) Effect of foliar spray of *Azotobacter chroococcum* on rice crop. Plant Soil 56:487–490
Laranjo M, Alexandre A, Oliveira S (2014) Legume growth-promoting rhizobia: an overview on the *Mesorhizobium* genus. Microbiol Res 169(1):2–17. https://doi.org/10.1016/j.micres.2013.09.012
Motsara MR, Bhattacharya P, Srivastava B (1995) Biofertilizer technology marketing and uses-a source book cum glossary. Fertilizer Development and Consultancy Organization, New Delhi, p l84
Nannipieri P, Ceccanti B, Grego S (1990) Ecological significance of biological activity in soil. In: Bollag GM, Stotzky G (eds) Soil biochemistry, vol 6. Marcel Dekker, New York, pp 293–354
National Mission for Sustainable Agriculture (2010) Department of Agriculture and Cooperation. Ministry of Agriculture, New Delhi
Odame H (1997) Biofertilizer in Kenya: research, production and extension dilemmas. Biotechnol Dev Monit 30:2023
Pindi PK (2012) Microbial consortium for soil health. J Biofertil Biopestici 3:3. https://doi.org/10.4172/2155-6202.1000e102
Rao DLN (1986) Nitrogen fixation in free living and associative symbiotic bacteria. In: Subba Rao NS (ed) Soil microorganisms and plant growth. Oxford and IBH Pub. Co., New Delhi
Reinhold B, Hurek T (1988) Location of diazotrophs in the root interior with special attention to the kallar grass association. Plant Soil 110:259–268

Roychowdhury R, Banerjee U, Sofkova S, Tah J (2013) Organic Farming for Crop Improvement and Sustainable Agriculture in the Era of Climate Change. Online J Biol Sci 13(2):50–65
Rüttimann JC, Rubio LM, Dean DR, Ludden PW (2003) VnfY is required for full activity of the vanadium-containing dinitrogenase in Azotobacter vinelandii. J Bacteriol 185:2383–2386
Shannon D, Sen AM, Johnson DB (2002) A comparative study of the microbiology of soils managed under organic and conventional regimes. Soil Use Manag 18:274–283
Singh MS, Devi RKT, Singh NI (1999a) Evaluation of methods for Azotobacter application on the yield of rice. Ind J Hill Farm 12:22–24
Singh T, Ghosh TK, Tyagi MK, Duhan JS (1999b) Survival of Rhizobia and level of contamination in charcoal and lignite. Ann Biol 15(2):155–158
Soni R, Sharma A (2016) Vermiculture technology: a novel approach in organic farming. Indian Hortic J 6(1):150–154
Stockdale EA, Shepherd MA, Fortune S, Cuttle SP (2002) Soil fertility in organic farming systems – fundamentally different? Soil Use Manag 18:301–308
Sturz AV, Christie BR, Nowak J (2000) Bacterial endophytes: potential role in developing sustainable Systems of Crop Production. Crit Rev Plant Sci 19(1):1–30
Sundaram S, Arunakumari A, Klucas RV (1988) Characterization of Azospirilla isolated from seeds and roots of turf grass. Can J Microbiol 34:212–217. https://doi.org/10.1139/m88-040
Vanhala P, Ahtiainen JH (1994) Soil respiration, ATP content and Photobacterium toxicity test as indicators of metal pollution in soil. Environ Toxicol Water Qual 9:115–121
Vessey JK (2003) Plant growth promoting rhizobacteria as bio-fertilizers. Plant Soil 255:571–586
Yadav A, Yadav K (2017) Exploring the potential of endophytes in agriculture: a minireview. Adv Plants Agric Res 6(4):00221. 10.15406/ apar.2017.06.00221

MIX
Papier aus verantwortungsvollen Quellen
Paper from responsible sources
FSC® C105338

If you have any concerns about our products,
you can contact us on
ProductSafety@springernature.com

In case Publisher is established outside the EU,
the EU authorized representative is:
Springer Nature Customer Service Center GmbH
Europaplatz 3, 69115 Heidelberg, Germany

Printed by Libri Plureos GmbH
in Hamburg, Germany